GENERAL RELATIVITY AND RELATIVISTIC ASTROPHYSICS

Cuvee 8ieme CCGRRA
Une Vraie Lueur Cosmique

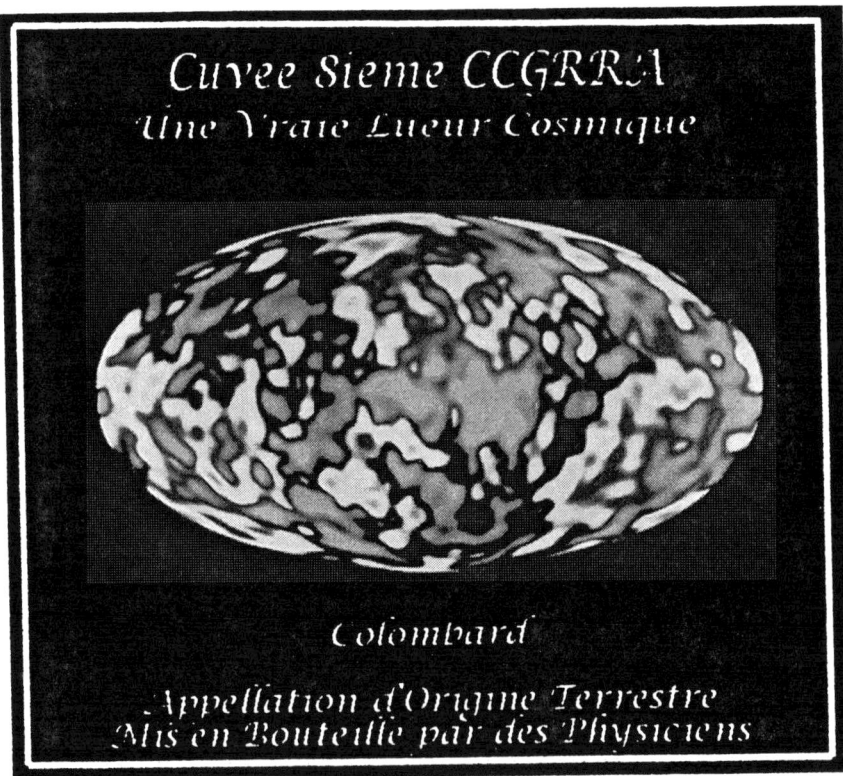

Colombard

Appellation d'Origine Terrestre
Mis en Bouteille par des Physiciens

GENERAL RELATIVITY AND RELATIVISTIC ASTROPHYSICS

Eighth Canadian Conference

Montréal, Québec   June 1999

EDITORS

C. P. Burgess
R. C. Myers
McGill University, Montréal

American Institute of Physics

AIP CONFERENCE
PROCEEDINGS 493

Melville, New York

Editors:

C. P. Burgess

R. C. Myers

Physics Department
McGill University
3600 University Street
Montréal, Québec H3A 2T8
CANADA

E-mail: cliff@physics.mcgill.ca
rcm@physics.mcgill.ca

L.C. Catalog Card No. 99-067693
ISBN 1-56396-905-X
ISSN 0094-243X
Printed in the United States of America

CONTENTS

III. COSMOLOGY

IV. QUANTUM GRAVITY

Plenary Sessions

Parallel Sessions

V. CLASSICAL GENERAL RELATIVITY

Parallel Sessions

viii

PREFACE

About one hundred participants closed the second millenium by celebrating the Eighth Canadian Conference on General Relativity and Relativistic Astrophysics during a hot, sunny week in June at McGill University in Montréal. Many came directly from the picturesque town of Val Morin, in the Laurentian hills north of the city, where the workshop *Black Holes II: Theory and Mathematical Aspects* had just been held.

As this volume indicates, this year's conference continued the tradition of previous meetings, with strong participation from graduate students and researchers from across the country, as well as from many physicists from Europe and the United States. The research presented during the meeting fell across many areas of gravitational physics and, although the sessions were full, there were nonetheless ample opportunities for enjoying Montréal's special summertime *je ne sais quoi*.

It was our good fortune to have this conference at the beginning of the Centre de Recherches Mathématiques' theme year devoted to mathematical physics, and so we had the luxury of both their financial and organizational support. Louis Pelletier of the CRM was particularly invaluable in ensuring the conference ran smoothly. The conference's success was in no small part also ensured by the contributions of our other co-sponsors, the Canadian Institute for Theoretical Astrophysics and McGill University's Faculty of Graduate Studies and Research.

Besides being the bedrock on which our local organizational efforts rested, Elizabeth Shearon left her mark through the imaginative flair with which she introduced numerous small welcoming touches. Our thanks are also due to Rim Dib, Ariel Edery, Steven Horwat, Julie Lainesse, Greg Mahlon, Jean-Sabin McEwen, Vincent Pelletier, Tanvir Rahman and David Winters, who helped without complaint in an infinite number of ways — from running errands to bottling and labelling our conference wine (it really was *mis en bouteille par des physiciens*).

We introduced our conference vintage, *Cuvée Huitième CCGRRA: Une Vraie Lueur Cosmique*, inspired by the introduction of *Black Hole Beer* during the seventh conference in this series, held in Calgary. We hope it is another of the fine traditions of these conferences which will continue into the new millenium!

Cliff Burgess and Rob Myers
McGill University

I. GRAVITATIONAL WAVES AND COMPACT OBJECTS

Relativistic Gravity and Binary Radio Pulsars

Victoria M. Kaspi

Department of Physics and Center for Space Research
Massachusetts Institute of Technology
Cambridge, MA 02139

Abstract.
Following a summary of the basic principles of pulsar timing, we present a review of recent results from timing observations of relativistic binary pulsars. In particular, we summarize the status of timing observations of the much celebrated original binary pulsar PSR B1913+16, draw attention to the recent confirmation of strong evidence for geodetic precession in this system, review the recent measurement of multiple post-Keplerian binary parameters for PSR B1534+12, and describe the Parkes Multibeam survey, a major survey of the Galactic Plane which promises to discover new relativistic binary pulsar systems.

INTRODUCTION

Not long after Einstein proposed his General Theory of Relativity, a variety of experimental tests to be done with solar system objects was suggested. These included the measurement of the perihelion advances of planets, the bending of light rays by the Sun, and radar echo delays from planets. However, such tests were limited by the fact that the effects to be measured were tiny perturbations on a classical description. They only verified the theory in the "weak-field" limit, akin to studying a function by only considering its Taylor expansion about zero. The "strong-field" regime, in which GR effects are more than a perturbation and a classical description is grossly violated, probably at first appeared inaccessible to Earth-bound observers.

The discovery of the first binary pulsar, PSR B1913+16, by Hulse & Taylor (1975) radically changed this situation. This binary system, consisting of two neutron stars in an eccentric 8 hr binary orbit, has permitted precise tests of GR predictions for the first time in the strong-field regime [Taylor & Weisberg 1982,Taylor & Weisberg 1989]. Thus far, GR has passed all tests with flying colours.

In this review, after an introduction to pulsars and pulsar timing, we present the most recent results of observations of PSR B1913+16, as well as of PSR B1534+12,

CP493, *General Relativity and Relativistic Astrophysics,* edited by C. P. Burgess and R. C. Myers
© 1999 American Institute of Physics 1-56396-905-X/99/$15.00

the second discovered binary pulsar system suitable for sensitive GR studies. We also describe a search for new pulsars that is currently underway, and which promises to find more such objects. For previous excellent reviews of relativistic binary pulsars and their experimental constraints on strong-field relativistic gravity see Taylor et al. (1992) and Damour & Taylor (1992).

RADIO PULSARS: SOME BACKGROUND

Pulsars are rotating, magnetized neutron stars. They exhibit beams of radio emission that can be observed, by a fortuitously located astronomer, as pulsations, once per rotation period. In the published literature there are 708 pulsars known (but see section "Parkes Multibeam Survey" below), all but a handful of which are in the Milky Way, the remainder being in the Magellanic Clouds. Known pulse periods range from a few seconds down to 1.5 ms. These pulse periods are observed to increase steadily, indicative of spin-down due to magnetic dipole radiation. From the observed pulse period and rate of spin-down of a pulsar, the magnitude of the dipole component of the stellar magnetic field, as well as an age estimate, can be deduced. See Lyne & Smith (1998) for a complete review of the properties of radio pulsars.

For our purposes here, we need highlight only two properties of radio pulsars: the stabilities of the radio pulse profile and the stellar rotation. By "pulse profile" we mean the result of the addition of many (typically thousands) of individual pulses, by folding the sampled radio telescope power output modulo the apparent pulse period. Two examples of such pulse profiles are shown in Figure 1. Average profiles are observed to be stable in that the summation of any few thousand consecutive pulses always results in the same pulse profile for a given radio pulsar at a given observing frequency, even though individual pulse morphologies vary greatly. Currently there is no theory to explain this observation; in radio pulsar timing it is simply accepted as fact. Less surprising perhaps is the observed rotational stability. In a reference frame not accelerating with respect to the pulsar, the observed times of pulsations (or TOAs, for times-of-arrival) are generally predictable with high precision, given only the pulse period and spin-down rate. This, we argue, is less surprising than the profile stability because of the large stellar moment of inertia and absence of external torques, in strong contrast to accreting neutron stars whose rotation is much less stable [e.g. Bildsten et al. 1997].

PULSAR TIMING

The combination of pulse profile and rotational stability makes a radio pulsar useful as an extremely precise clock; in some cases the stability of the pulsar-clock is comparable to those of the world's best atomic time standards [e.g. Kaspi, Taylor & Ryba 1994]. However, the realization of this stability can come only after effects extrinsic to the pulsar are accounted for. In particular, TOAs measured

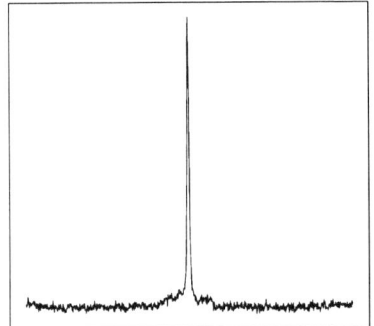 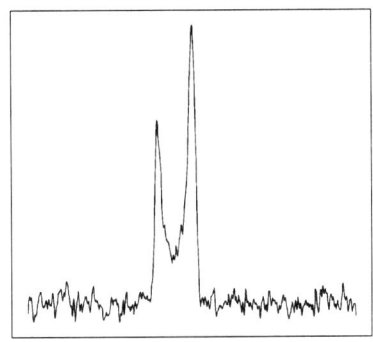

FIGURE 1. Average pulse profiles for PSR B1534+12 [after Kramer et al. 1998], and PSR B1913+16 (courtesy M. Kramer), both at radio frequencies near 1400 MHz.

at an Earth-bound radio telescope must be transformed to a reference frame that is not accelerating with respect to the pulsar. For this purpose, the solar system barycentre reference frame is generally used.

Standard pulsar timing thus consists of observing a pulsar at a radio telescope continuously over many cycles. The start time of the observations is recorded with high precision, and the sampled telescope power output is folded at the topocentric (i.e. apparent) pulse period. The resulting average pulse profile is cross-correlated with a high signal-to-noise template (e.g. Fig. 1) in order to determine the arrival time of the average pulse. That time is then transformed to the solar system barycentre. This transformation can be summarized by the expression

$$t_{\mathrm{SSB}} = t_{\mathrm{O}} + \Delta t_{\mathrm{C}} + \Delta t_{\mathrm{R}} + \Delta t_{\mathrm{E}} + \Delta t_{\mathrm{S}} + \Delta t_{\mathrm{D}}, \tag{1}$$

where t_{SSB} is the pulse arrival time at the solar system barycentre (typically in Baryncetic Dynamical Time), t_{O} is the arrival time as observed at an Earth-bound radio telescope, Δt_{C} is the difference between the observatory clock and a suitably stable atomic time standard (such as Terrestrial Dynamical Time), Δt_{R} is the Roemer delay, or the difference in arrival time of a pulse at the solar system barycentre and at the observatory due to the geometric path length difference, Δt_{E} is the Einstein delay due to (weak-field) GR effects in the solar system, and Δt_{S} is the so-called "Shapiro delay," which depends logarithmically on the impact parameter of the Earth-pulsar and Earth-Sun line of sights. Note that Δt_{R}, Δt_{E} and Δt_{S} require precise knowledge of the sky coordinates of the pulsar; this is turned around so that if observations of the source are available over at least one year, the known motion of the Earth in its orbit permits the measurement of the pulsar's coordinates with high precision. The last term, Δt_{D}, is an observing frequency-dependent term that accounts for the dispersion of radio waves in the ionized interstellar medium according to the cold plasma dispersion law. The delay term is proportional to DM/f^2, where DM is the dispersion measure, or integrated electron density along the line of sight, and f is the observing frequency. The measured DM, together with a model for the distribution of free electrons in the Galaxy [e.g. Taylor & Cordes 1993], provides an estimate of the distance to a pulsar. Details of all the above terms can be found in various references [e.g. Manchester & Taylor 1977].

The above procedure for timing a pulsar of interest is repeated typically on a bi-weekly or monthly basis, so that the spin and astrometric parameters are improved in an iterative fashion: the squares of the residual differences between the initial model-predicted TOAs and the observed TOAs are minimized by varying, and hence improving, the model parameters. The transformation and subsequent determination of the five optimal spin and astrometric parameters (the period P, its rate of change $\dot{P}$, two sky coordinates and DM) are done using a publically available software package, tempo, which consists of several thousand lines of Fortran code[1]. Note that by using TOAs, as opposed to measuring the pulse period at each observing epoch, the timing analysis is coherent in the sense that every rotation of the neutron star is accounted for.

TIMING BINARY PULSARS

If the pulsar is in a binary system, its motion about the binary centre of mass will cause regular delays and advances in observed TOAs just as the Earth's motion around the Sun does.[2] Classically, five additional parameters are required to describe and predict pulse arrival times for binary pulsars, in addition to the five spin and astrometric parameters. Conventionally the five Keplerian parameters are the orbital period P_b, the projected semi-major axis $a \sin i$, where i is the inclination angle of the orbit, the orbital eccentricity e, the longitude of periastron ω measured from the line defined by the intersection of the plane of the orbit and the plane of the sky, and an epoch of periastron T_0. Only the projected semi-major axis is measurable, as pulsar timing is only sensitive to the radial component of the pulsar's motion. Therefore, the component masses cannot be uniquely determined. Note that under certain circumstances, even in a classical system, the five Keplerian parameters may be insufficient to fully describe the orbit; for example, in the binary pulsar PSR J0045−7319, classical spin-orbit coupling induces post-Keplerian dynamical effects, a result of the quadrupole moment of the pulsar's rapidly rotating B-star companion [Lai, Bildsten, & Kaspi 1995,Kaspi et al. 1996].

In some binary systems, particularly double neutron star binaries, relativistic effects must also be taken into account in order to model the binary orbit and hence observed TOAs properly. A list of the known double neutron star binaries is

---

[1] http://pulsar.princeton.edu/tempo/index.html

[2] Although most non-degenerate stars are in binary systems, most pulsars are isolated because supernova explosions usually disrupt binaries. See Bhattacharya & van den Heuvel (1991) for a review of the circumstances under which binary pulsars form.

Table 1: Double Neutron Star Binaries[a,b,c]

PSR	P_b	e	measured PK parameter	Reference
J1518+4904	8.6 day	0.25	$\dot{\omega}$	Nice, Sayer & Taylor (1996)
B1534+12	10.1 hr	0.27	$\dot{\omega}, \gamma, \dot{P_b}, r, s$	see text
J1811−1736	18.8 day	0.83	$\dot{\omega}$	Lyne et al. (1999)
B1913+16	7.8 hr	0.62	$\dot{\omega}, \gamma, \dot{P_b}$	see text
B2127+11C	8.0 hr	0.68	$\dot{\omega}$	Prince et al. (1991)

[a]Sources suitable for tests of GR are indicated in bold.
[b]PSR B1820−11 is not included as the nature of its companion is uncertain
[Phinney & Verbunt 1991].
[c]PSR B2303+46, previously thought to have a neutron-star companion, has recently
been shown to have a white dwarf companion [van Kerkwijk & Kulkarni 1999].

given in Table 1. The only non-classical post-Keplerian (PK) effects to have been measured in a binary pulsar system thus far are: the rate of periastron advance $\dot{\omega}$, the combined effects of relativistic Doppler shift and time dilation γ (equivalent to the solar system Einstein delay – see Eq. 1), the rate of orbital decay $\dot{P_b}$, and r and s, the two parameters describing the Shapiro Delay, or the observed pulse time delay due to the bending of space-time near the pulsar companion, important for highly inclined orbits (equivalent to Δt_S in Eq. 1). The relativistic post-Keplerian parameters measured in each of the known double neutron star binaries are given in Table 1. The systems for which tests of theories of relativistic gravity are possible are indicated by bold type: these are binaries for which N post-Keplerian parameters are measurable, where $N > 2$. These systems permit $N - 2$ tests of gravity, as the first two parameters determine the masses of the two components.

Overall, the suitability of a binary pulsar system for tests of GR or other theories of gravity is determined by a number of factors, including orbital period, orbital eccentricity, orbital inclination angle, the morphology of the pulse profile (narrower pulses permit higher measurement precision) and of course, the pulsar's radio flux. For example, PSR B2127+11C, though in a binary system that is superb for testing GR [Prince et al. 1991], is faint (it was discovered in a deep search of the globular cluster M15) and has thus far not permitted any tests of GR.

PSR B1913+16

The results of long-term timing observations of the relativistic binary pulsar PSR B1913+16 are well-known; indeed they have been distinguished with the 1993 Nobel Prize in Physics awarded to the discoverers Joseph Taylor and Russell Hulse. Detailed descriptions and reviews of the results and implications of those timing observations can be found in a variety of references [Hulse & Taylor 1975,Taylor et al. 1976,Taylor & Weisberg 1982,Taylor 1987,Taylor & Weisberg 1989,Damour & Taylor 1991,Taylor et al. 1992,Damour & Taylor 1992,Taylor 1992,Taylor 1993]. Here we briefly summarize the status of those observations, and discuss the recently reported evidence for geodetic precession in this system.

Status of Timing Observations of PSR B1913+16

As reported by Taylor (1993), timing observations of the 59 ms PSR B1913+16 made at the 305 m radio telescope at Arecibo, Puerto Rico through 1993 (the Arecibo telescope became inoperable not long afterward in preparation for a major upgrade, which is nearly complete) have resulted in the determination of three post-Keplerian parameters: the rate of periastron advance $\dot{\omega} = 4°.226621 \pm 0°.000011$, the combined time dilation and gravitational redshift $\gamma = 4.295 \pm 0.002$ ms, and the observed orbital period derivative $\dot{P}_b = (-2.4225 \pm 0.0066) \times 10^{-12}$. The first two of these parameters determine the component masses to be 1.4411 ± 0.0007 $M_.$ and 1.3874 ± 0.0007 $M_.$. The third post-Keplerian parameter, $\dot{P}_b$, in principle allows for one test of GR (or other theory of gravity).

However, the observed value of $\dot{P}_b$ must first be corrected for the effect of acceleration in the Galactic potential. This correction follows from the simple first-order Doppler effect, where $P_b^{obs}/P_b^{int} = 1 + v_R/c$, where P_b^{obs} and P_b^{int} are the observed and intrinsic values, and v_R is the radial velocity of the pulsar relative to the solar system barycentre. A changing v_R leads to a Galactic term

$$\left(\frac{\dot{P}_b}{P_b}\right) = \frac{a_R}{c} + \frac{v_T^2}{cd}, \tag{2}$$

where a_R is the radial component of the acceleration, v_T is the transverse velocity, and d is the distance to the pulsar. The second term in this equation is the familiar transverse Doppler or "train-whistle" effect. The best estimate correction factor for PSR B1913+16, given its only approximately known location in the Galaxy, is $(-0.0124 \pm 0.0064) \times 10^{-12}$ [Damour & Taylor 1991,Taylor 1992]. With this correction applied to $\dot{P}_b^{obs}$, the comparison with the GR prediction can be made; the result [Taylor 1992] is that

$$\frac{\dot{P}_b^{obs}}{\dot{P}_b^{GR}} = 1.0032 \pm 0.0035. \tag{3}$$

Note that the uncertainty in this expression is dominated by the uncertainty in the Galactic acceleration term. Since a_R and d are unlikely to be known with much greater precision than is currently available, this particular test of GR will probably not improve much in the near future.

Additional tests of GR may still be possible with the PSR B1913+16 system if the parameters r and s can be measured. This may be possible given the recent major upgrade to the Arecibo telescope, as higher timing precision should now be available.

PSR B1913+16 and Geodetic Precession

Relativistic geodetic precession, the gravitational analogue of Thomas precession (the origin of fine structure in atomic spectra), is predicted to result in a changing

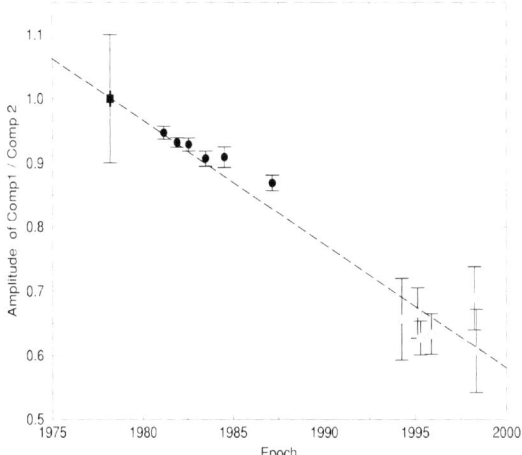

FIGURE 2. The variation in the ratio of the amplitudes of the components of the PSR B1913+16 average radio pulse profile, after Kramer (1998).

orientation of the pulsar spin axis. As the pulsar precesses, our line of sight should intersect different parts of the radio emission beam. Thus, the average pulse profile could vary significantly over time. The first evidence for this in the PSR B1913+16 system was presented by Weisberg, Romani & Taylor (1989) [but see also Cordes, Wasserman & Blaskiewicz 1990]. They reported a gradual, secular evolution in the ratio of the amplitudes of the two pulse peaks (see Fig. 1).

Recently, Kramer (1998) has clearly demonstrated that this trend continues. Figure 2 shows the ratio of the amplitudes of the two pulse components as a function of time; the variation is striking. If the emission results from a cone of radiation, then a secular change in the separation of the two peaks ought to be observed as well; strong evidence for this is also now seen [Kramer 1998]. Quantitative modeling of this variation depends on the unknown beam morphology. Under the assumption of a hollow, circular emission beam, if GR is correct, Kramer shows that the pulsar, sadly, will no longer grace the skies of our Earth after the year 2025. Happily however, it should reappear around the year 2220. The exact dates of disappearance, together with the form of the secular variation in average pulse morphology, will permit the first direct observation and study of the morphology of a radio pulsar emission beam.

PSR B1534+12

The binary pulsar PSR B1534+12 was discovered by Wolszczan (1991) using the Arecibo telescope. This 38 ms pulsar is in a 10 hr eccentric orbit with a second neutron star (see Table 1). PSR B1534+12 offers the hope of additional and more precise tests of GR for a number of reasons: first, the narrower pulse profile of PSR B1534+12 (Fig. 1) means higher timing precision. Second, the orbital plane of this system is more inclined than that of PSR B1913+16, which facilitates the measurements of two additional relativistic parameters r and s. Thus, in principle, five relativistic post-Keplerian parameters are measurable with high precision for PSR B1534+12, which allows two new additional tests of GR that have not been done for PSR B1913+16. This is particularly important for testing alternative theories of gravity, as it permits the separation of the radiative and strong-field components of the theory. This cannot be accomplished in the simple $\dot{\omega}$-γ-$\dot{P_b}$ test, as it mixes radiative and non-radiative effects [see Damour & Taylor 1992 for details].

Stairs et al. (1998) report on seven years of timing observations of PSR B1534+12 made at Arecibo, at the 43 m dish at Green Bank, as well as at the 76 m Lovell radio telescope at Jodrell Bank. As expected, they measure the five post-Keplerian relativistic parameters $\dot{\omega}, \gamma, \dot{P_b}, r$ and s. The results are nicely summarized in Figure 3 (Fig. 4 in Stairs et al. 1998), where the component masses are plotted on the axes. As each of the five post-Keplerian parameters has a different dependence on the masses, each parameter defines a curve in this plane. If GR holds, then the five curves, as calculated in GR, should meet at a single point. As can be seen in Figure 3, the curves for $\dot{\omega}, \gamma$ and s agree to better than 1% (though that for r is not yet precise enough to be very constraining). Surprisingly, their intersection implies that the pulsar and companion have exactly equal masses within uncertainties, $1.339 \pm 0.003 \, M_\cdot$.

As is clear in Figure 3, the curve for $\dot{P_b}$ just misses this intersection point. Note, however, that the value of $\dot{P_b}$ used to produce the curve in Figure 3 included a correction for Galactic acceleration (Eq. 2) that assumed a distance of 0.7 kpc to the pulsar, from its observed DM and the best model for the free electron distribution [Taylor & Cordes 1993]. The model is known to be only approximate, with uncertainties on inferred distance for anyone source optimistically 25%, and realistically considerably larger. Stairs et al. therefore argue that the discrepancy seen in Figure 3 can be removed by simply invoking a larger distance to the pulsar, 1.1 kpc. Put differently, by assuming GR is correct, the distance to this relativistic binary pulsar can be determined with greater precision than is otherwise available [Bell & Bailes 1996]. This demonstrates that the measurement of an improved $\dot{P_b}$ for PSR B1534+12 is unlikely to offer a useful test of GR unless the distance to the source can be determined independently (for example, via a timing or interferometric parallax measurement). However, the expected improved determination of the r parameter, following the Arecibo upgrade, could yield a useful test in addition to that from $\dot{\omega}$-γ-s.

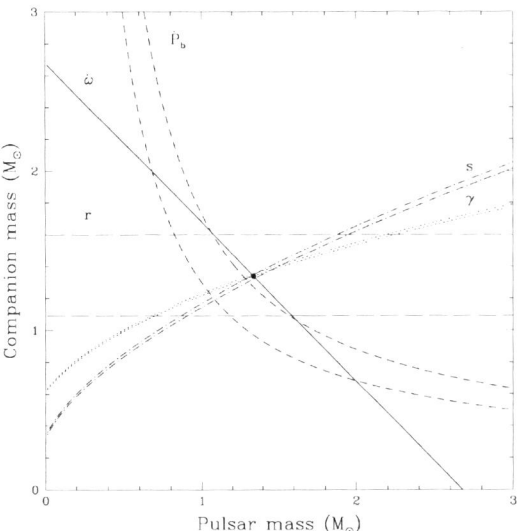

FIGURE 3. The constraints imposed by the observed values of the five measured post-Keplerian parameters on the masses of the components of PSR B1534+12, after Stairs et al. (1998).

The improved distance determination to PSR B1534+12 made by Stairs et al. (1998) has implications for estimates of the coalescence rate of double neutron star binaries. A larger distance implies a more intrinsically luminous pulsar, which in turn implies that there are fewer in the Galaxy, as otherwise more would be detected. Stairs et al. suggest that the expected rate must be reduced relative to previous estimates [Phinney 1991,Curran & Lorimer 1995,van den Heuvel & Lorimer 1996] by factors of 2.5–20. This rate is of considerable interest to the builders of gravitational wave detectors like LIGO (see paper by P. Saulson, this volume). Of course rates that vary greatly depending on the estimated distance to a single object should be regarded as crude estimates only.

FINDING MORE RELATIVISTIC BINARIES: THE PARKES MULTIBEAM SURVEY

A major survey of the Galactic Plane for radio pulsars is currently underway. This survey offers the hope of finding new examples of relativistic binary pulsars suitable for studying GR effects. The observations are being done using the Parkes 64 m radio telescope in Australia [Lyne et al. 1999]. The survey is planned to cover the inner Galactic Plane, in the Galactic longitude range $260° < l < 50°$ and Galactic latitude range $|b| < 5°$. The search is being carried out at radio frequencies near 1400 MHz and has roughly seven times the sensitivity of previous

1400 MHz surveys of the Galactic Plane [Clifton & Lyne 1986,Johnston et al. 1992], owing mainly to the longer integration time permitted by the use of the new multibeam receiver at Parkes. This new instrument consists of 13 independent, non-overlapping receivers in the telescope focal plane. This allows the Galaxy to be surveyed to much greater depth than was previously possible, without using a prohibitive amount of telescope time. Each beam pointing consists of a 35 min integration, with a total of 288 MHz bandwidth 1-bit sampled every 250 μs. Some 35,000 beams will be observed, and the data for each will be subject to a Fast Fourier Transform of 2^{23} points. The project is thus computer resource intensive. With approximately half of the survey complete, 405 previously unknown radio pulsars have been discovered, making this by far the most successful pulsar survey ever.

Among the first sources found in the survey is the very likely double neutron star binary PSR J1811−1736 (see Table 1) [Lyne et al. 1999]. Although this system is unlikely to be useful for tests of GR, its early discovery in the survey suggests there are many more such systems to be found. Indeed, not long after the conference for which these proceedings are a record, a third relativistic binary pulsar suitable for tests of GR was discovered among the new Parkes Multibeam sources. Detailed observations of this exciting source are just getting underway as this paper is being written.

CONCLUSIONS

The now famous technique of timing relativistic binary pulsars has yielded confirmation that GR is the correct theory of gravity at better than the 1% level. Future additional tests of GR, using the only two known sources well-suited to such tests, PSR B1534+12 and PSR B1913+16, are possible, from improved measurements of the Shapiro delay r and s parameters. The precision in the $\dot{\omega}$-γ-$\dot{P}_b$ test is limited by the uncertainty in our estimates for the Galactic acceleration of these objects. However, under the now justified assumption that GR is correct, observations of relativistic binary pulsars can yield unique astrophysical measurements that have never before been possible, including precise determination of neutron star masses, distances to these sources, LIGO source rates, and morphological studies of the pulsar radio emission beam. The ongoing Parkes Multibeam survey of the Galactic Plane promises (and indeed has already begun) to discover new examples of these fascinating objects.

VMK is an Alfred P. Sloan Research Fellow. She thanks Michael Kramer and Ingrid Stairs for sharing their figures, and the organizers of the 8th Canadian Conference on General Relativity and Relativistic Astrophysics for their hospitality and patience.

REFERENCES

Bell, J. F. & Bailes, M. 1996, ApJ, 456, L33

Bhattacharya, D. & van den Heuvel, E. P. J. 1991, Phys. Rep., 203, 1

Bildsten, L. et al. 1997, ApJS, 113, 367

Clifton, T. R. & Lyne, A. G. 1986, Nature, 320, 43

Cordes, J. M., Wasserman, I., & Blaskiewicz, M. 1990, ApJ, 349, 546

Curran, S. J. & Lorimer, D. R. 1995, MNRAS, 276, 347

Damour, T. & Taylor, J. H. 1991, ApJ, 366, 501

Damour, T. & Taylor, J. H. 1992, Phys. Rev. D, 45, 1840

Harrison, E. R. & Tademaru, E. 1975, ApJ, 201, 447

Hulse, R. A. & Taylor, J. H. 1975, ApJ, 195, L51

Johnston, S., Lyne, A. G., Manchester, R. N., Kniffen, D. A., D'Amico, N., Lim, J., & Ashworth, M. 1992, MNRAS, 255, 401

Kaspi, V. M., Bailes, M., Manchester, R. N., Stappers, B. W., & Bell, J. F. 1996, Nature, 381, 584

Kaspi, V. M., Taylor, J. H., & Ryba, M. 1994, ApJ, 428, 713

Kramer, M. 1998, ApJ, 509, 856

Kramer, M., Xilouris, K. M., Lorimer, D. R., Doroshenko, O., Jessner, A., Wielebinski, R., Wolszczan, A., & Camilo, F. 1998, ApJ, 501. in press

Lai, D., Bildsten, L., & Kaspi, V. M. 1995, ApJ, 452, 819

Lyne, A. G. et al. 1999, MNRAS. submitted

Lyne, A. G. & Smith, F. G. 1998, Pulsar Astronomy, (: Cambridge University Press)

Manchester, R. N. & Taylor, J. H. 1977, Pulsars, (San Francisco: Freeman)

Nice, D. J., Sayer, R. W., & Taylor, J. H. 1996, ApJ, 466, L87

Phinney, E. S. 1991, ApJ, 380, L17

Phinney, E. S. & Verbunt, F. 1991, MNRAS, 248, 21P

Prince, T. A., Anderson, S. B., Kulkarni, S. R., & Wolszczan, W. 1991, ApJ, 374, L41

Stairs, I. H., Arzoumanian, Z., Camilo, F., Lyne, A. G., Nice, D. J., Taylor, J. H., Thorsett, S. E., & Wolszczan, A. 1998, ApJ. submitted (astro-ph/9712296)

Taylor, J. H. 1987, in The Origin and Evolution of Neutron Stars, IAU Symposium No. 125, ed. D. J. Helfand & J.-H. Huang, (Dordrecht: Reidel), 383

13

Taylor1992 Taylor, J. H. 1992, Philos. Trans. Roy. Soc. London A, 341, 117

Taylor1993 Taylor, J. H. 1993, in Particle Astrophysics,IVth Rencontres de Blois, ed. G. Fontaine & J. Trân Thanh Vân, (Gif-sur-Yvette, France: Editions Frontieres), 367

Taylor&Cordes1993 Taylor, J. H. & Cordes, J. M. 1993, ApJ, 411, 674

Tayloretal.1976 Taylor, J. H., Hulse, R. A., Fowler, L. A., Gullahorn, G. E., & Rankin, J. M. 1976, ApJ, 206, L53

Taylor&Weisberg1982 Taylor, J. H. & Weisberg, J. M. 1982, ApJ, 253, 908

Taylor&Weisberg1989 Taylor, J. H. & Weisberg, J. M. 1989, ApJ, 345, 434

Tayloretal.1992 Taylor, J. H., Wolszczan, A., Damour, T., & Weisberg, J. M. 1992, Nature, 355, 132

van denHeuvel&Lorimer1996 van den Heuvel, E. P. J. & Lorimer, D. R. 1996, MNRAS, 283, L37

vanKerkwijk&Kulkarni1999 van Kerkwijk, M. H. & Kulkarni, S. R. 1999, ApJ, 516, 25

Weisberg,Romani,&Taylor1989 Weisberg, J. M., Romani, R. W., & Taylor, J. H. 1989, ApJ, 347, 1030

Wolszczan1991 Wolszczan, A. 1991, Nature, 350, 688

Surprises from Rotating Neutron Stars

Sharon M. Morsink

Department of Physics, University of Alberta, Edmonton, AB, T6G 2J1

Abstract. Gravitational radiation can drive a fluid instability in a rotating neutron star, causing it to spin down to slower rotation rates. This instability has been understood for the last quarter of a century. However, until recently is was never clear that the process would operate in astrophysical circumstances. The surprising discovery, due to Nils Andersson, that sluggish "ocean" currents (known to astronomers as *r*-modes) can be unstable at arbitrarily slow rotation rates has opened the door to a number of interesting applications in relativistic astrophysics. One exciting implication is that the instability may explain the spin rate of pulsars born with high angular velocity. It also may be possible to observe the gravitational radiation associated with these currents with an advanced version of LIGO.

I INTRODUCTION

Neutron stars encompass extreme conditions which make them exciting objects for practically any physicist to study. Many aspects of neutron stars make them fascinating objects: denser than atomic nuclei cores; superconducting, superfluid interiors (with $T_c \geq 10^9$ K); high magnetic fields (9−15 orders of magnitude stronger than the Earth's field) and fast rotation rates approaching the speed of light (see Shapiro & Teukolsky [35]). By far the most exciting aspect for the relativist is the strong gravitational field of a neutron star. The gravitational field of a neutron star is so strong that general relativity is needed for a correct description of its structure. The short timescales associated with neutron stars combined with their strong gravitational fields also imply that nonradial fluid motions could be a source of gravitational radiation.

This review will give an elementary introduction to the physics behind the production of gravitational waves from oscillations in a *rotating* neutron star's fluid. An excellent review (Stergioulas [37]) of the properties of rotating neutron stars already exists, so the present review will concentrate on the instability of whirlpool-like fluid perturbations known as Rossby waves or *r*-modes. Fluid instabilities occur when a perturbation can have negative energy. If the perturbation is radiative (say quadrupolar for general relativity) it will radiate away positive energy and the perturbation's energy will become more negative, causing an instability. Gravitational physicists probably will be more familiar with negative energy states in rotating

CP493, *General Relativity and Relativistic Astrophysics*, edited by C. P. Burgess and R. C. Myers
© 1999 American Institute of Physics 1-56396-905-X/99/$15.00

black holes than in rotating neutron stars, so section II will review superradiant scattering in the Kerr black hole. In section III the gravitational-radiation–driven fluid instability of rotating neutron stars will be reviewed with reference to black hole scattering. In section IV a brief classification of neutron star perturbations will be given, along with a review of their stability properties. Astrophysical implications of the instability will be discussed in section V. I will conclude in section VI with a list of questions for future research.

II SUPERRADIANT BLACK HOLE SCATTERING

Rotating objects in general relativity produce an effect known as the dragging of inertial frames. In the simplest manifestation of the frame-dragging effect, a test-particle at infinity with zero angular momentum will acquire an angular velocity as it moves along a geodesic towards a rotating object. In the rotating Kerr black hole an extreme version of the frame-dragging effect occurs in a region exterior to the event horizon but interior to the static-limit surface, known as the ergoregion. Outside of the static-limit surface, it is possible for test particles to be static. However once inside the static limit, all test particles must move around the black hole in the same sense as the black hole's rotation.

In more technical terms, the vector $t^a = (\partial/\partial t)^a$ which generates time translations at infinity is not timelike everywhere in the spacetime. At the static limit the time-translation vector's norm vanishes ($t^a t_a = 0$) and in the ergoregion t^a is spacelike ($t^a t_a > 0$). Recall (e.g. Wald [39]) that the energy, E, of a test particle is just the negative of the time-component of the particle's four-momentum, i.e., $E = -t^a p_a$. Since t^a is spacelike inside the static-limit surface, the energy can be negative in the ergoregion. The possibility of negative-energy orbits within the ergosphere can be exploited to extract energy from a rotating black hole, as first discussed by Penrose [32]. The Penrose process is essentially as follows: Send in a particle with energy $E_0 > 0$ into the ergoregion. Arrange for the particle to split in two pieces so that $E_1 < 0$, $E_2 > 0$ and $E_0 = E_1 + E_2$. The negative energy piece enters black hole while the positive energy piece escapes to infinity with energy $E_2 = E_0 + |E_1| > E_0$. Energy is extracted! The process continues until the angular momentum of the black hole is reduced to zero.

The process of superradiant scattering is just a wave version of the Penrose process. Consider a nonaxisymmetric scalar wave which has the form

$$\Phi = \Phi_0 e^{-i\omega t + im\phi} \tag{1}$$

in a frame rotating with the black hole. This wave is scattered off of a rotating black hole so that the transmitted wave is absorbed by the black hole and the reflected wave escapes to infinity. The incoming wave's energy (time averaged) is

$$< E_0 >= \frac{1}{2}(\omega + m\Omega_H)^2 |\Phi_0|^2 > 0. \tag{2}$$

The transmitted energy entering the black hole's horizon is [39]

$$< E_t >= \frac{1}{2}\omega(\omega + m\Omega_H)|\Phi_0|^2 \qquad (3)$$

where Ω_H is the angular velocity of the black hole.

Clearly, if the perturbation counter-rotates relative to the black hole in the rotating frame (corresponding to $\omega < 0$ and $m > 0$) and if frequency satisfies

$$0 < |\omega| < m\Omega_H \qquad (4)$$

then the energy entering the event horizon is negative. By conservation of energy, the energy of the reflected wave at infinity is positive and larger than the energy of the incident wave. Once again, energy is extracted from the black hole. All that is required is a perturbation which appears to counter-rotate in a frame rotating with the black hole but with phase velocity small enough that in the inertial frame, it appears to co-rotate with the black hole. Similar results hold for the scattering of electromagnetic and gravitational waves off of rotating black holes.

One important item to note is that the increase in energy of a mode is always finite: It has been proven by Whiting [40] that the mode perturbations of Kerr are stable and die away with time. Heuristically, one could think of the negative energy being swallowed by the black hole. Once it enters the event horizon, it is safely defused. If the negative energy state didn't get absorbed by the black hole, multiple scatterings could occur which would allow the negative energy state to become unbounded from below and an infinite amount of energy could be extracted from the black hole causing an instability. On this note, we should now consider the case of neutron stars.

III   GRAVITATIONAL-RADIATION–DRIVEN INSTABILITY OF ROTATING NEUTRON STARS

If a neutron star is very compact and rotates rapidly it can have an ergoregion. (However, the only known rigidly rotating stars with ergoregions also have constant density [8]. It may be possible for differentially rotating stars with a realistic equation of state to have an ergoregion.) For neutron stars with ergoregions, it is always possible to find a nonaxisymmetric perturbation with negative energy. These perturbations are the analogues of the black hole superradiant states. Radiation always acts to reduce the energy of a perturbation, so the perturbation's energy becomes *more* negative and an instability sets in. As a result stars with ergospheres are unstable to nonaxisymmetric perturbations [12]. However, the growth time of this instability is slow enough to render it uninteresting for astrophysics [8].

The key difference between black holes and neutron stars is the presence or absence of an event horizon. In a black hole the negative energy states are lost down the horizon, rendering them harmless. In the case of a neutron star, the negative energy "collects" in the ergosphere. One then expects that if the unstable mode could grow fast enough, a star with an ergoregion would be forced to slow down until the ergoregion disappeared.

Although stars with ergospheres are unstable to scalar, electromagnetic and gravitational perturbations, a much stronger result holds for gravitational perturbations. Since gravitational radiation couples to the source the nature of the matter source must be taken into account. In the case of a star composed of a perfect fluid, nonaxisymmetric perturbations are unstable for any value of the star's angular velocity. In other words, no ergosphere is required to produce an instability. Furthermore, gravitational radiation will be produced from the instability as the star spins down to a stable solution. This result is known as the CFS instability, named after Chandrsekhar, Friedman and Schutz. Chandrasekhar's original work [7] showed that perturbations with $m = 2$ are unstable. Friedman and Schutz [16] showed that the instability holds for any value of m for nearly Newtonian stars. Finally, Friedman [13] showed that the result holds for fully relativistic stars.

The CFS instability can be thought of as an analogue of black hole superradiant scattering. In this case, the unstable modes have a form similar to Eq (1) and counter-rotate with respect to the star in the rotating frame. If the star rotates with angular velocity $\Omega_\star$, then the unstable modes have

$$\omega < 0 \qquad \text{and} \qquad 0 < |\omega| < m\Omega_\star, \tag{5}$$

so that the mode co-rotates in the inertial frame. As long as these inequalities can be satisfied, the instability will occur. As long as we are free to choose the mode number m arbitrarily large, it is always possible to arrange for the inequalities to be satisfied even if the star is rotating very slowly. However, this analysis assumes that the fluid has no source of dissipation. If the viscosity of neutron star fluid is included, the viscosity would tend to damp out the perturbation. Hence the instability will only be of astrophysical interest if gravitational radiation leaks out energy faster than viscosity damps out the perturbation. Since the power radiated by gravitational radiation is inversely proportional to a function of m which grows very quickly with increasing m, the only physical values of the mode number lie in the range $2 \leq m \leq 5$ [9]. Until recently this was thought to limit the instability to stars which rotate very rapidly. The big surprise was the realization by Nils Andersson [1] that a previously overlooked type of mode has very different stability properties. In order to understand the surprise it will be important to understand something of the normal mode classification scheme and some of the properties of these modes.

IV NEUTRON STAR PERTURBATIONS

There are as many types of perturbations as there are types of restoring forces, and it would be impossible to list them all here. Instead I will make a short list of the most important modes for the purpose of gravitational radiation. There are two types of classifications of neutron star perturbation. The first classification divides the modes into two classes: those that either do or do not exist in the Newtonian limit (i.e. when gravitational radiation is turned off). The Newtonian limit modes rely mainly on the pulsation of the fluid, and the change in the star's gravitational potential is not as important. The other modes have no analogue in Newtonian physics and exist even in the limit of vanishing fluid pulsation.

The other classification scheme is based on the behaviour of the mode under parity (reflections through the star's centre). Recall that any scalar field on a 2-sphere can be decomposed in a series of spherical harmonics. Under the parity operator Π, the spherical harmonic function $Y_\ell^m(\theta, \phi)$ transforms as $\Pi(Y_\ell^m) = (-1)^\ell Y_\ell^m$. Similarly, vector fields on the 2-sphere can be decomposed in series of vector spherical harmonics. There are two types of vector spherical harmonics. One type, *polar*, transforms under parity with the same law as the spherical harmonics, $\Pi(_p\vec{Y}_\ell^m) = (-1)^\ell\, _p\vec{Y}_\ell^m$. The polar perturbations include radial motions and nonradial motions proportional to a gradient of a spherical harmonic. The second type of perturbation, *axial*, transforms as $\Pi(_a\vec{Y}_\ell^m) = (-1)^{\ell+1}\, _a\vec{Y}_\ell^m$. Axial motions consist only of nonradial motions which are proportional to the cross product of a radial vector with a polar perturbation. Similar decompositions can be done for higher rank tensor perturbations.

The main types of neutron star normal modes are:

Modes with a Newtonian Limit

> p-modes: The *pressure* modes are polar parity oscillations with the fluid pressure acting as the restoring force. These modes are sound waves and include both radial and nonradial motions.
>
> f-modes: The f-modes are the *fundamental* frequency pressure modes, dominated by nonradial motion.
>
> g-modes: The g-modes (or *gravity* modes) are polar parity oscillations driven by the buoyancy of the perturbations and have frequencies much smaller than the f-modes.
>
> r-modes: The r-modes are axial parity oscillations driven by the Coriolis force. Since their frequency is nonzero only if the star rotates, many people refer to them as rotational modes. However, the "r" refers to Rossby, a meteorologist who studied these modes of the Earth's oceans and atmosphere. These modes produce vortices on surfaces of constant radius.

Purely relativistic modes

> w-modes: The w-modes are oscillations in the gravitational field, the "w" standing for gravitational *waves*. The w-modes can have either polar or axial parity. Since the w-modes are dominated by gravitational (not fluid) perturbations it follows that w-modes are *not* unstable to the gravitational-wave driven fluid instability discussed in section III.

All work on the fluid instability prior 1997 concentrated on the stability properties of the f-modes in rapidly rotating neutron stars (for a review of this earlier period, see reference [14]). The f-modes of static stars have very high frequency, similar to the light-crossing frequency of the star, $\omega_{(f)} \sim \sqrt{\frac{M}{R^3}}$, which is close to the mass-shed limit on the angular velocity of a star. According to the instability criteria, the f-mode will only be unstable when a backward moving mode in the rotating frame appears to co-rotate in the inertial frame. From Eq (5) this implies that that $\omega_{(f)} \sim m\Omega_\star$. Assuming that rotation does not change the frequency of the f-modes greatly, this implies that for $m = 2$, the star will only be unstable if the angular velocity is very close to the mass-shed limit. The stellar angular velocity for the onset of the instability has been calculated exactly in general relativity [38,30]. The result depends on the assumed equation of state for dense nuclear matter, but typically stars spinning at a rate of at least 83% - 93% of the star's mass-shed limit are unstable. The fact that the instability only sets in if the star rotates very rapidly meant that its application to astrophysics was limited. The limitation arises since it there are no known newly-born neutron stars with such high rotation rates. The viscosity of young neutron stars is very low, so it is thought that the instability will be most important in the first year of a neutron star's life, whether the neutron star is born in the core collapse supernova of a main sequence star or the merger of two neutron stars.

The new twist appears with the r-modes, which are axial parity Newtonian solutions of the fluid pulsation equations in the slow rotation limit ($\Omega_\star \ll \sqrt{\frac{M}{R^3}}$). It is fairly simple to see that frequency of the r-modes must vanish if the star does not rotate. This vanishing occurs since the pressure, density and gravitational field perturbations are all scalars and must have polar parity. In a static star, polar and axial parity perturbations decouple. Hence the velocity perturbations produced by these perturbations must also be polar and no axial solutions can exist, except for zero-frequency trivial solutions. However, when the star rotates, axial and polar perturbations couple, and axial parity velocity perturbations with nonvanishing frequency exist. The frequency of these modes in the slow-rotation limit are linear in the star's angular velocity. In the rotating frame, the r-mode frequency is

$$\omega_{(r)} = -\frac{2m\Omega_\star}{\ell(\ell+1)} < 0 \qquad \text{for} \qquad m\Omega_\star > 0 \tag{6}$$

while in the inertial frame, the frequency is measured to be

$$\omega_{(r)\infty} = \omega_{(r)} + m\Omega_\star = \frac{m\Omega_\star}{\ell(\ell+1)}(\ell+2)(\ell-1) > 0. \tag{7}$$

Clearly the r-modes satisfy the CFS instability criterion (5) for *any* value of the star's angular velocity and any value of the spherical harmonic index m if $\ell > 1$. This Newtonian argument is due to Andersson [1] who invoked it to explain the instability of a family of modes which he computed using general relativity. Unfortunately the numerical approximation scheme he employed truncated some terms in the equations, so that it correctly showed the presence of the instability, but it did not accurately calculate the relativistic modes. After learning of Andersson's results, an analytic method was used [15] to prove the existence of unstable perturbations with axial parity in general relativity. Although existence of perturbations was proved, relativistic normal mode solutions were not found. The calculation of the frequencies and wavefunctions of the relativistic r-modes still remains as one of the more important problems in this field.

V   ASTROPHYSICAL IMPLICATIONS

The discussion of the previous sections has shown that rotating perfect fluid balls are unstable to the emission of gravitational radiation. What does this have to do with neutron stars?

Neutron stars are not made of a perfect fluid. Real neutron star fluid is viscous and will oppose the motion created by the r-modes. The CFS instability will only be of interest if the growth time of the instability due to gravitational radiation reaction is shorter than the viscous damping timescale. These timescales have been calculated in the slowly rotating, nearly Newtonian limit [23,2]. The results of these calculations were surprising: for a temperature of 10^9 K, neutron stars would be unstable at angular velocities above 7% of the mass-shed limit, corresponding to rather slow speeds. This corresponds to a spin-period of about 19 ms, which remarkably is the inferred "birth"-period of the Crab pulsar. (The present day spin period of the Crab pulsar is 33 ms.) The temperature of 10^9 K is approximately the critical temperature below which the fluid will make a transition to a superfluid state. Once the star has changed to the superfluid state, the fluid's viscosity increases greatly and it is generally thought that fluid perturbations will be damped.

The scenario in which the r-mode instability may be of interest is as follows. A neutron star is born in the supernova death of a main sequence star (or the collision of two neutron stars). The birth temperature of neutron stars is approximately 10^{11} K. If the neutron star was born with a very high angular velocity, the f-mode instability will occur and gravitational radiation will be emitted. This will cause the neutron star to spin down to more moderate angular velocities. At these lower angular velocities the r-mode instability will dominate and cause the neutron star to spin down further. During this spin-down period, the neutron star will also be cooling, say via the standard modified URCA process (beta decay with spectator neutrons) [35]. In the standard cooling scenarios neutron stars take about 1 year to cool to a temperature of 10^9 K, at which point the instability would presumably turn off leaving behind a star which is rotating relatively slowly. This star will have a changing dipole magnetic field and will continue to slowly spin down due to the emission of magnetic dipole radiation as is observed today. If the neutron star were born with a lower angular velocity, the scenario would be similar, but the f-modes would not be unstable. Note that the time during which gravitational radiation is emitted is very short (1 year) so the chance of observing a gravitational radiation induced spin-down by randomly pointing a detector is rather slim. However, if a supernova is observed optically, the region could be monitored by a detector such as an advanced version of LIGO with some hope of detecting the radiation. Since the frequency of the r-mode is linear in angular velocity, the signal would be an inverse chirp. Methods to detect such a signal are being developed [31,6,10].

Although it was originally assumed that superfluidity would kill off the instability, so little is known about superfluids that this may not be the case. If the r-mode instability can occur in older cold stars, it could limit the angular velocity of neutron stars in low mass x-ray binaries [5,3,20] or in white dwarfs [17,21].

VI QUESTIONS FOR FUTURE RESEARCH

The original calculations which indicated that the r-mode version of the gravitational-radiation–driven instability is of interest depend on a number of simplifying assumptions. This leaves open many questions which should be answered in order to assess the true astrophysical implications. I will conclude this review with a list of questions which I feel are important and include some comments on what progress has been made in each case.

1. *What is the form of the r-modes for rapidly rotating stars?* The form of the r-modes for rapidly rotating uniform density Newtonian stars (Maclaurin spheroids) has been found [22]. Another approach [27] has found the modes for more general equations of state in the slow rotation case and promises to provide the correct modes for rapid rotation. These two papers show that the correct form of the generalized r-modes for rapid rotation is a hybrid of polar and axial contributions.

2. *What is the generalization of the r-modes in general relativity?* Some confusion about relativistic r-modes exists, since it has been shown that if the leading order contribution is axial, the eigenvalue problem is singular [18,4]. The resolution to the problem is that the correct modes are hybrids similar to the Newtonian hybrid solution [27]. The method developed in reference [27] has been extended to relativistic stars with $M/R \leq 1/10$ and an algorithm to find the modes for stars of arbitrary compactness has been developed [26]. In addition, the modes have been found in the Cowling approximation for stars with a buoyant restoring force [19].

3. *What happens if the star's rotation is differential?* The r-modes are solutions assuming the star rotates rigidly. However, in a short period after birth, neutron stars rotate differentially.

4. *What is the effect of nonlinear fluid interactions?* The r-modes are solutions of linear perturbation theory, but the fluid perturbation equations are nonlinear. The nonlinear effects should become important when the amplitude of the

velocity perturbation is similar in size to the angular velocity of the star. If turbulence sets in, one would expect energy from long wavelength modes to be channelled to short wavelength modes and be dissipated by viscosity, effectively killing off the instability. Work is in progress on a perturbative approach to this problem, both when buoyancy forces are [29] and aren't [11] present. The other more direct way to consider this problem is to do a direct numerical evolution of the equations. The main problem with the direct approach is the numerical viscosity inherent to the usual schemes which would artificially obscure the physics. Some new approaches to the numerics [34] should provide answers soon.

5. *What is the effect of magnetic fields?* All work discussed so far has assumed that the neutron star has a negligible magnetic field. This assumption is good as long as the energy in the star's rotation is larger than the energy stored in the magnetic field. The possibility that the motion of the r-modes could distort and amplify the star's magnetic field has been explored [36,33]. In particular one calculation [33] suggests that if the r-modes develop in a rapidly rotating neutron star, then the magnetic field would be amplified to the point where the oscillations could not occur. This suggests that stars born rotating very fast might turn into magnetars (highly magnetic neutron stars), while those born with more moderate speeds would spin down via the r-mode gravitational radiation emission route. The key point would be to calculate the critical angular velocity separating the two regimes.

6. *What are the effects of superfluidity?* Work is in progress on this question [24]. One paper [20] suggests that runaway heating due to r-modes in superfluid neutron stars might occur.

7. *How fast do neutron stars cool?* The standard modified URCA cooling process was used in the simplest spin-down model [23]. Different cooling rates would change the details of the spin-down.

8. *What is the dependence on equation of state?* The r-modes have very small pressure and density perturbations, so their frequencies are almost equation of state independent. However, the composition of the star can change the viscosity of the fluid above the superfluid transition temperature. In the case of a strange star [28], (composed of up, down and strange quarks) the coefficient of bulk viscosity is large enough to damp out the instability.

9. *Can the r-mode instability trigger other fluid or plasma instabilities?* In plasma and fluid studies one type of instability usually triggers another.

Clearly, many questions remain to be answered. The resolution of these questions will lead to a better understanding of the spin evolution of neutron stars and of the gravitational radiation that rotating neutron stars emit.

REFERENCES

1. Andersson, N., 1998, ApJ 502, 708
2. Andersson, N., Kokkotas, K., & Schutz, B. 1999, ApJ 510, 846
3. Andersson, N., Kokkotas, K., & Stergioulas, N. 1999 ApJ 516, 307
4. Beyer, H. & Kokkotas, K. 1999, gr-qc/9903019
5. Bildsten, L. 1998, ApJ 501, L89
6. Brady, P.R. & Creighton, T. 1998, gr-qc/9812014
7. Chandrasekhar, S. 1970, Phys. Rev. Lett., 24, 611
8. Comins, N. & Schutz, B.F. 1978, Proc. R. Soc. Lond. A. 364, 211
9. Cutler, C. & Lindblom, L. 1987, ApJ 314, 234
10. Ferrari, V., Matarrese, S., & Schneider, R. 1999, MNRAS 303, 258
11. Flanagan, E.E., Schenk, K., & Wasserman, I. 1999, paper in preparation
12. Friedman, J.L. 1978, Commun. math. Phys. 63, 243
13. Friedman, J.L. 1978, Commun. math. Phys. 62, 247
14. Friedman, J.L. 1996, J. Astrophys. Astr. 17, 199
15. Friedman, J.L. & Morsink, S.M. 1998, ApJ 502, 714
16. Friedman, J. L. & Schutz, B. F. 1978, ApJ, 222, 281
17. Hiscock, W. 1998, gr-qc/9807036
18. Kojima, Y. 1998, MNRAS 293, 49
19. Kojima, Y. & Hosonuma, M. 1999, ApJ 520, 788
20. Levin, Y. 1999, ApJ 517, 328
21. Lindblom, L. gr-qc/9903042
22. Lindblom, L. & Ipser, J. 1999, PRD 59, 044009
23. Lindblom, L., Owen, B.J., & Morsink, S.M. 1998, PRL 80, 4843
24. Lindblom, L. & Mendell, G. 1999, paper in preparation
25. Lindblom, L., Mendell, G., & Owen, B.J. 1999, gr-qc/9902052
26. Lockitch, K.H. 1999, PhD Thesis, University of Wisconsin-Milwaukee
27. Lockitch, K.H., & Friedman, J.L. 1998, gr-qc/9812019
28. Madsen, J. 1998, PRL 81, 3311
29. Morsink, S.M. 1999, paper in preparation
30. Morsink, S.M., Stergioulas, N., & Blattnig, S.R. 1999, ApJ 510, 854
31. Owen, B.J. et al. 1998, PRD 58, 084020
32. Penrose, R. 1969, Rev. del Nuovo Cimento, 1, 252
33. Rezzolla, L., Lamb, F.K., & Shapiro, S.L. 1999, preprint
34. Rezzolla, L. et al. 1999, gr-qc/9905027
35. Shapiro, S.L., & Teukolsky, S.A. 1983, *Black Holes, White Dwarfs, and Neutron Stars*, (John Wiley & Sons, New York)
36. Spruit, H. 1999, A&A 341, L1
37. Stergioulas, N. 1998, Living Reviews in Relativity 1988-8, http://www.livingreviews.org/Articles/Volume1/1998-8stergio
38. Stergioulas, N & Friedman, J.L. 1998, ApJ 492, 301
39. Wald, R.M. 1984, *General Relativity*, (University of Chicago Press, Chicago, IL)
40. Whiting, B.F. 1989, J. Math. Phys. 30, 1301

What Will We Learn from the Detection of Gravitational Waves?

Peter R. Saulson

Department of Physics, Syracuse University, Syracuse, NY 13244-1130, USA

Abstract. Gravitational waves are the most relativistic of the predictions of general relativity, and yet are one of the last of its experimental predictions to be fulfilled. In this article, I review the nature of gravitational waves, the methods by which we hope to detect them, what we will learn from their detection, and the prospects for the field.

THE NATURE OF GRAVITATIONAL WAVES

The history of our understanding of gravitation is marked by concern with questions about whether gravitational effects can be transmitted instantaneously across the distance separating two massive bodies. Newton's theory appeared to require instantaneous action at a distance, but this feature caused him great anguish. When he could not expunge it from his view of gravitation, he took refuge in the idea that it was not his job to "make hypotheses" how gravity *worked*, only to explicate how it appeared to *behave*.

Einstein's relativistic worldview revived concern with gravitational action at a distance, since the relativistic picture of causality required that no signal of any kind could travel faster than the speed of light. If Newton's theory were literally true, then a sufficiently sensitive measurement of the gravitational field could be used as the receiver in a gravitational radio system with no propagation delays. All that the sender would have to do would be to make a change in the distribution of matter in her vicinity, and others throughout the universe could get the news without having to wait.

This, as much as any other feature of gravitation, pointed out the need to bring gravity explicitly into the framework of relativistic physics. General relativity was the result of this effort. This tremendously rich and beautiful theory was built on the basis of other physical insights (e.g. that bodies move on geodesics through curved space-time), so it must have been a wonderful confirmation that gravitational waves appear in it in a completely natural way; the weak field version of the vacuum field equations (in the transverse traceless gauge) is just the wave equation.

The list of classic tests of general relativity didn't include gravitational waves, however. There is a good reason for this, of course, no matter how central grav-

CP493, *General Relativity and Relativistic Astrophysics*, edited by C. P. Burgess and R. C. Myers
© 1999 American Institute of Physics 1-56396-905-X/99/$15.00

itational waves are to the intellectual structure of the theory. Even though all of the famous tests involved small effects, the feebleness of gravitational waves in our universe is of another order altogether. I will discuss below a nice way to estimate the amplitude of gravitational waves likely to arrive at our detectors.

As gravitational waves got lost in the excitement over other dramatic confirmations of general relativity, a curious thing happened. Many began to doubt the existence of gravitational waves, even in principle. (For a time, their number even included Einstein himself!) This doubt was part of a broader confusion over one of the subtlest aspects of G.R., the question of how to identify which predictions of the theory were genuinely physical (i.e. gauge invariant), as opposed to coordinate artifacts. This idea was expressed in Eddington's quip that some people thought that gravitational waves "traveled at the speed of thought." [1] The doubts were finally laid to rest only by Bondi's 1957 paper that showed that a suitable detector could absorb energy from a gravitational wave. [2]

We can shake our heads in puzzlement at our forebears, and see that in many ways gravitational waves are like their much more familiar cousins, electromagnetic waves. Both are transverse disturbances that travel at the speed of light. The "disturbance" of the gravitational wave is a quadrupolar strain field $h(t)$ that affects all freely-falling bodies equally, as demanded by the Equivalence Principle. (See Figure 1.) The observable effect of the wave is the change x in the distance L between two such free bodies,

$$x(t) = \frac{1}{2}Lh(t). \tag{1}$$

Detection could involve measuring the strain in an extended body, or more ideally the relative motion of widely separated bodies (since the greater is L, the greater will be the relative motion for a given wave amplitude h.)

GENERATORS OF GRAVITATIONAL WAVES

It remains to be seen whether the waves which are observable in principle are observable in practice. That depends in part on the strength of gravitational wave sources. The Equivalence Principle (combined with the conservation of linear and angular momentum) ensures that dipoles cannot radiate, but time-varying mass quadropole moments do exist, and radiate much as do their electromagnetic counterparts. The connection between source and wave in this case is given by the Quadrupole Formula

$$h_{\mu\nu} = \frac{2G}{c^4 R}\ddot{I}_{\mu\nu}. \tag{2}$$

This bears a strong resemblance to the Larmor formula of electromagnetism, with the field proportional to the time derivative of the moment and inversely proportional to the distance R from source to measurment point. But the dimensionful

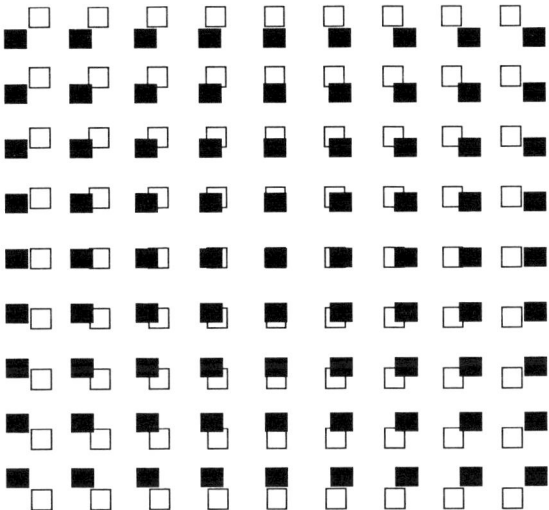

FIGURE 1. Displacement of a set of freely-falling masses by a gravitational wave.

prefactor $2G/c^4$ has the distressingly small SI value of 10^{-44}. This is an indication that gravitational waves are very weak. The strongest laboratory quadrupoles one can imagine making (multi-ton steel bars rotating at break-up speed) yield strains of order 10^{-38}, weak indeed.

Evidently, if gravitational waves are to be detectable, they must be generated by stronger sources than we can build in the lab. Astronomical objects may play this role. Consider a binary star system, where the two stars orbit one another under the influence of gravity. As long as they are not too close to one another, then Newtonian gravity works fine, and we can derive a simple expression for the amplitude of the quasi-sinusoidal $h(t)$ generated by the time-varying quadrupole moment of the system. [3] We find

$$h = \frac{r_{S1} r_{S2}}{aR},\qquad(3)$$

where r_{S1} and r_{S2} are the Schwarzschild radii of each of the stars, and a is their separation. (A nice feature of this expression is that as a ratio of products of lengths it is dimensionless on its face, unlike the more general Quadrupole Formula.)

Under what circumstances will h be large? Clearly, when massive stars orbit close to one another, as close to us as we can find them. As an example, let's make the number for a neutron star binary in the Virgo Cluster. Toward the end of the lifetime of the system as the stars spiral together due to gravitational radiation reaction, the two stars can come so close that $r_{S1} r_{S2}/a$ is of order 1 km. The distance to the Virgo Cluster is of order 10^{21} km, so we might hope for a signal of amplitude 10^{-21}. This is much better than any terrestrial generator, by 17 orders of magnitude in wave amplitude (34 in power flux!), even across astronomical distances. Just as remarkably, the peak of this signal will come at a frequency between 100 Hz and 1 kHz (in the middle of the audio band), a frequency range at which it is possible to carry out sensitive measurments with terrestrial gravitational wave detectors.

There is of course the question of how often neutron star binary coalescences occur as close as the Virgo Cluster. Most likely, we'll have to see farther, perhaps another factor of 10 in distance (and thus a factor of 10 lower in amplitude) in order to register one event or more per year.

The lesson of this exercise should be clear. No one should aspire to mimic what Hertz did for electromagnetic waves: build a wave generator on one side of his lab, set up a receiver at the other, and then demonstrate that waves with the expected properties traveled from generator to receiver. [4] The sensitivity required for such an experiment would be impossible to achieve. Instead, using astronomical generators, the sensitivity required is challenging but possible to achieve, as I will show below.

A corollary is that physics issues and astronomical questions tend to get mixed up together in the business of gravitational wave detection. While astronomy rescues us from our inability to make strong enough generators in the lab, this means that the strengths of waves arriving at our receiver (as well as other properties like arrival time and rate, gravitational waveform, and polarization) are out of the experimenters' control. It is impossible to be simply a gravitational wave physicist; one must also be a gravitational wave astronomer. But then again, this gives a depth of interest to the quest for gravitational waves that it might not otherwise have. Even if all of the strictly gravitational physics questions should be answered exhaustively, the enterprise will continue to pay dividends as a branch of astronomy.

WHAT WE MIGHT LEARN

Thus there is a large list of things we might learn, once we succeed in detecting gravitational waves from cosmic sources. It includes:

1. Final demonstration that gravitational waves propagate away from sources and interact with detectors as expected,

2. Verification of a wide range of wave properties, such as waveform, speed and polarization,

3. Demonstration of the existence of astrophysical black holes,

4. Exploration of strong gravity phenomena in the orbits of neutron stars and the oscillations of black holes,

5. Obtain astronomical information concerning the abundance of neutron stars and black holes,

6. Gain insight into the process of stellar core collapse,

7. Test models for the production of gamma-ray bursts, and

8. Search for Planck-era information in the form of a gravitational wave background.

The first two items are classic wave physics tests of the Hertzian style, albeit displaced outside the lab for the reasons discussed above. Since the discovery of the Hulse-Taylor Binary Pulsar, [5] there can be little doubt about the existence of gravitational waves. Still, simply as a matter of good scientific craftsmanship, it is incumbent upon us to see gravitational waves as fully time-dependent waves, and not merely as a time-averaged flux of energy and angular momentum from a generator. Similarly, while it would be astonishing to find departures from theoretical predictions about the detailed properties of the waves, we have to check nevertheless.

A useful analogy from the history of physics is the long interval between the discovery of the neutrino's existence in β-decay and the discovery of individual neutrinos themselves. While no one doubted their existence, they were more "ghostly" than they ought to have been until seen directly. And, once it became possible to detect neutrinos, a vast new area of physics became experimentally accessible.

The question of the existence of black holes has a somewhat analogous character. The astrophysical evidence for the existence of black holes is strong and growing stronger, for black holes in several different mass ranges. But no one would claim that the proof is as strong as that of the Hulse-Taylor pulsar for the existence of gravitational waves. With the possible exception of the new arguments about advection-dominated accretion flows, [6] the astrophysical case for black holes has only to do with highly concentrated non-luminous mass, and nothing to do with the spacetime structure that makes a black hole a black hole. Detection of the characteristic gravitational wave signature of quasi-normal mode ringing would not only clinch the case for black holes' existence, but it would immediately make possible an investigation into strong-gravity phenomena that is completely impossible by any other means. The properties of the strongly damped sinusoidal gravitational waveform are inextricably linked to the character of the strongly curved spacetime immediately outside the black hole horizon, and to the existence of the horizon itself.

This is just one example of the way in which successful detection of gravitational waves will open vistas onto regimes of strong-gravity phenomena. As the orbit of a neutron star binary shrinks, it will cross over from a quasi-Newtonian regime into one where general relativity is essential. The final collision and ringdown of these systems will provide a rich laboratory of relativistic physics, whether or not a black hole is the final state.

Beyond measurements related to gravitational physics, gravitational wave detection will provide the foundation for a whole new branch of astronomy, yielding information about the abundance of collapsed objects, glimpses into the process of stellar core collapse, and tests of various models of the generation of gamma ray bursts. As with other new branches of astronomy, it is impossible to predict what might be the most interesting discoveries, only that it is likely that among the first events there will be some surprises.

A long shot possibility with tremendously high payoff would be a detection of a cosmic background of gravitational waves. There are a variety of possible mechanisms by which such a background might be generated. [7] Most of the predictions have large uncertainties. Regardless, such a signal will be eagerly sought. As good a reason as any is that gravitational waves hold out the promise of peering through the otherwise impenetrable wall of the "surface of last scattering" that serves as the source of the Cosmic Background Radiation. Before decoupling, the Universe was opaque to electromagnetic radiation, making extension of ordinary (electromagnetic) astronomy to earlier times impossible even in principle. Whatever information might be revealed by gravitational waves emerging from that hidden epoch would be of tremendous value. This includes the possibility that one might be able to observe the Universe at (or not long after) the Planck era.

DETECTING GRAVITATIONAL WAVES

As Figure 1 illustrates, the measurable effect of a gravitational wave on a set of freely-falling masses is the change in their separation in one direction, accompanied by the opposite change in separation in the perpendicular direction. This pattern of motion is well suited to measurement by the beautiful and versatile device known as the Michelson interferometer. An idealized version is shown in Figure 2.

A Michelson interferometer can be thought of as a transducer whose input is the difference in the length of its two arms, and whose output is the amount of light power which exits from one side of the beam splitter. Depending on whether the arms are precisely matched in length or have a slight mismatch, the light power leaving the output port can vary from the full power entering the interferometer to zero; the full range is swept out over a change in path length difference of one half the wavelength of the light in the interferometer. Since light has such a small wavelength, an interferometer is a ruler whose "tick marks" are very closely spaced indeed.

Imagine a gravitational wave incident on a Michelson interferometer with 4 kilometer arms, and ask what effect that would have on light that traveled once down its arms and back. It would take fractional length changes of about a part in 10^{10} to change the output from bright to dark. But we aspire to sensitivities to changes of only 10^{-21}, or even finer.

Evidently, more cleverness is required. A couple of orders of magnitude can be gained by causing the light to make multiple round trips in the arms, thus amplifying the effect of the wave by making the light take a longer path. In most gravitational wave detectors, this is achieved by replacing each interferometer arm with a resonant Fabry-Perot cavity. This is great, but can't improve things without limit; once the light spends more than half a gravitational wave period in the arms, there is no more increase in signal.

30

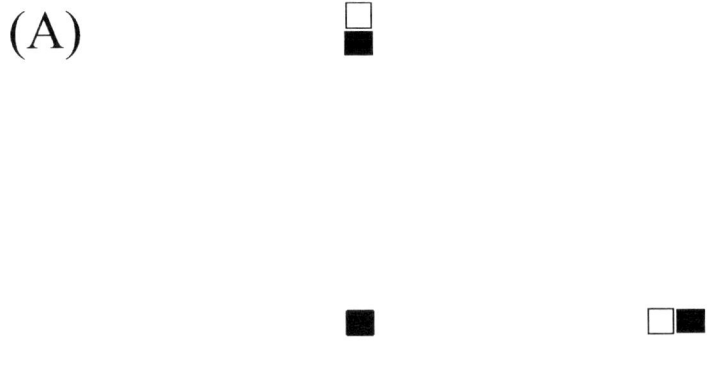

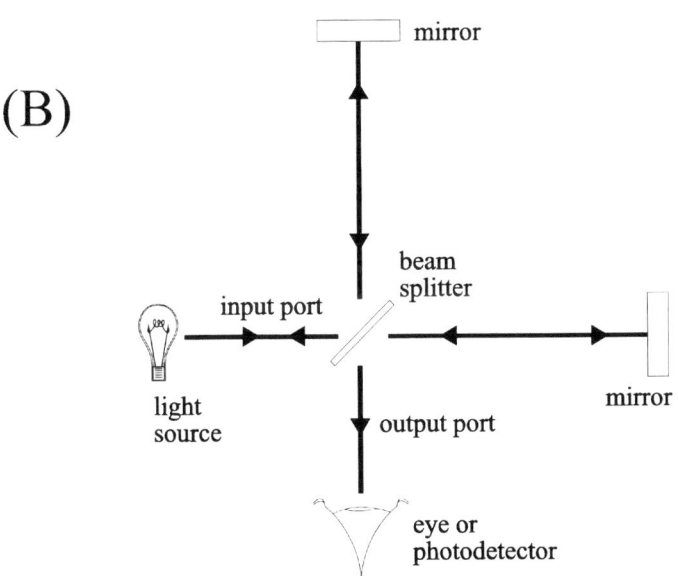

FIGURE 2. (A) An 'L' of three freely-falling masses, with their positions distorted by a gravitational wave. (B)The vertex mass becomes a beam splitter, and the other two masses become mirrors, thus allowing a measurement of the shift in positions by means of a Michelson interferometer.

Most of the sensitivity gap is made up by making very precise measurements of the optical output power, to determine where exactly "on the fringe" one is sitting. In round numbers, we need to determine the arm length difference to an amount equivalent 10^{-9} radians of optical phase difference in the light from the two arms. This corresponds to a power measurement of a part in 10^9. Is this possible? Yes. If a power measurement is made by accumulating 10^{18} photons (corresponding to of order 10 Watts of light), then Poisson statistics will allow sufficient precision. It is still a technological *tour de force* to achieve this level of sensitivity, but such precision has now been demonstrated. [8]

While this demonstrates that the required sensitivity is allowed, making gravitational wave measurements with the required precision involves other challenges as well. The mirrors that define the arms of the Michelson interferometer must also play the role of the freely-falling masses of general relativity. But other effects besides gravitational waves can cause objects to move. The interferometer's test masses must be rigorously isolated from the effects of the ambient vibrations of the laboratory environment (seismic waves and acoustic vibration). Even Brownian motion might easily swamp the 10^{-19} meter motions that are sought, unless extraordinary precautions are taken to limit its magnitude.

LIGO, VIRGO, GEO, TAMA

The considerations above are embodied in a new generation of gravitational wave detectors that will come on line in the next few years. In the United States, a pair of 4 km long interferometer facilities has been built by the Laser Interferometer Gravitational-Wave Observatory (LIGO) Project. [9] Located at Hanford, Washington and Livingston, Louisiana, these sites will hold a total of three interferometers (the Hanford vacuum system will house a 2 km long interferometer in addition to its 4 km interferometer.) Coincident operation of these instruments will allow a search for gravitational wave impulses, periodic signals, or a stochastic background. Its first observation run, at a sensitivity of 10^{-21}, is planned to start in 2002.

A single 3 km long interferometer is now under construction at Cascina, Italy by the French-Italian VIRGO collaboration. [10] Its design includes special features to make low frequency noise as low as possible, by the construction of elaborate seismic filters. It is expected to come on line not much later than LIGO.

The British-German GEO Project has built a 600 meter interferometer facility near Hannover. [11] While its shorter length is undoubtedly a handicap, GEO will remain competitive for some time by aggressive application of the most advanced ideas for optical sensitivity enhancement and for reduction of mechanical noise.

A 300 meter interferometer has been built in the outskirts of Tokyo by the TAMA project. [12] In addition to an astronomical role for this instrument, it is hoped that it will serve as a prototype for a future 3 km long interferometer in Japan.

These instruments will soon be scanning the skies with a precision nearly 3 orders of magnitude in amplitude (5 to 6 orders of magnitude in power) more sensitive than the best previous gravitational wave observations. Their sensitivity is just at the level at which simple order-of-magnitude estimates (like the one given above) say one should expect signals to be detected.

Whether or not success comes immediately will depend on details of the composition of our universe that are not entirely certain. If the universe contains no objects more exotic than the known neutron star binaries, then detections may not come until sensitivity is improved by another order of magnitude. But there is much room for surprise, perhaps from as-yet-undetected black hole binaries, or from whatever is behind the enigmatic gamma-ray bursts. The recent history of astronomy gives courage to the optimists, since new observing techniques have typically been blessed with surprisingly strong signals from unsuspected objects.

But just to be certain, the developers of the present generation of interferometers have already begun work on upgraded versions to be retrofitted into the new facilities. For example, LIGO is putting together a plan for a new set of interferometers for installation at mid-decade, which would add enough sensitivity to reach the range at which neutron star binary signals are expected. Thus, it seems very likely that gravitational waves from astronomical sources will be detected by the end of the next decade. When that happens, a new era in gravitational physics and astrophysics will have begun.

REFERENCES

1. Eddington, A. S., *The Mathematical Theory of Relativity*, Cambridge: Cambridge University Press, 1923, p. 130.
2. Bondi, H., *Nature* **179**,1072–73 (1957). This article is illuminatingly candid about the confusion that surrounded this subject in 1957.
3. Kafka, P., in *Space Science and Fundamental Physics* Proceedings of Summer School held at Alpbach, Austria, 28 July – 7 August 1987, ESA SP-283, May 1988.
4. Hertz, H. *Electric Waves*, London: Macmillan, 1893.
5. Hulse, R. A., and Taylor, J. H., *Ap. J. Lett.* **195**, L51–L53 (1975).
6. Narayan, R., Garcia, M. R., and McClintock, J. E., *Ap. J. Lett.* **478**, L79–L82 (1997).
7. Allen, B., and Brustein, R., *Phys. Rev. D* **55**, 3260–64 (1997); Allen, B., and Ottewill, A. *Phys. Rev. D* **56**, 545–63 (1997).
8. Fritschel, P., Gonzalez, G., Lantz, B., Saha, P., and Zucker, M., *Phys. Rev. Lett.* **80**, 3181–84 (1998).
9. Abramovici, A., Althouse, W. E., Drever, R. W. P., Gürsel, Y., Kawamura, S., Raab, F. J., Shoemaker, D., Sievers, L., Spero, R. E., Thorne, K. S., Vogt, R. E., Weiss, R., Whitcomb, S. E., and Zucker, M. E., *Science* **256**, 325–33 (1992).
10. Bradaschia, C., Del Fabbro, R., Di Virgilio, A., Giazotto, A., Kautzky, H., Montelatici, V., Passuello, D., Brillet, A., Cregut, O., Hello, P., Man, C. N., Manh, P. T., Marraud, A., Shoemaker, D., Vinet, J.-Y., Barone, F., Di Fiore, L., Milano, L., Russo, G., Solimeno, S., Aguirregabiria, J. M., Bel, H., Duruisseau, J.-P., Le Denmat, G., Tourrenc, P., Capozzi, M., Longo, M., Lops, M., Pinto, I., Rotoli, G.,

Damour, T., Bonazzola, S., Marck, J. A., Gourghoulon, Y., Holloway, L. E., Fuligni, F., Iafolla, V., and Natale, G., *Nucl. Inst. Meth. Phys. Res.* bf A289, 518–25, (1990).

11. Danzmann, K., Chen, J., Nelson, P. G., Niebauer, T. M., Rüdiger, A., Schilling, R., Schnupp, L., Strain, K. A., Walther, H., Winkler, W., Hough, J., Campbell, A. M., Cantley, C. A., Logan, J. E., Meers, B. J., Morrison, E., Newton, G. P., Robertson, D. I., Robertson, N. A., Rowan, S., Skeldon, K. D., Veitch, P. J., Ward, H., Welling, H., Aufmuth, P., Kröpke, I., Ristau, D., Hall, J. E., Bennett, J. R. J., Corbett, I. F., Edwards, B. W. H., Elsey, R. J., Greenhalgh, J. R. S., Schutz, B. F., Nicholson, D., Shuttleworth, J., Ehlers, J., Kafka, P., Schäfer, G., Braun, H., and Kose, V. *Lecture Notes in Physics* **410**, 184–209, (1992).

12. Tsubono, K., and the TAMA Collaboration, in *Gravitational Wave Detection*, K. Tsubono, M.-K. Fujimoto, and K. Kuroda, eds., Tokyo: Universal Academy Press, 1997, pp. 183–92.

The Characteristics of Colliding Black Holes

Jeffrey Winicour

Department of Physics and Astronomy,
University of Pittsburgh, Pittsburgh, PA 15260

Abstract. I review the characteristic initial value problem, its implementation as a robust computational algorithm for a 4-dimensional vacuum space-time (the PITT NULL CODE) and its application to the calculation of gravitational waveforms emitted by black holes.

I describe the potential applications of the code to the binary black hole problem, via Cauchy-characteristic matching or pure characteristic evolution. In particular, the event horizon is itself a characteristic hypersurface and can be treated by characteristic methods as a stand-alone object. This allows an analytic treatment of the intrinsic geometry of the event horizon for colliding black holes which produces the pair-of-pants horizon found in the numerical simulation of the head-on-collision of black holes and the initially toroidal event horizon found in the simulation of a collapsing, rotating cluster.

Most previous studies of black hole formation and merger are restricted to the axisymmetric case. However, axisymmetric horizons, like the Schwarzschild horizon, are non-generic. When applied to a *non*-axisymmetric horizon, the characteristic approach reveals substantially new features. In particular, coalescing black holes generically go through a toroidal phase before they become spherical. In a bigger picture, this analytic model of the event horizon provides part of the data for a simulation of the space-time exterior to a binary merger by means of the null code.

NULL CONE EVOLUTION

An effort started many years ago [1] to develop an evolution code based upon the characteristic initial value problem has recently culminated in a highly accurate, efficient and robust code - the PITT NULL CODE [2]. The code implements the pioneering work of Bondi [3] and Penrose [4] as a new computational algorithm for general relativity based upon null cones. Because null cones are generated by the characteristic rays of the theory this initial value problem offers several advantages for numerical investigations. Some of the fruits of these investigations relevant to black hole physics will be described here.

In null cone coordinates, Einstein's equations contain no elliptic constraints and reduce to propagation equations along the radial light rays, which can be inte-

CP493, *General Relativity and Relativistic Astrophysics*, edited by C. P. Burgess and R. C. Myers
© 1999 American Institute of Physics 1-56396-905-X/99/$15.00

grated in hierarchical order for one variable at a time. We have turned this into a highly efficient characteristic marching algorithm in which evolution to the next grid point is carried out explicitly. Also, because no second time derivatives of the metric enter in the equations for characteristic evolution the number of variables is half that for the Cauchy problem. There is a single complex evolution variable which encodes the free degrees of freedom of the gravitational field and four auxiliary variables. The use of a compactified grid, based upon Penrose's conformal description of null infinity, removes the necessity of an artificial outgoing radiation condition and makes possible a rigorous description of geometrical quantities such as the Bondi mass and news function. The news function supplies both the true waveform and polarization of the gravitational radiation incident on a distant antenna. Furthermore the grid domain is exactly the region in which waves propagate, which is ideally efficient for the purpose of radiation studies. Since each null cone extends from the source to null infinity, the radiation appears immediately with no need for numerical evolution to propagate it across the grid. Because the null cones extend to infinity, the growth of a large redshift offers the bonus of forecasting event horizon formation.

Although the technique of shooting along characteristics is a standard computational tool in one spatial dimension, use of characteristic hypersurfaces as the underlying foliation for numerical evolution in higher dimensions is exclusive to relativity [5]. The basic approach is applicable to any of the hyperbolic systems occurring in physics, such as the wave equation, electromagnetic theory and hydrodynamics.

For general relativity, the computational grid is based on coordinates constructed from a family of outgoing null hypersurfaces emanating from a worldtube of topology $S^2 \times R$. Let u label these hypersurfaces, x^A ($A = 2,3$) be angular coordinates for the null rays and r be a surface area distance. In the resulting $x^\alpha = (u, r, x^A)$ coordinates, the metric takes the Bondi-Sachs form [3,6]

$$
ds^2 = -\left(e^{2\beta}\frac{V}{r} - r^2 h_{AB}U^A U^B\right)du^2 - 2e^{2\beta}dudr - 2r^2 h_{AB}U^B dudx^A
$$
$$
+ r^2 h_{AB}dx^A dx^B, \tag{1}
$$

where $h^{AB}h_{BC} = \delta^A_C$ and $det(h_{AB}) = det(q_{AB}) = q(x^A)$, with q_{AB} a unit sphere metric. For purposes of including null infinity as finite grid points, the code uses a compactified radial coordinate.

Note that the traditional $3 + 1$ decomposition of space-time used in the Cauchy formalism is not applicable here because the foliation by null hypersurfaces of constant u has a degenerate 3-metric and null normal. However, an analogous $2 + 1$ decomposition can be made on the timelike worldtube of constant r, which has intrinsic metric

$$
{}^{(3)}ds^2 = -e^{2\beta}\frac{V}{r}du^2 + r^2 h_{AB}(dx^A - U^A du)(dx^B - U^B du). \tag{2}
$$

In this form, we can identify $r^2 h_{AB}$ as the metric of the surfaces of constant u which foliate the worldtube, $e^{2\beta}V/r$ as the square of the lapse function and $(-U^A)$ as the shift vector.

A Schwarzschild geometry in outgoing Eddington-Finklestein coordinates is given by the choice $\beta = 0$, $V = r - 2m$, $U_A = 0$ and $h_{AB} = q_{AB}$. In the nonspherically symmetric case, in order to computationally treat derivatives of tensor fields on the sphere, we introduce two stereographic coordinate patches with complex unit sphere dyad vector satisfying

$$q_{AB} = \frac{1}{2} \left(q_A \bar{q}_B + \bar{q}_A q_B \right), \tag{3}$$

This allows use of the spin weight operators $\eth$ and $\bar{\eth}$ [7] to express the covariant derivative ∇_A (associated with q_{AB}) of a tensor field on the sphere as spin-weighted fields. Our computational eth formalism [8] allows robust, accurate use of covariant derivatives in spherical coordinates.

The conformal 2-metric h_{AB} can be represented by its dyad component $J = h_{AB} q^A q^B/2$. The role of h_{AB} as the null hypersurface data for the characteristic initial value problem can thus be transferred to J. The Einstein equations impose no constraints so that the complex function J encodes the two free degrees of freedom of the gravitational field of the null hypersurface.

The evolution algorithm is a computational version of the mixed initial value problem based upon a worldtube and a null hypersurface [9]. Consider a convex worldtube whose interior contains the sources and whose exterior is an asymptotically flat region of spacetime. If such a worldtube is sufficiently large it admits a slicing whose outgoing normal null hypersurfaces extend to infinity without developing caustics. Given J on the initial null hypersurface, Einstein's equations propagate data on the world tube along the outgoing characteristics to determine the exterior space-time. The required worldtube data consists of the conformal 2-geometry of a foliation of the the worldtube and the mass and angular momentum aspects of the initial slice. There are constraint equations on the worldtube which are hyperbolic versions of the elliptic constraints of the Cauchy problem. These constraints are generalized mass and angular momentum conservation laws. In addition, the lapse and shift associated with the foliation of the world tube represents gauge freedom which must be specified.

The worldtube can be eliminated by shrinking it to a nonsingular worldline (in which case the conservation equations reduce to regularity conditions). The code was first implemented this way in its original axisymmetric form [10]. However, this has high computational cost due to the Courant-Friedrich-Lewy condition which requires that the physical domain of dependence be smaller than the domain of dependence determined by the numerical algorithm. For a null cone algorithm, this forces a small time step near the vertex of the null cone, which makes 3-dimensional evolution computationally unfeasible on a uniform grid. This perhaps can be circumvented by an adaptive grid but at present, 3-dimensional null cone evolution is feasible only exterior to a worldtube. The world tube can be null, as well as timelike, and this allows important application to black holes.

Given worldtube data and initial data, the evolution provides the waveform at future null infinity $\mathcal{I}^+$ - first in a super-accelerated frame determined by the gauge conditions adopted on the world tube and then converted to an asymptotic inertial frame to give the Bondi news function, whose real and imaginary parts are the standard plus and cross polarization modes. This has been tested to be second order accurate in grid size in a wide number of test beds including linearized waves and nonlinear waves propagating outside a black hole (which are independently constructed by solving the Robinson-Trautman equation) [2].

APPLICATION TO BLACK HOLES

The Pitt Null Code is a robust, accurate new tool for the simulation of highly curved space-times. As historically asked with the discovery of new solutions to Einstein's equations: "What do you do with them?" The subtle part of this question is the determination of physically relevant worldtube data.

Nonlinear scattering off a black hole

One simple choice of inner worldtube is supplied by the ingoing $r = 2m$ hypersurface in a Schwarzschild spacetime (the white hole horizon). The associated worldtube data automatically satisfies the conservation conditions. With this worldtube data, we pose the nonlinear problem of gravitational wave scattering off a Schwarzschild black hole by setting initial data on an outgoing null hypersurface extending to $\mathcal{I}^+$ consisting of an ingoing pulse of compact radial support with various angular modes.

We have calculated the news function radiated from this system [2]. As expected, for small pulse amplitude the news scales linearly but, as the amplitude increases it shows a stronger than linear dependence. In the perturbative regime, the news results from the backscattering of the incoming pulse off the effective potential of the interior Schwarzschild black hole [11]. However, in the nonlinear regime, not only does the news scale much stronger than linearly with amplitude but the waveform also reveals nonlinear generation of additional modes. In this regime, the mass of the system is dominated by the incoming pulse, which essentially backscatters off itself in a nonlinear way. In the extreme nonlinear regime the total back scattered energy radiated to $\mathcal{I}^+$ is itself much larger than the mass of the interior black hole hole. In dimensionless units the Bondi news = 400, corresponding to a radiation power of 10^{13} solar masses/sec.

Waveforms from matter falling into a black hole

We have incorporated a crude hydrodynamic code for matter, in the form of a perfect fluid into the null code. The combined 3-dimensional null code has been tested for stability and accuracy to verify that nothing breaks, at least in the regime that hydrodynamical shocks do not form. The results establish the feasibility of a combined characteristic gravity and matter evolution [12].

The tests include a localized blob of matter falling radially into a black hole. The peak of the blob follows a geodesic of the background spacetime, with the waveform of the emitted gravitational radiation monitored at $\mathcal{I}^+$. This simulation is a prototype of a neutron star orbiting a black hole. Thus a refined characteristic hydrodynamic code would open the way to explore important problems in black hole astrophysics. Recently, a high resolution shock-capturing code has been successfully implemented in the null formulation in the case of spherical symmetry [13] and the code is being generalized to 3-dimensions.

Horizon tracking and dynamical singularity excision

Characteristic evolution can also be based upon a family of *ingoing* null cones with data given on a worldtube at their *outer* boundary and on an initial *ingoing* null cone. We use this system to evolve black holes in the region interior to the worldtube by locating a marginally trapped surface (MTS) on the ingoing cones and excising the singular region inside it [14]. This is a characteristic version of the conventional strategy for evolving black holes by excising the interior of an apparent horizon, as initially suggested by W. Unruh [15]. The ingoing null code locates the MTS, tracks it during the evolution and stably excises its interior from the numerical grid.

We used this code to simulate a distorted "black hole in a box" [16]. Worldtube data was induced from a Schwarzschild or Kerr spacetime but the worldtube was allowed to move relative to the stationary trajectories; i.e. with respect to the grid the worldtube is fixed but the black hole moves. The initial null data consisted of a pulse of radiation which subsequently travels outward to the worltube where it is reflected back toward the black hole. The approach of the system to equilibrium was monitored by the area of the MTS, which is equivalent to its Hawking mass. When the worldtube is stationary (static or rotating in place), the distorted black hole inside evolves to equilibrium with the boundary. A boost or other motion of the worldtube with respect to the black hole does not affect this result. The marginally trapped surface always reaches equilibrium with the outer boundary, confirming that the motion of the boundary is "pure gauge".

The code essentially runs "forever" even when the worldtube wobbles with respect to the black hole to produce artificial periodic time dependence. "Forever" cannot be rigorously attained in any finite simulation but we appeal to the characteristic time necessary to obtain accurate waveforms for the inspiral and merger of two black holes. The inspiral from a a post-Newtonian orbit at $r = 20M$ to the innermost stable orbit at $6M$ lasts $\approx 10,000M$. We have successfully evolved an initially distorted, wobbling black hole for a time of $60,000M$, clearly as long as needed for a smooth transition from the post-Newtonian regime to merger, if this success could be duplicated in the ultimate binary black hole code [17]. This capability is important because the coordinates used to simulate a binary may not become exactly stationary after merger and ring-down to final equilibrium.

This is the most demanding black hole simulation by any code to date. Results can be viewed at http://artemis.phyast.pitt.edu/animations.

Cauchy-Characteristic-Matching (CCM)

The sole weakness of characteristic evolution is its limitation to regions admitting a nonsingular foliation by null cones. This breaks down when focusing effects introduce caustics or crossovers. Caustics result not just from the lensing effect of strong curvature but also the location of the lens with respect to the vertex. In a spacetime with negligible curvature containing, say, two peanuts, no global null cones exist if the peanuts are sufficiently far apart (approximately 10^{10} light years). Yet in quasispherical spacetimes with strong curvature, global null cones with point vertices exist (or end on physical singularities) and they exist for a binary neutron star with orbital separation less than 5 neutron star radii.

Given the appropriate worltube data for a binary system in its interior, characteristic evolution can supply the exterior spacetime and radiated waveform. But determination of the worldtube data for the complete evolution of a binary system requires the Cauchy evolution of the interior. CCM is a matched Cauchy-characteristic evolution designed to tackle such radiation problems [18]. The two evolutions are matched across a worldtube, with the Cauchy domain supplying the boundary values for characteristic evolution and vice versa. Just as several coordinate patches are necessary to describe a spacetime with nontrivial topology, an effective attack on the binary black hole problem is to use CCM to patch together regions of spacetime handled by different algorithms.

Because of the singular time dependence of the compactified version of spatial infinity, a globally compactified approach to the Cauchy problem is not feasible numerically. Instead, the grid is terminated at a finite boundary, where the necessity of an artificial boundary condition, such as the Sommerfeld condition, leads to strong back reflection in the case of high asymmetry. Here the strengths and weaknesses of the Cauchy and characteristic approaches complement themselves in a fortuitous way.

The potential advantages of CCM over traditional artificial boundary conditions are: (1) Accurate waveform and polarization properties at infinity; (2) Computational efficiency for radiation problems in terms of both the grid domain and algorithm; (3) Elimination of an artificial outer boundary condition on the Cauchy problem, which eliminates contamination from back reflection and clarifies the global initial value problem; and (4) A global picture of the spacetime exterior to the horizon. These advantages have been realized in the following two important test cases.

First, CCM has been successfully implemented for nonlinear scalar waves propagating in 3-dimensional Euclidean space, matching a spherical null grid to a Cartesian Cauchy grid [19]. Performance of the matching algorithm was compared with the prime examples of both local and nonlocal radiation boundary conditions proposed in the computational physics literature. For linear problems, CCM outperformed all local boundary conditions and was about as accurate (for the same grid resolution) as the best nonlocal conditions. However, since the computational cost of conventional nonlocal conditions is many times that of matching, the matching algorithm can be used with finer grids, yielding a significantly higher final accuracy. For strongly nonlinear problems, matching was *significantly more accurate* than all other methods tested. This is because all other currently available outer boundary conditions are based on linearizing the equations in the far field, while CCM consistently takes nonlinearity into account in both interior and exterior regions.

Second, in the context of black hole spacetimes, a two-fold version of CCM has been used to evolve globally the spherical collapse of a self-gravitating scalar field onto a black hole [14]. Null evolution on ingoing light cones, bounded on the inside by a dynamically tracked MTS, are matched at their outer boundary to the inner boundary of a Cauchy evolution. In turn, the outer boundary of the Cauchy evolution is matched to an exterior characteristic evolution extending to infinity. This study reveals further advantages of CCM for dealing with black holes. Because the null evolution is extended to infinity, the appearance of infinite redshifts acts as an automatic mechanism for stopping the evolution when the event horizon is reached. In addition, the null approach provides a mechanism for eliminating extraneous incoming radiation from the initial data, thus sharpening the physical model behind the emitted waveforms.

The success of these scalar models suggests the ultimate application of CCM to the binary black hole problem. Two disjoint characteristic evolutions based upon ingoing light cones are matched across worldtubes Γ_1 and Γ_2 to a Cauchy evolution of the region between them, as illustrated in Fig. 1. The outer boundary Γ of the Cauchy region is matched to an exterior characteristic evolution based upon outgoing light cones extending to infinity, where the waveform is calculated. There are major computational advantages in posing the Cauchy evolution in a frame co-rotating with the orbiting black holes. Indeed, such a description may be necessary in order to keep the numerical grid from being intrinsically twisted. In co-orbiting coordinates the individual holes would be tracked as they wobble inside the inner matching worldtubes.

All the pieces for such an attack on the binary black hole problem are presently in place except for the long term stability of 3-dimensional CCM for general relativity. A CCM module incorporating all the necessary geometric transformations between curved space versions of a Cartesian Cauchy grid and a spherical null grid has been written and thoroughly debugged [20]. However, at present there are instabilities

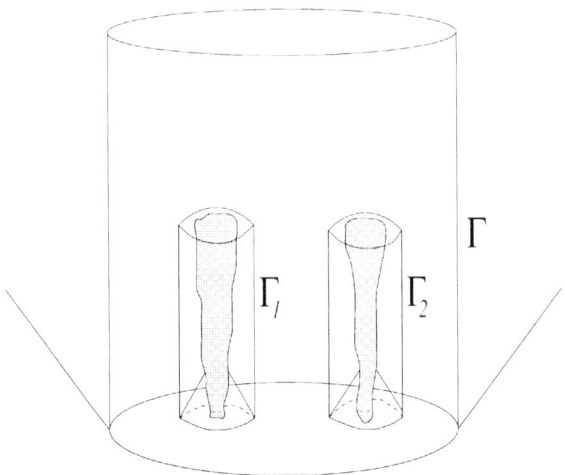

FIGURE 1. A matching scheme for two orbiting black holes (in a co-rotating frame which eliminates the major source of time dependence). This global strategy has been successfully implemented for spherically symmetric self-gravitating scalar waves evolving in a single black hole spacetime.

of a type not found in the simpler test problems.

THE CHARACTERISTICS OF COLLIDING BLACK HOLES

An event horizon is a special type of null hypersurface whose light rays emerge from an initial caustic-crossover region, then expand and asymptotically "hover" at a finite surface area. The elementary caustics have been classified in terms of catastrophe optics and, equivalently, in terms of singular maps of dynamical systems. Because of its inherent structural stability, this classification is unchanged by the tidal distortions of the light rays produced by spacetime curvature.

Axisymmetric Horizons

We have utilized this basic theory to identify the caustic structure of the horizon found in simulations of the axisymmetric head-on collision of two black holes [21]. In the case of spherical symmetry, the light rays generating the horizon originate at the vertex of a light cone. However, this special pointlike origin of the horizon is unstable and is nongeneric when spherical symmetry is broken. Only cusps and folds are structurally stable caustics in the axisymmetric case. Although the theory

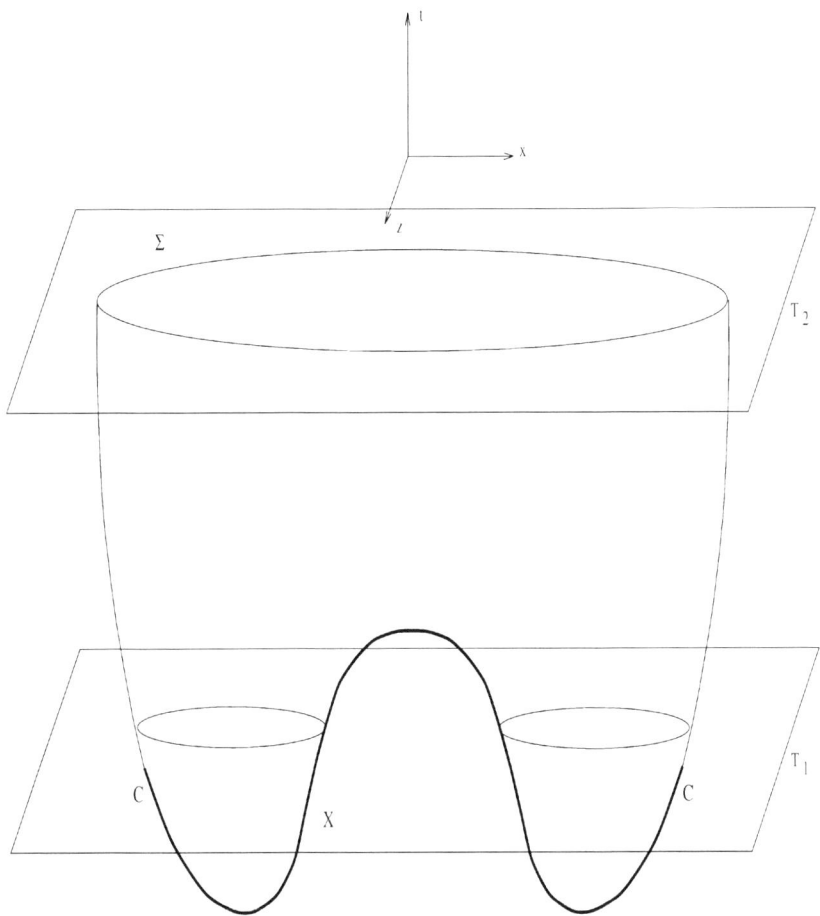

FIGURE 2. Spacetime picture of a pair-of-pants horizon emerging from the (thicker shaded) crossover points X bounded by cusps C.

of elementary caustics is local, we have used it to piece together a global model of the "trousers" shaped horizon obtained in the simulations. Some surprising spacetime features emerge. As schematically illustrated in Fig. 2, the light rays generating the horizon originate on a spacelike seam (a set of crossover points) running along the inside of the trouser legs. The trousers are bowlegged with no special sharpness at the crotch. For black holes formed at a finite time, the crossover seam extends around the bottom of each trouser and slightly up the outer side where it becomes asymptotically null and terminates at a cusp.

Surprisingly, the same trousers shape appears in the horizon found in the axisymmetric gravitational collapse of a rotating cluster [22,23], except now the rotational symmetry axis in Fig. 2 must be identified differently. In the head-on-collision, the

43

symmetry axis lies in the x-direction of Fig. 2. After replacing the t-direction in the figure by the suppressed spatial dimension (the y-direction), a rotation about the x-axis produces two spherical black holes at time T_1. Also, in the head-on collision, the crossover points X in Fig. 2 lie on the rotation axis so that they generate a crossover line under rotation.

For the horizon formed by the rotating cluster, the symmetry axis in Fig. 2 lies the z-direction. Now replacing the t-direction in the figure by the suppressed y-direction, a rotation about the symmetry axis produces a single toroidal black hole at time T_1. Furthermore, the crossover points generate a disc with a circular boundary of cusps. In both models, at the later time T_2 the horizon consists of a single spherical black hole Σ, which in one case results from the collision of two black holes and in the other from the hole in the torus closing up.

The discovery of temporarily toroidal black holes raised the concern [24] of a potential mechanism for violating topological censorship. Under assumptions of asymptotic flatness, global hyperbolicity and a suitable energy condition on the matter fields, the topological censorship theorem requires that any two causal curves extending from past to future null infinity be deformable into each other [25]. The key issue is whether two twins in the spacetime outside the horizon could depart and travel to the same spacetime point by homotopically inequivalent causal paths. A light ray traveling from the infinite past through the hole in the torus and back out to future null infinity would not be causally deformable to a light ray that altogether skirts the horizon if the hole in the torus were too long lived. However, because the intersection of two null hypersurfaces is spacelike, the crossover set X in Fig. 2 must also be spacelike so that the the hole closes up superluminally. Consequently, a causal curve passing inside the hole at a given time can be deformed below the pants leg to a causal curve entirely outside the hole in the torus.

Exact solution for the intrinsic geometry of generic black hole coalescence

Remarkably, the horizon geometry found in the axisymmetric numerical simulations can be independently constructed from a purely analytic approach based upon the conformal rescaling of a flat space null hypersurface [26]. For a null hypersurface emanating from a prolate spheroid, it reproduces the pair-of-pants found in the head-on-collision of black holes. For an oblate spheroid, it yields the temporarily toroidal horizon found in the collapse of a rotating cluster. But the approach is not confined to axisymmetry and yields a generic picture of black hole coalescence in which there is always a toroidal phase [27]. This is illustrated in Fig. 3 which shows the generic tidal distortion of two black holes just prior to merger. At merger, the pincers on the individual holes join to form a single (temporarily) toroidal black hole. In the axisymmetric limit, the pincers close up so that the individual holes have teardrop shape and merge without a toroidal phase. Details of the merger can be viewed at http://artemis.phyast.pitt.edu/animations.

FIGURE 3. Prelude to a generic collision of black holes.

The conformal model treats the horizon in stand-alone fashion as a 3-dimensional manifold with degenerate metric γ_{ab} and affine parameter t along the null rays. The metric is obtained from the conformal mapping $\gamma_{ab} = \Omega^2 \hat{\gamma}_{ab}$ of the metric $\hat{\gamma}_{ab}$ of a flat space null hypersurface with affine parameter $\hat{t}$ emanating from a convex surface. The flat space null hypersurface expands forever but the conformal factor is chosen to stop the expansion so that the cross-sectional area of the horizon approaches a finite limit in the future. At the same time, the Raychauduri equation (which governs the growth of surface area) forces a nonlinear relation between the affine parameters t and $\hat{t}$ which introduces non-trivial topology to the affine slices of the black hole horizon.

The *number* of black holes or the topology of a black hole at a given time is not conventionally defined in terms of such a stand-alone model but in terms of the number of disjoint intersections with a Cauchy hypersurface. In the conformal model of a black hole collision the notion of "two holes" and their merger arises intrinsically from the affine foliation of the horizon. The analytic nature of the model provides further insight into the saddle shape geometry at the crotch of the trousers and implications for the hoop conjecture.

Waveforms from coalescing black holes

A prime application of the analytic horizon model is the calculation of the waveform emitted by coalescing black holes using the null code. The horizon geometry supplies part of the data necessary for an evolution of the exterior spacetime. The evolution is carried out along a family of ingoing null hypersurfaces intersecting the horizon and restricted to the period from merger to ringdown in order to avoid caustics. The other necessary piece of data is the conformal geometry of the final ingoing null hypersurface in the family, which is taken to approximate $\mathcal{I}^+$; i.e. the missing data is essentially the outgoing waveform! The evolution then proceeds backward in time to determine the exterior spacetime in the post-merger regime.

Since the outgoing waveform appears as part of the data, it would seem that this is a circular approach. However, we can proceed by setting the outgoing waveform to zero and evolve backwards in time to calculate the incoming wave entering from $\mathcal{I}^-$ (which is eventually absorbed by the black holes). If the calculation were carried out as a perturbation of a Schwarzschild or Kerr geometry, as in the close approximation [28], a simple application of the time reflection symmetry would then supply the outgoing waveform under the physically appropriate condition of no ingoing radiation. More generally, the appropriate outgoing waveform can be obtained numerically by a more complicated inverse scattering procedure.

Thus the horizon data for a coalescing binary black hole offers a new way to calculate the post-merger waveform. Because this is an unexplored area of binary black hole physics we are beginning this study in the simple case of a head-on collision, where the close approximation waveform has been calculated. This will provide some preliminary physical checks for extending the work into the nonlinear and nonaxisymmetric case where inspiraling black holes can be treated. Preliminary calculations for the head-on-collision show that at late times the waveform is entirely quadrupole ($\ell = 2$) in agreement with the close approximation, but that a strong $\ell = 4$ mode exists just after merger.

ACKNOWLEDGMENTS

The results reported here are cumulative of years of productive work by members of the Pittsburgh numerical relativity group. I am especially grateful to Roberto Gomez who supplied the continuity and guidance critical to bringing the PITT NULL CODE to fruition. Support has been provided by NSF grants PHY 9510895, PHY 9800731 and NSF INT 9515257. Computer time has been provided by the Pittsburgh Supercomputing Center and the San Diego Supercomputing Center.

REFERENCES

1. R. A. Isaacson, J. S. Welling, and J. Winicour, *J. Math. Phys.* **24**, 1824 (1983).
2. N. T. Bishop, R. Gómez, L. Lehner, M. Maharaj and J. Winicour, *Phys. Rev. D* **56** 6298 (1997).
3. M. van der Burg, H. Bondi, and A. Metzner, *Proc. R. Soc. London Ser. A* **269**, 21 (1962).
4. R. Penrose, *Phys. Rev. Letters* **10**, 66 (1963).
5. J. Winicour, *Living Reviews* (1998) http://www.livingreviews.org/
6. R. Sachs, *Proc. R. Soc. Ser. A* **270**, 103 (1962).
7. E. T. Newman and R. Penrose, *J. Math. Phys.* **7**, 863 (1966).
8. R. Gómez, L. Lehner, P. Papadopoulos and J. Winicour, *Class. and Quantum Gravity* **14**, 977 (1997).

9. L. A. Tamburino and J. Winicour, Phys. Rev. **150**, 1039 (1966).

10. R. Gómez, P. Papadopoulos, and J. Winicour, *J. Math. Phys.* **35**, 4184 (1994).

11. R.H. Price, *Phys. Rev.* **D5**, 2419 (1972).

12. N. T. Bishop, R. Gómez, L. Lehner, M. Maharaj and J. Winicour, *Phys. Rev.* **D60**, 24005 (1999).

13. P. Papadopoulos and J. Font, "Relativistic hydrodynamics on spacelike and null surfaces: Formalism and computation of spherically symmetric spacetimes", gr-qc/9902018 (1999).

14. R. Gómez, R. Marsa and J. Winicour, *Phys. Rev.*, **D56**, 6310 (1997).

15. See J. Thornburg, *Class. Quantum Grav.* **4**, 1119 (1987).

16. R. Gómez, L. Lehner, R. Marsa, and J. Winicour, *Phys. Rev.* **D57**, 4778 (1997).

17. R. Gómez, L. Lehner, R. Marsa, J. Winicour and the BBH Grand Challenge, *Phys. Rev. Letters* **80**, 3915 (1998).

18. N. T.Bishop, C. Clarke, and R. d'Inverno, *Class. Quantum. Grav.* **7**, L23 (1990).

19. N. T. Bishop, R. Gómez, P. R. Holvorcem, R. A. Matzner, P. Papadopoulos and J. Winicour, *J. Comp. Phys.* **136**, 140 (1997).

20. N. T. Bishop, R. Gomez, L. Lehner, R. Isaacson, B. Szilágyi and J. Winicour, in *Black Holes, Gravitational Radiation and the Universe*, eds. B. R. Iyer and B. Bhawal (Kluwer, Dordrecht, 1998).

21. R. A. Matzner, H. E. Seidel, S. L. Shapiro, L. Smarr, W-M Suen, S. A. Teukolsky, and J. Winicour, *Science* **270**, 941 (1995).

22. S. A. Hughes, C. R. Keeton, P. Walker, K. Walsh, S. L. Shapiro, and S. A. Teukolsky, *Phys. Rev.* **D49**, 4004 (1994).

23. S. Shapiro, S. Teukolsky and J. Winicour, *Phys. Rev.* **D52**, 6982 (1995).

24. T. Jacobson and S. Venkataramani, *Class. Quantum Grav.* **12**, 1055 (1995).

25. J. L. Friedman, K. Schleich, and D. M. Witt, *Phys. Rev. Letters* **71**, 1486 (1993).

26. L. Lehner, N. T. Bishop, R. Gómez, B. Szilágyi, and J. Winicour, *Phys. Rev.* **D60**, 44005 (1999).

27. "The Asymmetric Merger of Black Holes", S. Husa and J. Winicour, *Phys. Rev.*, to appear (1999), gr-qc/9905039.

28. R. H. Price and J. Pullin, *Phys. Rev. Letters*, **72**, 3297 (1994).

Detection of Gravitational Waves from Eccentric Compact Binaries

Karl Martel

Department of physics, University of Guelph, Guelph, Ont., N1G 3E1

Abstract. Coalescing compact binaries have been pointed out as the most promising source of gravitational waves for kilometer-size interferometers such as LIGO. Gravitational wave signals are extracted from the noise in the detectors by matched filtering. This technique performs really well if an *a priori* theoretical knowledge of the signal is available. The information known about the possible sources is used to construct a model of the expected waveforms (templates). A common assumption made when constructing templates for coalescing compact binaries is that the companions move in a quasi-circular orbit. Some scenarios, however, predict the existence of eccentric binaries. We investigate the loss in signal-to-noise ratio due to non-optimal filtering of eccentric signals and calculate the fitting factor associated with the matched filtering of eccentric signals by circular templates.

I INTRODUCTION

Coalescing compact binaries have been pointed out as the most promising source of gravitational waves for the LIGO/VIRGO/TAMA/GEO interferometers [1,2]. These binaries typically have formed a long time ago, giving them time to radiate most of their eccentricity away. The templates (model of the radiation) needed for matched filtering are thus constructed according to this assumption. Gravitational waves emitted by a circular binary will be explicitly searched for in the output of the detectors, but not gravitational waves emitted by eccentric binaries. The scenario we have in mind allows for the formation of young eccentric binary systems, young enough that they did not have had time to be fully circularized by the radiation reaction. For example, the collapse of a dense Newtonian globular cluster can lead to the formation of a copious number of eccentric binaries via two- and three- body encounters [3,4].

These eccentric binaries will emit strongly in the frequency band of the LIGO interferometers. It may seem that these eccentric binaries can be dealt with by incorporating adequate templates in the bank of templates already available, but this may prove to be inefficient. The addition of new templates has two undesirable effects: It adds to the already heavy computational burden associated with data processing and it increases the probability of false detection (mistaking the noise in the detector for a signal). A better solution might be to search for these eccentric

CP493, *General Relativity and Relativistic Astrophysics*, edited by C. P. Burgess and R. C. Myers
© 1999 American Institute of Physics 1-56396-905-X/99/$15.00

signals with the circular templates, and once a signal is concluded to be present, to extract the information using eccentric templates.

For this to be possible, the circular templates have to follow the phase of the eccentric signals very well. To assess the quality of the circular templates at modeling eccentric signals, special detection tools are needed.

II MATCHED FILTERING AS A DETECTION METHOD

Gravitational wave signals are very weak and at best they will be of the same order of magnitude as the noise in the detectors. This motivates the general belief that matched filtering will be needed to extract the signals from the noisy output of the detectors [1]. When the signals are of known shape, this technique produces the highest signal-to-noise ratio [5]. Suppose a gravitational wave $h(t)$ reaches the detector. The output of the detector $o(t)$ is then a superposition of the useful signal $h(t)$ and the noise $n(t)$. In matched filtering, the signal is extracted by using a theoretical template (theoretical model) that mimics the signal as well as possible; we call this template $m(t, \Omega)$. The vector Ω denotes the parameters that characterize the template. If the template were a perfect copy of the signal, the parameters would represent the real parameters of the source, such as its mass and distance from earth. If the templates are not a perfect approximation to the real signal, the parameters Ω represent phenomenological parameters.

If instead of working in the time domain we work in frequency space, we can introduce the natural inner product of matched filtering. For two functions $a(t)$ and $b(t)$ with Fourier transforms $\tilde{a}(f)$ and $\tilde{b}(f)$, the inner product is defined as [6]

$$(a|b) = 2 \int_0^\infty \mathrm{d}f \frac{\tilde{a}^*(f)\tilde{b}(f) + \tilde{a}(f)\tilde{b}^*(f)}{S_n(f)} , \tag{1}$$

where a " * " denotes complex conjugation and $S_n(f)$ is the one-sided spectral density of the detector's noise. In terms of this inner product, the average signal to noise ratio is [7]

$$\langle \rho \rangle = \frac{(m(\Omega)|h)}{\sqrt{(m(\Omega)|m(\Omega))}} . \tag{2}$$

In practice, the set of parameters Ω is varied until a maximum of the signal-to-noise ratio is found. This maximum is the signal-to-noise ratio achievable by using $\tilde{m}(f, \Omega)$ as a template. The signal-to-noise ratio of equation (2) does not give any information about the quality of the templates or, equivalently, how well the template models the signal. The Schwartz inequality provides an answer to this question [7]. The absolute maximum the signal-to noise ratio can take is achieved when the template is a *perfect* match of the signal, and the parameters Ω correspond to the parameters of the source ($\tilde{m}(t, \Omega) \equiv \tilde{h}(f)$). The optimal SNR is [7]

$$\langle \rho \rangle_{max} = \sqrt{(h|h)} \,. \tag{3}$$

By dividing the signal-to-noise ratio (equation (2)) with the value achieved by optimal filtering (equation (3)), we construct the ambiguity function $\mathcal{A}(\mathbf{\Omega})$:

$$\mathcal{A}(\mathbf{\Omega}) = \frac{(m(\mathbf{\Omega})|h)}{\sqrt{(m(\mathbf{\Omega})|m(\mathbf{\Omega}))(h|h)}} \,. \tag{4}$$

This function takes values between 0 and 1. It is equal to 1 when the optimal template is used.

The value of the parameters $\mathbf{\Omega}$ can be varied until $\mathcal{A}(\mathbf{\Omega})$ is maximized. The maximum value of the ambiguity function is the fitting factor:

$$FF = \max_{\mathbf{\Omega}} \mathcal{A}(\mathbf{\Omega}) \,. \tag{5}$$

The fitting factor is a direct measure of the template's quality since it can be related to the loss of event rate, i.e. the number of events missed by using an inappropriate set of templates. This loss is calculated according to $1 - FF^3$ [6]. For example, if the fitting factor is 0.8, then 48.8% of the events would be mistaken for noise. We adopt a threshold of $FF = 0.9$ for the present work. This corresponds to a loss in event rate of 27%.

III   THE GRAVITATIONAL WAVEFORMS

We calculate the waveforms for both circular and eccentric binaries in the quadrupole approximation. In this approximation the waveforms are given by [8]

$$h_{ij}^{TT} = \frac{2}{R} \frac{d^2}{dt^2} \left(I_{ij} - \frac{1}{3} \delta_{ij} I^k{}_k \right)^{TT} \,, \tag{6}$$

where R is the distance between the source and the observer, I_{ij} is the source's quadrupole moment and the superscript TT reminds us that gravitational waves are traceless and live in the plane transverse to the direction of propagation.

For eccentric binary systems, the waveforms are [9]

$$
\begin{aligned}
\mathrm{h}_+ = h_{xx} = -h_{yy} = \frac{1}{R} \frac{\mu}{p} &\Bigg\{ 2(1 + \cos^2 \theta_o) \cos 2(\varphi - \varphi_o) \\
&+ e \left[(1 + \cos^2 \theta_o) \left(\frac{5}{2} \cos(\varphi - 2\varphi_o) + \frac{1}{2} \cos(3\varphi - 2\varphi_o) \right) + \sin^2 \theta_o \cos \varphi \right] \\
&+ e^2 (1 + \cos^2 \theta_o) \cos 2\varphi_o + \sin^2 \theta_o \Bigg\} \,,
\end{aligned}
\tag{7}
$$

$$
\begin{aligned}
\mathrm{h}_\times = h_{xy} = h_{yx} = -\frac{1}{R} \frac{\mu}{p} \cos \theta_o &\Bigg\{ 4 \sin 2(\varphi - 2\varphi_o) \\
&+ e \left[5 \sin(\varphi - 2\varphi_o) + \sin(3\varphi - 2\varphi_o) \right] - 2e^2 \sin 2\varphi_o \Bigg\} \,,
\end{aligned}
\tag{8}
$$

50

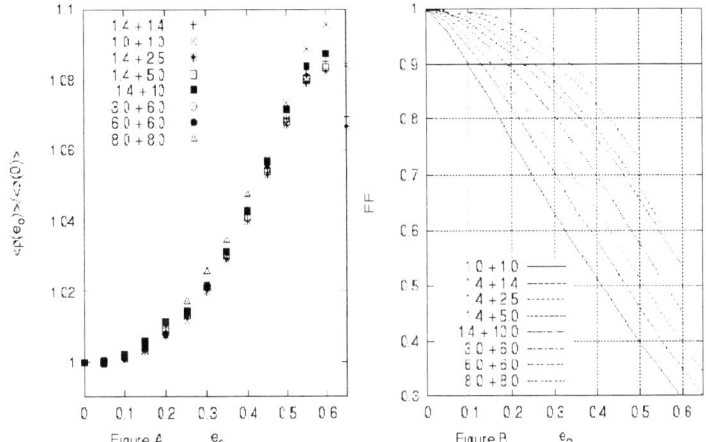

FIGURE 1. Figure A: The ratio of the optimal signal-to-noise ratio $\langle \rho(e_o) \rangle$ to the signal-to-noise ratio $(\langle \rho(0) \rangle)$. The figure shows that eccentric binary systems will be easier to detect if they are explicitly searched for at the output of the interferometers.

Figure B: The fitting factor as a function of e_o for various binary systems. Two trends are apparent. The first one is the net decrease in the fitting factor as e_o increases, while the total mass of the binary is held fixed. The second one is the increase in the detection probability when the total mass of the binary increases. The various binaries studied are labeled by the two masses of the companions; they are given in units of the solar mass.

where θ_o and φ_o are the two angles defining the location of the observer with respect to the orbital plane, μ is the reduced mass, and p and e are defined in terms of the turning points $(r_\pm)$ of the Newtonian orbit as

$$r_\pm = \frac{Mp}{1 \pm e}.$$

The eccentric waveforms oscillate at once, twice and thrice the orbital frequency, whereas the circular waveforms oscillate only at twice the orbital frequency. We parameterize our binaries by specifying the two masses, and the eccentricity e_o they have when they first enter the LIGO frequency band (at 40 Hz).

The Fourier transform of the waveforms is calculated numerically for the eccentric signal, and obtained through the stationary phase approximation for the circular templates [10]. Once these Fourier transforms are known, it is straightforward to calculate the optimal SNR (equation (3)), build the ambiguity function (equation (4)), and maximize it over the different parameters of the templates to get the fitting factor (equation (5)).

The results for the signal-to noise ratio and the fitting factor are displayed in figure (1). The signal-to-noise ratio for an eccentric signal is higher than the ratio obtained for an equivalent circular binary. This means that if both binaries are located at the same distance R, the eccentric binary will emit stronger radiation and will be easier to detect if optimal filters are used. On the other hand, the fitting factor decreases as the initial eccentricity is increased. The circular templates fail to model the eccentric signal properly. The good news is that circular templates are still accurate enough to detect some eccentric signals. For example, a neutron star binary system will be detected as long as its initial eccentricity does not exceed 0.13. If the total mass of the system is increased, the detection probability increases as well. For example, for a system of two 8.0 $M_\odot$ black holes, the initial eccentricity can be as high as 0.33. This trend is explained in the following way. As the total mass of the system increases, the radiation it emits is stronger and the system coalesces in a shorter time. The shorter the signal, the less opportunity the circular templates have to go out of phase with it. This is a good news, because the higher the mass is, the easier it is to detect the signal because it is stronger. Thus, the LIGO interferometers should be able to detect radiation from some eccentric binaries, those with large masses and relatively low eccentricities [11].

Acknowledgments: This work was carried out with É. Poisson. It was supported by NSERC.

REFERENCES

1. K.S. Thorne. *Three Hundred Years of Gravitation*, pages 330–458. Cambridge University Press, New York(1987).
2. B.F. Schutz. Gravitational waves. Nuclear Physics B**35**(Proc. Suppl.), 44(1994).
3. S.L. Shapiro and S.A. Teukolsky. The collapse of dense star clusters to supermassive black holes: The origin of quasars and AGN's. Astrophys. J. **292**, L41(1985).
4. S.L. Shapiro and G.D. Quinlan. The collapse of dense star clusters to supermassive black holes: Binaries and gravitational radiation. Astrophys. J. **321**, 199(1987).
5. É. É. Flanagan and S. A. Hughes. Measuring gravitational waves from binary black hole coalescences: I. signal to noise for inspiral, merger, ringdown. Phys.Rev. D**57**, 4535(1998).
6. T.A. Apostolatos. Construction of a template family for the detection of gravitational waves from coalescing binaries. Phys.Rev.D**54**, 2421(1996).
7. L.A. Wainstein and V.D. Zubakov. *Extraction of Signals from Noise*. Prentice-Hall, Englewood Cliffs (1962).
8. C.W. Misner, K.S. Thorne, and J.A. Wheeler. *Gravitation*. Freemann, San Francisco(1973).
9. H. Wahlquist. The doppler response to gravitational waves from a binary star source. General Relativity and Gravitation, vol.19, no 11, 1101(1987).
10. L.S. Finn and D.F. Chernoff. Observing binary inspiral in gravitational radiation: One interferometer. Phys. Rev. D**47**, 2198(1993).
11. K. Martel and É. Poisson. http://xxx.lanl.gov/archive/gr-qc, 9907006.

Implementing Fully Relativistic Hydrodynamics in Three Dimensions

T. W. Baumgarte[1], S. A. Hughes[1], L. Rezzolla[1], S. L. Shapiro[1,2] and M. Shibata[1,3]

[1] *Department of Physics, University of Illinois at Urbana-Champaign, Urbana, Il 61801*
[2] *Department of Astronomy and NCSA, University of Illinois at Urbana-Champaign, Urbana, Il 61801*
[3] *Department of Earth and Space Science, Osaka University, Toyonaka, Osaka 560-0043, Japan*

Abstract. We report on our numerical implementation of fully relativistic hydrodynamics coupled to Einstein's field equations in three spatial dimensions. We briefly review several steps in our code development, including our recasting of Einstein's equations and several tests which demonstrate its advantages for numerical integrations. We outline our implementation of relativistic hydrodynamics, and present numerical results for the evolution of both stable and unstable Oppenheimer-Volkov equilibrium stars, which represent a very promising first test of our code.

INTRODUCTION

The physics of compact objects is entering a particularly exciting phase. New instruments, including X-Ray and Gamma-Ray satellites and the new neutrino observatories SNO and Super-Kamiokande, can now yield unprecedented observations of neutron stars and black holes. Perhaps most excitingly, the new gravitational wave detectors LIGO, TAMA, GEO and VIRGO promise to open a gravitational wave window to the Universe and make gravitational wave astronomy a reality (see, e.g., [1]).

Simultaneously, the availability of computational resources at modern supercomputers makes the simulation of realistic astrophysical scenarios involving relativistic compact objects feasible. Several groups, including two "Grand Challenge Alliances" [2], have launched efforts to construct numerical codes capable of solving Einstein's equations with or without matter sources in three dimensions, and simulating the merger of black hole or neutron star binaries (see, e.g., [3–9]). Numerical simulations will be necessary to predict the gravitational wave form from such processes to increase the likelihood of detections and, ultimately, to extract physical information from observations. Even though much progress has recently been made (see [9] for the most recent developments), significant obstacles still remain.

CP493, *General Relativity and Relativistic Astrophysics,* edited by C. P. Burgess and R. C. Myers

In this contribution, we report on our systematic approach towards constructing such a numerical code. We first review our formulation of Einstein's equations and describe several tests, both for vacuum spacetimes and analytical matter sources, which demonstrate its advantages for numerical integrations. We then outline our implementation of hydrodynamics, and present promising numerical test results. We adopt geometrized units (G = c = 1) and the convention that Greek indices run from 0 to 3, while Latin indices only run from 1 to 3.

EVOLUTION OF THE GRAVITATIONAL FIELDS

Most numerical implementations of Einstein's equations adopt a Cauchy formulation based on the 3+1 formulation by Arnowitt, Deser and Misner ([10], see, e.g., [11] for an alternative characteristic formulation). However, a straightforward implementation of these "undressed" ADM equations tends to develop instabilities even for the evolution of small amplitude gravitational waves on a flat background (compare [12]). Following Shibata and Nakamura [13] and the spirit of many earlier one and two-dimensional codes [14], we have recently developed a modification of the ADM equations which proves to be much more suitable for numerical implementations [15].

More specifically, we modify the ADM equations in two ways. First, we split the spatial metric γ_{ij} into a conformal factor $\exp(\phi)$ and a conformally related metric $\tilde{\gamma}_{ij}$ according to

$$\gamma_{ij} = e^{4\phi}\tilde{\gamma}_{ij} \tag{1}$$

and evolve ϕ and $\tilde{\gamma}_{ij}$ separately. We choose ϕ such that $\det(\tilde{\gamma}_{ij}) = 1$, and similarly split the extrinsic curvature into its trace and its trace-free part. This split separates the "radiative" variables from the "non-radiative" variables in the spirit of the "York-Lichnerowicz" decomposition [16].

In the second stage we introduce "conformal connection functions"

$$\tilde{\Gamma}^i \equiv \tilde{\gamma}^{lm}\tilde{\Gamma}^i_{lm} = -\tilde{\gamma}^{il}{}_{,l} \tag{2}$$

as independent functions (compare [17]). Some of the mixed second derivatives of $\tilde{\gamma}_{ij}$ in the conformal, spatial Ricci tensor $\tilde{R}_{ij}$ can now be written as first derivatives of the $\tilde{\Gamma}^i$. As a result, $\tilde{R}_{ij}$ becomes a manifestly elliptic operator on the metric $\tilde{\gamma}_{ij}$. The analogous technique was used for the four-dimensional Ricci tensor $R_{\alpha\beta}$ as early as in the 1920's to make Einstein's equations manifestly hyperbolic [18]. For more details of our formulation, including an evolution equation for the $\tilde{\Gamma}^i$, the reader is referred to [15] (see also the mathematical analysis in [19], and further numerical applications in [20]).

In [15], we tested this form of Einstein's equations for small amplitude gravitational waves, and found that it performs far better than a similar implementation of the original ADM formulation. In particular, we found that we could evolve such waves with harmonic slicing without encountering growing instabilities, whereas the original ADM code crashed after about 35 light-crossing times for the same initial data. In [21], we inserted analytic matter sources on the right hand side of Einstein's equations and evolved the fields in their presence. This approach allows us to study the numerical properties of the field evolution in the presence of highly relativistic matter sources without having to solve the equations of hydrodynamics: "hydro-without-hydro". We inserted the Oppenheimer-Snyder solution for a relativistic, static star to test the long term stability of the evolution code, and the Oppenheimer-Volkov solution for the collapse of a sphere of dust to a Schwarzschild black hole. These simulations focus on the highly relativistic, longitudinal fields, and complement our earlier tests involving dynamical transverse fields in [15]. With the code having passed these tests, we are now implementing both collisionless matter, which will be described elsewhere, and hydrodynamics, as described below, to evolve the matter self-consistently with the fields.

RELATIVISTIC HYDRODYNAMICS

For a perfect fluid, the stress-energy tensor can be written

$$T^{\alpha\beta} = (\rho_0 + \rho_0\epsilon + P)u^\alpha u^\beta + Pg^{\alpha\beta}, \tag{3}$$

where ρ_0 is the rest mass density, ϵ the specific internal density, P the pressure, u^α the fluid four velocity, and $g_{\alpha\beta}$ the four dimensional spacetime metric. We construct constant entropy initial data with a polytropic equation of state

$$P = K\rho_0^\Gamma, \tag{4}$$

where $\Gamma = 1 + 1/n$ and n is the polytropic index, and where we assume the polytropic constant K to be unity without loss of generality. During the evolution, we adopt the gamma-law relation appropriate for adiabatic flow,

$$P = (\Gamma - 1)\rho_0\epsilon. \tag{5}$$

Following [9,22], we write the equation of continuity

$$(\rho_0 u^\alpha)_{;\alpha} = 0 \tag{6}$$

and the equations of motion

$$T^{\alpha\beta}_{\;\;;\beta} = 0 \tag{7}$$

in the form

$$\rho_{*,t} + (\rho_* v^i)_{,i} = 0, \tag{8}$$

$$e_{*,t} + (e_* v^i)_{,i} = 0, \tag{9}$$

$$(\rho_* \tilde{u}_i)_{,t} + (\rho_* \tilde{u}_i v^j)_{,j} = -\alpha e^{6\phi} P_{,i} - \alpha\rho_* \tilde{u}^0 \alpha_{,i} + \rho_* \tilde{u}_l \beta^l_{\;,i} +$$

$$\frac{\rho_* \tilde{u}_l \tilde{u}_m}{e^{4\phi}\tilde{u}^0}\left(2\tilde{\gamma}^{lm}\phi_{,i} + \frac{1}{2}\left(\tilde{\Gamma}^l_{ki}\tilde{\gamma}^{km} + \tilde{\Gamma}^m_{ki}\tilde{\gamma}^{kl}\right)\right). \tag{10}$$

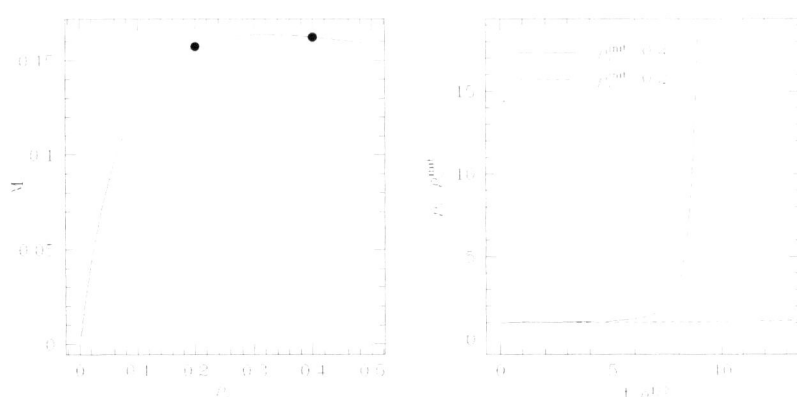

FIGURE 1. Mass versus central density for a $n = 1$ polytrope (left panel). We chose the two configurations marked by the filled circles ($\rho_c = 0.2$ and $\rho_c = 0.4$) as initial data for our dynamical simulations. The central density as a function of time for the evolution of the two initial data sets (right panel). As expected, the $\rho_c = 0.4$ configuration soon collapses, whereas the $\rho_c = 0.2$ configuration remains stable for several dynamical timescales.

Here we have defined the auxiliary quantities

$$\rho_* \equiv \alpha e^{6\phi} u^0 \rho_0, \tag{11}$$

$$e_* \equiv \alpha e^{6\phi} u^0 (\rho_0 \epsilon)^{1/\Gamma}, \tag{12}$$

$$\tilde{u}_i \equiv (1 + \Gamma\epsilon) u_i, \tag{13}$$

$$\tilde{u}^0 \equiv (1 + \Gamma\epsilon) u^0, \tag{14}$$

$$v^i \equiv u^i / u^0 = -\beta^i + \gamma^{ij} u_j / u^0. \tag{15}$$

Similar equations have been used by many other groups (e.g. [23] and references therein). We integrate equations (8) - (10) with an artificial viscosity scheme suggested in [3] (see [8] for implementations of more elaborate shock capturing schemes). Given ρ_*, e_* and $\tilde{u}_i$ on a new timelevel, u^0 can be found iteratively from the normalization relation $u^\alpha u_\alpha = -1$, which yields

$$(\alpha u^0)^2 = 1 + \frac{\tilde{\gamma}^{ij} \tilde{u}_i \tilde{u}_j}{e^{4\phi}} \left(1 + \Gamma \frac{e_*^\Gamma}{\rho_* (\alpha u^0 e^{6\phi})^{\Gamma-1}} \right)^{-2}. \tag{16}$$

The matter sources for the right hand sides of Einstein's equations can then be constructed from these variables.

NUMERICAL RESULTS

As a first test of our implementation of hydrodynamics, we adopt the Oppenheimer-Volkov solution describing equilibrium neutron stars in spherical symmetry as initial data and evolve these dynamically. Constructing a sequence of such Oppenheimer-Volkov solutions for increasing central rest mass densities ρ_c, the ADM mass M of the star takes a maximum $M_{\max}$ at ρ_c^{crit} (see the left panel in Figure 1). For central densities smaller than ρ_c^{crit}, the star is in stable equilibrium, while for $\rho_c > \rho_c^{\mathrm{crit}}$ it is unstable and will collapse to a black hole.

We adopt a polytropic equation of state with $\Gamma = 2$ ($n = 1$), for which $\rho_c^{\mathrm{crit}} = 0.32$ and $M_{\max} = 0.164$ when $K = 1$. We choose as initial data the two configurations marked with filled circles in Figure 1; a stable configuration with $\rho_c = 0.2$, $M = 0.157$ and $R/M = 5.5$, and an unstable configuration with $\rho_c = 0.4$, $M = 0.162$ and $R/M = 4.4$.

In the right panel of Figure 1, we show the central density as a function of time for the two initial configurations. We performed these runs on quite modest numerical grids with $(64)^3$ gridpoints using cartesian coordinates and imposing outgoing wave boundary conditions at $x, y, z = 2$. We employ harmonic slicing and zero shift. As expected, the unstable configuration soon collapses, while the stable configuration remains stable for several dynamical timescales (the period of the fundamental radial oscillation for a $\rho_c = 0.2$ configuration is approximately $7\rho_c^{-1/2}$, compare [9]). Ultimately, accumulation of numerical error causes this configuration to collapse too, but this can be delayed by increasing the grid resolution.

In Figure 2 we show density contours at the beginning and towards the end of the simulation for the unstable configuration with $\rho_c = 0.4$ initially. We also include arrows indicating the fluid flow $\bar{u}_i$. The star is rapidly contracting and collapsing to a black hole. Up to the late stage of the collapse, the mass M is conserved to about 5 %. We can follow the collapse to about a 18 fold increase of the central density. By this time, the central lapse has decreased from 0.4 initially to about 0.03.

SUMMARY AND DISCUSSION

We report on our systematic approach towards constructing a fully relativistic hydrodynamics code in three spatial dimensions. As part of this program, we have developed a new formulation of Einstein's equations which in several tests and applications has proved to be much more suitable for numerical implementations than the traditional ADM formulation [15,20]. Mathematical properties of this formulation have been analyzed in [19]. We have studied the evolution of small amplitude gravitational waves to test dynamical transverse fields and have inserted analytical solutions as matter sources (hydro-without-hydro) to test highly relativistic longitudinal fields.

We outline our implementation of the relativistic equations of hydrodynamics and present preliminary test results for spherical neutron stars in hydrostatic equilibrium (Oppenheimer-Volkov stars). As expected, we find that stable stars remain stable for several dynamical timescales, while unstable stars soon collapse to black holes. We conclude that our method seems like a very promising approach towards

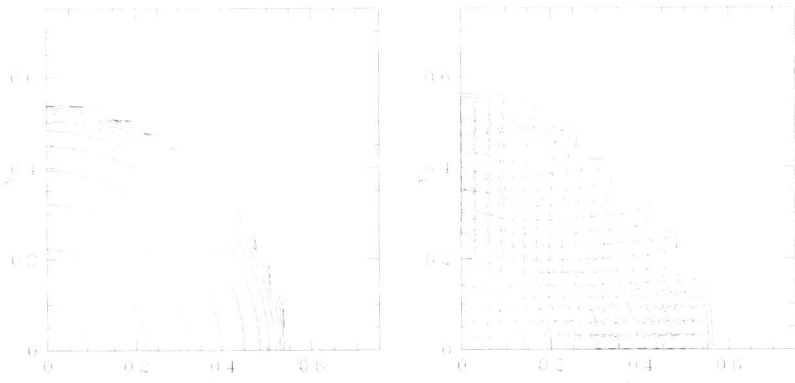

FIGURE 2. Rest mass density contours at $t = 0$ (left panel) and $t = 13$ (right panel) in the $z = 0$ (equatorial) plane for the unstable configuration with $\rho_c = 0.4$ initially. The contours logarithmically span densities between ρ_c and $10^{-3}\rho_c$. We also include arrows indicating the fluid flow $\tilde{u}_i$.

simulating the final plunge and merger of binary neutron stars.

On a more speculative note, we point out that fully self-consistent hydrodynamics may not be a feasable approach towards simulating the coalescence and gravitational wave emission from neutron stars in the intermediate inspiral phase. In this epoch, the stars are very close and interact through a strongly relativistic tidal field, but reside outside the innermost stable circular orbit and hence move on a nearly circular orbit. Simulating the slow inspiral would require evolving the stars for hundreds or thousands of orbits, which is presently impossible. It may be possible, however, to insert known quasi-equilibrium binary configurations (e.g. [6,7]) into the field evolution code to get the transverse wave components approximately. Decreasing the orbital separation (and increasing the binding energy) at the rate consistent with the computed outflow of gravitational-wave energy would generate an approximate strong-field wave inspiral pattern. Such a "hydro-without-hydro" calculation may yield an approximate gravitational waveform from inspiraling neutron stars, without having to couple the matter and field integrations.

Calculations were performed on SGI CRAY Origin2000 computer systems at the National Center for Supercomputing Applications, University of Illinois at Urbana-Champaign. This work was supported by NSF Grants AST 96-18524 and PHY 99-02833 and NASA Grant NAG 5-7152 at Illinois.

REFERENCES

1. See article by P. Saulson in this Volume.

2. Information about the Binary Black Hole Grand Challenge can be found at www.npac.syr.edu/projects/bh/, and about the Binary Neutron Star Grand Challenge at wugrav.wustl.edu/Relativ/nsgc.html.

3. K. Oohara and T. Nakamura, *Prog. Theor. Phys.* **82**, 535 (1989).

4. K. Oohara and T. Nakamura, in *Relativistic Gravitation and Gravitational Radiation*, edited by J.-A. Marck and J.-P. Lasota (Cambridge University Press, Cambridge, 1997).

5. J. R. Wilson and G. J. Mathews, *Phys. Rev. Lett.* **75**, 4161 (1995); J. R. Wilson, G. J. Mathews and P. Marronetti, *Phys. Rev. D* **54**, 1317 (1996).

6. T. W. Baumgarte, G. B. Cook, M. A. Scheel, S. L. Shapiro, and S. A. Teukolsky, *Phys. Rev. Lett.* **79**, 1182 (1997); *Phys. Rev. D* **57**, 6181 (1998); *Phys. Rev. D* **57**, 7292 (1998).

7. S. Bonazzola, E. Gourgoulhon and J.-A. Marck, *Phys. Rev. Lett.* **82**, 892 (1999)

8. J. A. Font, M. Miller, W. Suen and M. Tobias, submitted (also gr-qc/9811015).

9. M. Shibata, submitted (1999).

10. R. Arnowitt, S. Deser and C. W. Misner, in *Gravitation: An Introduction to Current Research*, edited by L. Witten (Wiley, New York, 1962).

11. See article by J. Winicour in this Volume.

12. A. M. Abrahams *et. al.* (The Binary Black Hole Grand Challenge Alliance) *Phys. Rev. Lett.* **80**, 1812 (1998).

13. M. Shibata and T. Nakamura, *Phys. Rev. D* **52**, 5428 (1995).

14. For example: J. M. Bardeen and T. Piran, *Phys. Rep.* **96**, 205 (1983); C. R. Evans, PhD thesis, University of Texas at Austin (1984); A. M. Abrahams and C. R. Evans, *Phys. Rev. D* **37**, 318 (1988); A. M. Abrahams, G. B. Cook, S. L. Shapiro and S. A. Teukolsky, *Phys. Rev. D* **49**, 5153 (1994).

15. T. W. Baumgarte and S. L. Shapiro, *Phys. Rev. D* **59**, 024007 (1999).

16. A. Lichnerowicz, *J. Math. Pure Appl.* **23**, 37 (1944); J. W. York, Jr., *Phys. Rev. Lett.* **26**, 1656 (1971).

17. T. Nakamura, K. Oohara and Y. Kojima, *Prog. Theor. Phys. Suppl.* **90**, 76 (1987).

18. T. De Donder, *La gravifique einsteinienne* (Gauthier-Villars, Paris, 1921); C. Lanczos, *Phys. Z.* **23**, 537 (1922).

19. S. Frittelli and O. Reula, submitted (1999) (also gr-qc/9904048).

20. Miguel Alcubierre *et.al*, submitted (1999) (also gr-qc/9904013).

21. T. W. Baumgarte, S. A. Hughes and S. L. Shapiro, *Phys. Rev. D*, in press.

22. M. Shibata, T. W. Baumgarte and S. L. Shapiro, *Phys. Rev. D* **58**, 023002 (1998).

23. J. F. Hawley, L. L. Smarr and J. R. Wilson, *Astrophys. J. Suppl.* **55**, 211 (1984).

On the Maximum Mass of Neutron Stars

Chris Vuille* and Jim Ipser†

*Department of Physical Sciences, Embry-Riddle University
Daytona Beach, FL 32114
†Department of Physics, University of Florida, Gainesville, FL 32611

Abstract. The equation of state for matter with energy density above $2x10^{14}g/cm^3$ is parametrized with three regions, two with adiabatic equations of state and the third with a causal equation of state. Searching the parameter space under a stability constraint, we find the maximum mass for a non-rotating relativistic neutron star depends on the maximum allowed values of the adiabatic index parameters. The maximum mass as a function of adiabatic index can be rendered to good approximation by a rational quadratic. Selecting a generous upper bound on the adiabatic index, it appears very likely that the maximum mass is less than three solar masses.

INTRODUCTION

Computing the maximum mass of a neutron star requires knowledge of the equation of state (EOS) of matter, especially in the core region, where it is likely to be quite stiff. Chandrasekhar [1] derived the maximum mass for white dwarfs, and it would be of interest to be able to obtain a similar result for neutron stars, which would be of value in differentiating neutron stars from candidate black holes.

Rhoades and Ruffini [2] developed an estimate of the maximum mass in terms of a variational principle, matching a causal EOS,

$$E = \frac{1}{2}\left(E_f - P_f/c^2 + \left(E_f + P_f/c^2\right)\left(\frac{N}{N_f}\right)^2\right) \tag{1}$$

$$P = Ec^2 - a \tag{2}$$

to the BPS/BBP EOS [3,4]. Here E is the energy density, N the number density, and P is the pressure, while $a = E_f c^2 - P_f$, with P_f and N_f the pressure and baryon number density corresponding to the fiducial matching energy density E_f, all taken from a tabulation of the BPS/BBP EOS. The mass calculated with the Tolman-Oppenheimer-Volkoff equation depends on the parameter a. Choosing $a/c^2 = 2.5x10^{14}g/cm^3$, the value of the saturation density, results in 4.25 sola

CP493, *General Relativity and Relativistic Astrophysics*, edited by C. P. Burgess and R. C. Myers
© 1999 American Institute of Physics 1-56396-905-X/99/$15.00

masses, whereas choosing $a/c^2 = 4.569x10^{14}g/cm^3$, the value below which the BPS/BBP equation of state is often assumed valid, results in 3.14 solar masses. Glendenning [5]prefers this latter value, since the EOS is unlikely to be causal at the lower density, and believes that further constraints, in particular a more realistic equation of state, can only lower the estimate. In this paper, we show evidence that, indeed, he is correct.

PARAMETRIZATION OF THE CORE EOS

For our purposes, the BPS/BBP EOS is assumed valid up to $E = 2.004x10^{14}g/cm^3$, corresponding to a number density $N_{f1} = 1.182x10^{38}cm^{-3}$, and pressure of $P_{f1} = 1.286x10^{33}dynes/cm^2$, taken from a tabulation of the BPS/BBP EOS. At densities above this, two adiabatic regions are defined in tandem, with fiducial number density N_{f2} representing the point at which a phase transition takes place, from the first adiabatic region to the second. The pressure in each adiabatic region is given by an equation of the form

$$P = kN^\Gamma \tag{3}$$

where Γ is the adiabatic index and k a constant which is determined by continuity. From the first law, the energy density in such a region is

$$E = \frac{P}{c^2(\Gamma - 1)} + DN \tag{4}$$

where D is a constant of integration, which, like k, is established by continuity. Finally, when conditions are such that the speed of sound equals the speed of light, a parametrized causal EOS similar to equations 1 and 2 is used for all densities above that point, as in [6], with constants again chosen so that the pressure and energy density join continuously to the underlying adiabatic region.

It is very likely that $4/3 \leq \Gamma_1, \Gamma_2 \leq 4$, since many derived equations of state can be accurately rendered with two adiabatic core regions, and in all cases studied the largest adiabatic index required was about 2.8, which is already quite stiff. Models were rejected when found on unstable regions of the mass versus central energy density graph. Systematically varying the parameters Γ_1, Γ_2, and N_{f2}, the Tolman-Oppenheimer-Volkoff equations were repeatedly integrated to obtain the masses of a large number of stellar models.

RESULTS

In terms of finding stars with the greatest possible mass, two adiabatic regions rendered no apparent advantage over one such region. Stable models with the largest allowed adiabatic index in both regions generally yielded the largest masses, though occasional slight deviations from this were found, depending on the resolution of the search through the parameter space. It was also found that the highest mass stable stars could be obtained with central number densities below about $1x10^{39}cm^{-3}$. Converting the code to a single adiabatic region in tandem with the causal EOS, the maximum mass was generally obtained with the maximum value of Γ. As the upper bound on the value of Γ was increased, the number density of the transition to the causal equation of state decreased, and the maximum calculated mass increased, asymptotically approaching the maximum given by using the causal equation of state, alone. To good accuracy for $\Gamma \geq 4/3$, the maximum mass as a function of core adiabatic index can be given by a rational quadratic,

$$M = \frac{4.8\Gamma^2 - 10.07\Gamma + 5.37}{\Gamma^2 - 1.62\Gamma + 4.88} \tag{5}$$

For $\Gamma = 4$, which is very stiff, the maximum is about 2.9 solar masses, while for $\Gamma = 3$ the maximum is 2 solar masses. Given the distribution of known neutron star masses, the former value is likely to be a conservative estimate of the upper bound on the non-rotating neutron star mass.

ACKNOWLEDGEMENTS

This work was supported in part by the NASA-JOVE program, a joint venture of NASA and Embry-Riddle Aeronautical University, and by the research fund instituted by Dean Andres Zellweger, ERAU.

REFERENCES

1. Chandrasekhar, S., *Astrophys. J. Phys.* **74**, 81 (1931).
2. Rhoades, C., and Ruffini, R., *Phys. Rev. Lett. Phys.* **32**, 324 (1974).
3. Baym, G., Pethick, C., and Sutherland, P., *Astrophys. J.* **170** 299 (1971).
4. Baym, G., Bethe, H., and Pethick, C., *Nucl. Phys. A* **175** 225(1971).
5. Glendenning, N., *Compact Stars*, New York, Springer-Verlag, 1996, ch. 3, pp. 110-112.
6. Koranda, S., Stergioulas, N., and Friedman, J.,*Ap. J.*, **488**,799-806 (1997).

Roche Lobe Overflows in Mass-Conservative Circular Binary Systems: Constraints on Rates of Mass Exchange and Orbital Angular Momentum Loss

T.T. Chia[1]

Department of Physics, National University of Singapore, Singapore
Canadian Institute of Theoretical Astrophysics, Toronto, Ontario, Canada

Abstract. Limits on the rates of mass-exchange and orbital angular momentum loss in a mass-conservative circular binary system evolving by loss of orbital angular momentum and Roche lobe overflow are derived.

THE MODEL

In a mass-conservative circular binary system, mass transfer can occur via Roche lobe overflow: matter falls toward the mass-gaining star with mass M_1 from its companion with mass M_2, whenever the latter's stellar radius R_2 exceeds or equals to its Roche lobe radius R_L. Further, the binary system may lose orbital angular momentum via gravitational waves emission, tidal interaction and magnetic braking [1-5]. Here, we shall examine analytically a simple model of such a system.

We assume that the orbital angular momentum J decreases with time. This could occur, for example, when the stars are spherically symmetrical and angular momentum is being carried away by the emission of gravitational waves. As the torque due to the gravitational waves does not affect the spins of spherical stars, gravitational waves must remove J from the system if the accretion disc is in a steady state [6]. The orbital angular momentum could also be lost through conversion to spin angular momentum via tidal interaction when the spin angular orbital velocities are less than the orbital angular velocity [3]. Moreover, J could also be lost from the binary system via magnetic braking [5].

Regardless of how J is lost from the system, we can write

$$\frac{dJ}{dt} = -JW \; , \tag{1}$$

[1] On leave from the National University of Singapore. Now at the Max-Planck-Institut für Astrophysik, Postfach 1523, D-85740 Garching, Germany.

where

$$J = \left(\frac{Ga}{M}\right)^{1/2} M_1 M_2 , \tag{2}$$

$M = M_1 + M_2$, a is the orbital radius, and W is a positive-definite function dependent on how J is lost.

Using (2), the angular momentum loss equation can be expressed as

$$\frac{\dot{a}}{2a} = -W + (1 - q)\frac{\Delta}{M_2} , \tag{3}$$

where $q = \frac{M_2}{M_1}$, and $\Delta = \dot{M_1} = -\dot{M_2}$ which is a positive quantity.

We assume that the mass-losing star obeys the usual mass-radius relation

$$R_2 = K M_2^n , \tag{4}$$

where K is a constant, and n is a constant dependent on the structure of the mass-losing star [2]. We take the radius of its Roche lobe R_L to be [7]

$$R_L = 0.46224a \left(\frac{M_2}{M}\right)^{1/3} \quad \text{for } 0 < q < 0.8 \tag{5}$$

$$R_L = a(0.38 + 0.2 \, log_{10} \, q) \quad \text{for } 0.3 < q < 20 \tag{6}$$

By differentiating (4) to (6) with respect to time, we get

$$\frac{\dot{R_2}}{R_2} = -\frac{n\Delta}{M_2} , \tag{7}$$

$$\frac{\dot{R_L}}{R_L} = \frac{\dot{a}}{a} - \frac{\alpha\Delta}{M_2} , \tag{8}$$

$$\alpha = \frac{1}{3} \quad \text{for } 0 < q < 0.8 , \tag{9}$$

$$\alpha = \frac{1 + q}{4.3749 + log_e \, q} \quad \text{for } 0.3 < q < 20 . \tag{10}$$

When mass exchange commences, $R_2 = R_L$. Then from (7)-(8), we get

$$\frac{\dot{a}}{a} = (\alpha - n)\frac{\Delta}{M_2} + X , \tag{11}$$

$$X = \frac{\dot{R_L} - \dot{R_2}}{R_L} . \tag{12}$$

which in general is not zero as R_L and R_2 have different functional dependence on M_2; hence they need not have the same slope when these two radii are equal [8].

By using (3) and (11), we obtain

$$\frac{f\Delta}{M_2} = W + \frac{X}{2} \; , \tag{13}$$

$$\frac{f\dot{a}}{a} = (\alpha - n)W + (1 - q)X \; , \tag{14}$$

$$f = 1 - q - \left(\frac{\alpha - n}{2}\right) = h - q \; . \tag{15}$$

CONSTRAINTS

Though four unknowns Δ, W, X and $\dot{a}$ appear in (13) and (14), constraints on them can be deduced in many cases; but we shall not dwell on the last variable here because of space constraint.

Note that even though the exact value of X may be unknown for a given binary system undergoing Roche lobe overflow, in principle, observational data could be used to determine the sign of X. This is possible as M_2 is related to the sign of X. A negative X at the moment of commencement of mass exchange means that R_2 grows faster than R_L, that is, the mass-exchange rate is fast. However, a positive X implies the opposite: the mass-exchange rate is slow [7,9].

It is possible to use (13), the Δ-equation, to classify such binary systems into three classes via the sign of X and the value of q. Further, within each class, inequalities relate Δ, W and X. Hence, if we know either the sign of X or the value of q, or both of them, and in addition, know any one of the three variables Δ, W and X, we will be able to derive constraints on the other two.

(i) The classes and constraints

From the Δ-equation, we see that the ranges of q and X fall into three separate classes A, B and C. In class A, q is larger than h, a quantity defined in (15), while in class B, the mass exchange rate is slow. Finally in class C, M_2 is fast and $q < h$.

Since W, M_2 and Δ are all positive quantities, a negative f implies that X must be negative giving rise to class A. If X is positive, (13) implies that $f > 0$ corresponding to class B. Finally, when $f > 0$ with $X < 0$, we have class C.

In each class, an inequality relating Δ, W and X follows from (13). These inequalities are given in Table 1.

Class	Bounds on Δ	Bounds on X	Bounds on W
A	$\frac{M_2 X}{2f} > \Delta > 0$	$X < \min\left(\frac{2f\Delta}{M_2}, -2W\right)$	$-\frac{X}{2} > W > 0$
B	$\Delta > \max\left(\frac{M_2 X}{2f}, \frac{M_2 W}{f}\right)$	$\frac{2f\Delta}{M_2} > X > 0$	$\frac{f\Delta}{M_2} > W > 0$
C	$\frac{M_2 W}{f} > \Delta > 0$	$0 > X > -2W$	$W > \max\left(-\frac{X}{2}, \frac{f\Delta}{M_2}\right)$

(ii) Checks on system parameters

For a binary system undergoing Roche lobe overflow with known or assumed q and n, one can calculate h. If $q > h$, the system belongs to class A with a fast mass-exchange rate. However, should observation indicate a slow mass-exchange rate, then one or more items in the following list must be incorrect: observed value of q, assumed value of n, circular orbit assumption, zero mass loss assumption, Roche lobe overflow assumption, and orbital angular momentum loss mechanisms.

If the system is observed to undergo slow mass-exchange, it belongs to class B, so that $q < h$. If the last inequality is wrong, then one or more of the items given in the list in the last paragraph must be incorrect.

(iii) Upper and lower bounds with a known variable

(α) Known W

If a mechanism for orbital angular momentum loss is known or assumed, then W is known. From Table 1, limits on Δ and X can now be obtained:

$$\text{Class A} \qquad\qquad X < -2W . \tag{16}$$

$$\text{Class B} \qquad\qquad \Delta > \frac{M_2 W}{f} . \tag{17}$$

$$\text{Class C} \qquad 0 > X > -2W , \qquad \frac{M_2 W}{f} > \Delta > 0 . \tag{18}$$

These bounds to the mass-exchange rate and X could be used for comparison with observational data or computational results.

(β) Known Δ

In this case of known Δ, bounds on X and W can be deduced from Table 1:

$$\text{Class A} \qquad\qquad X < \frac{2f\Delta}{M_2} . \tag{19}$$

$$\text{Class B} \qquad \frac{2f\Delta}{M_2} > X > 0 , \qquad \frac{f\Delta}{M_2} > W > 0 . \tag{20}$$

$$\text{Class C} \qquad\qquad W > \frac{f\Delta}{M_2} . \tag{21}$$

If Δ is known from observation, the bounds on W given in (20) and (21) could shed some light on the angular momentum loss mechanisms in that system. However, if Δ is obtained from evolutionary model computations, (19)-(21) can be used as consistency checks for the values of W and X computed.

(γ) Known X

If an expression for X is adopted in a model, then we have bounds on Δ and W:

$$\text{Class A} \qquad \frac{M_2 X}{2f} > \Delta > 0 \ , \qquad -\frac{X}{2} > W > 0 \ . \tag{22}$$

$$\text{Class B} \qquad \Delta > \frac{M_2 X}{2f} \ . \tag{23}$$

$$\text{Class C} \qquad W > -\frac{X}{2} \ . \tag{24}$$

The bounds in (22) and (23) can be used to check the adopted expression for X in the model when Δ is available from observation. Further, (22)-(24) provide consistency checks for the computed values of Δ and W from the model.

SUMMARY

In a mass-conservative circular binary system evolving by Roche lobe overflow and by orbital angular momentum loss, two equations relate the variables Δ, W, a and X. If the sign of X or the value of q or both are known, the binary system can be classified as in Table 1. For class A or B systems, some consistency checks on the data and model assumptions can be made.

In each class, there are inequalities governing the variables. If one of the variables Δ, W, and X is also known, bounds to the other variables can be obtained. These are given in (16)-(24) and can be used to test models and for comparison with observational data.

REFERENCES

1. Paczynski, B., *Acta.Astr.* **17**, 287 (1967).
2. Faulkner, J., *Ap.J.* **170**, L99 (1971).
3. Kopal, Z., *Dynamics of Close Binary Systems*, Dordrecht: Reidel, (1977).
4. Zahn, J.P., *Astr. Ap.* **57**, 383 (1977).
5. Verbunt, F., and Zwaan, C., *Astr. Ap.* **100**, L7 (1981).
6. Chia, T.T., *Ap.Sp. Sci.* **51**, 159 (1977).
7. Paczynski, B., *Ann.Rev.Astr.Ap.* **9**, 183 (1971).
8. Chia, T.T., This proceeding
9. Paczynski, B., Ziolkowski, J., and Zytkow, A., *Mass Loss from Stars*, ed. M. Hack, Dordrecht: Reidel, (1969) pp. 237.

On a Model for the Evolution of a Mass-Conservative Circular Binary System by Roche Lobe Overflow and by Emission of Gravitational Waves

T.T. Chia[1]

Department of Physics, National University of Singapore, Singapore
Canadian Institute of Theoretical Astrophysics, Toronto, Ontario, Canada

Abstract. In some ranges of mass ratio and mass-radius exponent, it is shown that there are inconsistencies in Faulkner's (1971) analytic model of a mass-conservative circular binary system that is evolving via the combined processes of mass exchange through Roche lobe overflow and emission of gravitational waves. These inconsistencies occur because of the use of an assumption that the time derivative of the Roche lobe radius is equal to that of the stellar radius. Consequently, his equations for the rate of mass exchange and orbital period variations are generally invalid.

INTRODUCTION

In one mechanism of mass transfer in a circular binary system, mass transfer occurs whenever the radius of the mass-losing star exceeds that of its critical potential surface known as the Roche lobe. This mass exchange rate is enhanced and or induced by the emission of gravitational radiation since the latter, by robbing angular momentum from the system, causes the orbital radius to decrease thereby shrinking the Roche lobe surface [1-3].

Since these early works which are responsible for arousing much interest in the subject, the evolution of circular binary systems by Roche lobe overflows and emission of gravitational waves has been studied computationally by many (for example see [4] and references therein). However, the original simple analytical model, "the standard model" which gives both the rates of mass-exchange and change of orbital period [3] has not been reexamined. Here, we shall show that this simple model is not self-consistent as it does give rise to contradictions in some ranges of mass-ratios for both degenerate and lower main-sequence stars, with the consequence that the model-dependent expressions for these rates are generally incorrect.

---

[1] On leave from the National University of Singapore. Now at the Max-Planck-Institut für Astrophysik, Postfach 1523, D-85740 Garching, Germany.

CP493, *General Relativity and Relativistic Astrophysics,* edited by C. P. Burgess and R. C. Myers
© 1999 American Institute of Physics 1-56396-905-X/99/$15.00

FORMULATION OF THE MODEL

We set up the model by using Faulkner's assumptions [3]. It consists of a mass-conservative circular binary system in which the mass-losing star, with mass M_2 and stellar radius R_2, is losing matter by Roche lobe overflow to the companion star whose mass is M_1.

The radius of the mass-losing star is assumed to obey the equation

$$R_2 = K M_2^n , \tag{1}$$

where K is a constant. The other constant n can be taken to be $-1/3$ for degenerate stars and 1 for lower main-sequence stars [3].

Though a single expression for the Roche lobe radius R_L was used in the original model [3], we use the more general expressions due to Paczynski [5]:

$$R_L = 0.46224a \left(\frac{M_2}{M} \right)^{1/3} \quad \text{for } 0 < q < 0.8 \tag{2}$$

$$R_L = a(0.38 + 0.2 \, log_{10} \, q) \quad \text{for } 0.3 < q < 20 \tag{3}$$

where

$$q = \frac{M_2}{M_1} , \tag{4}$$

$$M = M_1 + M_2 , \tag{5}$$

and a is the orbital radius.

The angular momentum of the binary system decreases with time owing to the emission of gravitational waves. Assuming that the disk is in a steady state and either the rotational angular momenta of the stars are constant as in the case of spherical stars [6] or negligibly small, it follows that the orbital angular momentum J must decrease. In the weak field limit of the general theory of relativity [7], we can write

$$\frac{dJ}{dt} = -\frac{32G}{5c^5} \left(\frac{M_1 M_2}{M} \right)^2 a^4 \omega^5 , \tag{6}$$

where ω is the angular velocity of the binary system and

$$J = \left(\frac{Ga}{M} \right)^{1/2} M_1 M_2 . \tag{7}$$

69

DEDUCTIONS FROM THE STANDARD FORMULATION

As in the original model, we shall assume that mass exchanges occurs when the radius of the mass-losing star is equal to that of its Roche lobe. Using (1)-(3) and by differentiating them with respect to time, we get

$$\frac{\dot{a}}{a} = (\alpha - n)\frac{\Delta}{M_2} \ , \tag{8}$$

where

$$\alpha = \frac{1}{3} \quad \text{for} \ \ 0 < q < 0.8 \ , \tag{9}$$

$$\alpha = \eta = \frac{1+q}{4.3749 + log_e \ q} \quad \text{for} \ \ 0.3 < q < 20 \ , \tag{10}$$

and

$$\Delta = \dot{M}_1 = -\dot{M}_2 \ , \tag{11}$$

which is a positive quantity.

When $\alpha = 1/3$, it can be shown that (6)-(8) are equivalent to the equations of the standard model [3].

Assuming that α can be taken to be equal to $1/3$, (8) implies that the orbital radius increases with time as long as $n < 1/3$; in particular, for a degenerate star with $n = -1/3$, this inequality is satisfied.

However if $\alpha = \eta$ applies, then $\dot{a} > 0$ when $n < \eta$; e.g., this holds in the special cases:

(a) $q \geq 1$ with $n < 0.4572$. This includes the class of degenerate stars.
(b) $q > 4.985$ with $n = 1$. This is for the class of lower main-sequence stars.

CONTRADICTIONS

From the last deduction, namely that the orbital radius *increases* when $n < \alpha$, we see that in particular this condition holds for stars which are degenerate or not very degenerate with an effective $n < 0.4572$ with $q \geq 1$, and also for lower main-sequence stars with $n = 1$ and $q > 4.985$. Here we shall deduce an opposite result from the rate of change of angular momentum equation (6) which does not involve the details of Roche lobe overflows such as the nature of the star, and approximations for both the stellar and Roche lobe radii. This equation can therefore be regarded as being more fundamental.

The angular momentum equation (6) can be written as

$$\frac{\dot{a}}{2a} = -\frac{32}{5c^5}\left(\frac{Ga^7}{M^3}\right)^{1/2} M_1 M_2 \omega^5 - (q-1)\frac{\Delta}{M_2}.$$ (12)

In this form, it follows immediately that whenever $q \geq 1$, the orbital radius *decreases* instead, a result which contradicts the deduction from the model that the orbital radius should *increase* as long as $n < \alpha$.

CONCLUSIONS

Though we have deduced contradictions for the cases above, we shall show elsewhere that this contradiction will occur for other cases such as for the combination $n = -1/3, q \geq 0.6253$.

We attribute the above contradictions to the assumption in the standard model that the time derivatives of the Roche lobe and the stellar radii are always equal to each other at all times. In general, they are different since the two radii are not identical functions.

Because of this flaw in the model, the deductions from the standard model are generally incorrect. In principle, one should take into account the fact that $\dot{R}_L - \dot{R}_2$ does not vanish generally [8].

REFERENCES

1. Paczynski, B., *Acta.Astr.* **17**, 287 (1967).
2. Vila, S.C., *Ap.J.* **168**, 217 (1971).
3. Faulkner, J., *Ap.J.* **170**, L99 (1971).
4. McDermott, P.N., Taam, R.E., and Ringwald, F.A. *Ap.J.* **328**, 617 (1988).
5. Paczynski, B., *Ann.Rev.Astr.Ap.* **9**, 183 (1971).
6. Chia, T.T., *Ap.Sp. Sci.* **51**, 159 (1977).
7. Landau, L.D., and Lifshitz, E.M., *The Classical Theory of Fields*, 4th revised English edition, Oxford: Pergamon, (1975).
8. Chia, T.T., This proceeding

Asymptotically Schwarzschild Spacetimes[1]

Uchida Gen[*] and Tetsuya Shiromizu[†]

[*]*Department of Earth and Space Science, Osaka University, Toyonaka 560-0043, Japan*
[†]*Department of Applied Mathematics and Theoretical Physics, University of Cambridge, Silver Street, Cambridge CB3 9EW, Untited Kingdom*

Abstract. It is shown that if an asymptotically flat spacetime is asymptotically stationary, in the sense that $\mathcal{L}_\xi g_{ab}$ vanishes at the rate $\sim t^{-3}$ for asymptotically timelike vector field ξ^a, and the energy-momentum tensor vanishes at the rate $\sim t^{-4}$, then the spacetime is an asymptotically Schwarzschild spacetime.

There are many astrophysical phenomena that are best explained by black holes. The analysis on the phenomena are done assuming that those black holes are described by the Kerr spacetimes. This is because the uniqueness theorem of a black hole guarantees that a spacetime which is stationary, vacuum and asymptotically flat is uniquely the Kerr spacetime, and we believe that when gravitational collapse takes place and a black hole is formed, the spacetime becomes vacuum and accordingly stationary. However, one may argue that such a spacetime does not become *exactly* vacuum nor *exactly* stationary: the spacetime becomes *asymptotically* vacuum, and accordingly *asymptotically* stationary at a certain rate of the time. Then, a required uniqueness theorem is one that states asymptotically stationary, vacuum and flat spacetime is uniquely an asymptotically Schwarzschild spacetime. This is what we show in this paper.

To show the theorem, we use the notion of the asymptotic flat spacetime, first introduced by Ashtekar and Romano [2] at spacelike infinity and succeedingly developed in our previous study [3] at timelike infinity. We investigate the asymptotic behaviour of the gravitational field at the future timelike infinity, because we would like to know whether the gravitational field approaches asymptotically that of the Schwarzschild spacetime at the *late time* when the spacetime becomes asymptotically stationary.

The completion method introduced by [2] for the definition of asymptotic flatness leaves the infinity as a 3-manifold. As a result, the complicated differential structure needed in the former treatment can be avoided. Subsequently, in our previous study [3], we clarified that the method leads to a definite picture of hierarchy in the asymptotic behaviour of the gravitational field and the symmetry, and that it is

CP493, *General Relativity and Relativistic Astrophysics*, edited by C. P. Burgess and R. C. Myers
© 1999 American Institute of Physics 1-56396-905-X/99/$15.00

suitable to discuss such a notion as an "asymptotically Schwarzschild spacetime". In this presentation, we further investigate the hierarchy and prove the theorem that states asymptotically stationary, vacuum and flat spacetime is uniquely an asymptotically Schwarzschild spacetime.

Let us recall the definition of asymptotic flatness of [2,3] that will be used in the main proof and fix the notation.

DEFINITION: A physical spacetime $(.\hat{\mathcal{M}}, \hat{g}_{ab})$ is said to possess an *asymptote at future timelike infinity* $\breve{i}^+$ *to order* n (ATI-n) for a non-negative integer n, if there exists a manifold $\mathcal{M}$ with boundary $\mathcal{H}$, a smooth function Ω defined on $\mathcal{M}$, and an imbedding Ψ of an open subset $\hat{\mathcal{F}}$ in $.\hat{\mathcal{M}}$ to $\mathcal{M} - \mathcal{H}$ satisfying the following conditions:

(1) $\breve{i}^+ := \partial\mathcal{F} \cap (.\mathcal{M} - \Psi(.\hat{\mathcal{M}}))$ is not empty and $\breve{i}^+ \subset I^+(\mathcal{F})$ where $\mathcal{F} := \Psi(\hat{\mathcal{F}})$;

(2) $\Omega \overset{\vee}{=} 0$ and $\nabla_a\Omega \overset{\vee}{\neq} 0$, where $\overset{\vee}{=}$ denotes the equality evaluated on $\breve{i}^+$;

(3) $n^a := \Omega^{-4}\Psi^*\hat{g}^{ab}\nabla_b\Omega$ and $q_{ab} := \Omega^2(\Psi^*\hat{g}_{ab} + \Omega^{-4}F^{-1}\nabla_a\Omega\nabla_b\Omega)$ admit smooth limits to $\breve{i}^+$ with q_{ab} having signature($+++$) on $\breve{i}^+$, where $F := -\mathcal{L}_n\Omega$; and

(4) $\lim_{-\breve{i}^+}\Omega^{-(2+n)}T_{\hat{\mu}\hat{\nu}} \overset{\vee}{=} 0$ where $\hat{T}_{\hat{\mu}\hat{\nu}} := \Psi^*[(\hat{e}_\mu)^a(\hat{e}_\nu)^b\hat{T}_{ab}]$ in which $\{(\hat{e}_\mu)^a\}$ and $\hat{T}_{ab}$ are a tetrad and the physical energy-momentum tensor of $(.\hat{\mathcal{M}}, \hat{g}_{ab})$, respectively.

Since all the equations appearing in the following discussion are those on $\mathcal{M}$, unless it may cause ambiguity, we omit hereafter Ψ^* in front of the tensors defined on $.\hat{\mathcal{M}}$ for brevity.

Now we explore the consequence of the definition of an ATI-n spacetime for $n = 0$. Solving the Einstein equation under the fall-off condition on the energy-momentum tensor, $\lim_{-\breve{i}^+}\Omega^{-2}\hat{T}_{\hat{\mu}\hat{\nu}} = 0$, it is found that

$$\boldsymbol{F} \overset{\vee}{=} 1 \quad \text{and} \quad \boldsymbol{q}_{ab} \overset{\vee}{=} \boldsymbol{h}_{ab} \tag{1}$$

in an ATI-0 spacetime, where h_{ab} is the 3-metric of the unit timelike 3-hyperboloid. (For the details, see [3].) These results imply

$$\hat{g}_{ab} = {}^{(0)}\hat{g}_{ab} + \Omega^{(1)}\hat{g}_{ab} + O(\Omega^2) \quad \text{where} \quad {}^{(0)}\hat{g}_{ab} = (e^{-\eta})^2[-(d\eta)_a(d\eta)_b + h_{ab}] \tag{2}$$

in which ${}^{(n)}\hat{g}_{ab}$ is defined by

$$ {}^{(n)}\hat{g}_{ab} := \sum_{\mu,\nu}(\hat{e}_\mu)_a(\hat{e}_\nu)_b\,{}^{(n)}g_{\hat{\mu}\hat{\nu}} \tag{3}$$

with a function $g_{\hat{\mu}\hat{\nu}} := (\hat{e}_\mu)^a(\hat{e}_\nu)^b\hat{g}_{ab}$ that admits a smooth limit to $\breve{i}^+$. ${}^{(0)}\hat{g}_{ab}$ is a metric of the Milne universe and is equivalent to the metric of a Minkowski spacetime, $\hat{g}_{ab}^{\text{MIN}}$. In other words, eq.(2) tell us that an ATI-0 spacetime is an asymptotically Minkowski spacetime:

$$\hat{g}_{ab} = \hat{g}_{ab}^{\text{MIN}} + O(\Omega) \tag{4}$$

Hence, it is no surprise that the Riemann tensor asymptotically vanishes in such a spacetime. The trace part of the Riemann tensor asymptotically vanishes by virtue

73

of the fall-off condition on the energy-momentum tensor. The traceless part, or the Weyl tensor $\hat{C}_{ambn}$, can be best investigated by decomposing the tensor into the electric part $\hat{E}_{ab} := \hat{C}_{ambn}\hat{n}^m\hat{n}^n$ and the magnetic part $\hat{B}_{ab} := {}^*\hat{C}_{ambn}\hat{n}^m\hat{n}^n$. Then, it can be shown that both admits a smooth limit to $\breve{\imath}^+$. Their leading terms are

$$
{}^{(0)}\boldsymbol{E}_{ab} \overset{\smile}{=} 0 \qquad \text{and} \qquad {}^{(0)}\boldsymbol{B}_{ab} \overset{\smile}{=} 0. \tag{5}
$$

Simple calculation shows that the behaviours of ${}^{(n)}\boldsymbol{E}_{ab}$ and ${}^{(n)}\boldsymbol{B}_{ab}$ for $n \geq 1$ depends on that of the higher order energy-momentum tensor, which is arbitrary in an ATI-0 spacetime. (See [3] for the derivations.)

Now we derive the first-order asymptotic structure, that is, a structure which is possessed by all the ATI-1 spacetimes but not by ATI-0 spacetimes.

The energy-momentum tensor of an ATI-1 spacetime satisfies a stronger fall-off condition, $\lim_{\to\breve{\imath}^+}\Omega^{-3}\hat{T}_{\bar{\nu}\bar{\mu}} = 0$, than that of an ATI-0 and thus the behaviour of asymptotic gravitational fields is constrained stronger. In other words, ATI-1 spacetimes possess more asymptotic gravitational structure in common. The structure can be derived by solving the Einstein equation under the condition $\lim_{\to\breve{\imath}^+}\Omega^{-3}\hat{T}_{\bar{\nu}\bar{\mu}} = 0$. To obtain the equation, we first decompose the first-order metric ${}^{(1)}\hat{g}_{ab}$ as

$$
{}^{(1)}\hat{g}_{ab} = (e^{-\eta})^2 \left[{}^{(1)}F(d\eta)_a(d\eta)_b - 2{}^{(1)}\beta_{(a}(d\eta)_{b)} - 2{}^{(1)}\psi h_{ab} + 2{}^{(1)}\chi_{ab} \right] \tag{6}
$$

where ${}^{(1)}\beta_a$ and ${}^{(1)}\chi_{ab}$ are tangential to the Ω-const. surfaces, i.e., $q_a{}^b{}^{(1)}\beta_b = {}^{(1)}\beta_a$ and $q_a{}^c q_b{}^d{}^{(1)}\chi_{cd} = {}^{(1)}\chi_{ab}$; and ${}^{(1)}\chi_{ab}$ is traceless, i.e., $q^{ab}{}^{(1)}\chi_{ab} = 0$. With these quantities, the Einstein equation induces on $\breve{\imath}^+$ the following set of differential equations in an ATI-1 spacetime satisfying $\lim_{\to\breve{\imath}^+}\Omega^{-3}\hat{T}_{\bar{\nu}\bar{\mu}} = 0$:

$$
3{}^{(1)}\boldsymbol{F} + 2\triangle{}^{(1)}\boldsymbol{\psi} \overset{\smile}{=} 0
$$
$$
\boldsymbol{D}_a({}^{(1)}\boldsymbol{F} + 2{}^{(1)}\boldsymbol{\psi}) + \tfrac{1}{2}(\triangle - 2){}^{(1)}\boldsymbol{\beta}_a^T \overset{\smile}{=} 0
$$
$$
(h_{ab}\triangle - \boldsymbol{D}_a\boldsymbol{D}_b)({}^{(1)}\boldsymbol{F} + 2{}^{(1)}\boldsymbol{\psi}) + 2\boldsymbol{D}_{(a}{}^{(1)}\boldsymbol{\beta}_{b)}^T + 2(\triangle + 3){}^{(1)}\chi_{ab}^{TT} \overset{\smile}{=} 0 \tag{7}
$$

where the over-bar is omitted which shows that the quantity is transformed and the subscripts T and TT denote that the quantities are transverse and transverse-traceless, respectively. In the above we adopted Poisson gauge. The solution is given by

$$
{}^{(1)}\boldsymbol{F} \overset{\smile}{=} \sum \boldsymbol{a}_{\ell m}{}^{(1)}\boldsymbol{F}^{\ell m}, \qquad {}^{(1)}\boldsymbol{\beta}_a \overset{\smile}{=} 0,
$$
$$
{}^{(1)}\boldsymbol{\psi} \overset{\smile}{=} -\frac{1}{2}\sum \boldsymbol{a}_{\ell m}{}^{(1)}\boldsymbol{F}^{\ell m} \qquad {}^{(1)}\chi_{ab} \overset{\smile}{=} \sum_{\ell \neq 0} \boldsymbol{b}_{\ell m}^{(-)}{}^{(1)}\chi_{ab}^{TT(-)\ell m}, \tag{8}
$$

where

$$
{}^{(1)}\boldsymbol{F}^{\ell m}(\chi, \theta, \phi) := T_0^\ell(\chi)\boldsymbol{Y}^{\ell m}(\theta, \phi)
$$
$$
{}^{(1)}\chi_{\chi\chi}^{TT(-)\ell m} :\overset{\smile}{=} 0, \qquad {}^{(1)}\chi_{\chi A}^{TT(-)\ell m} :\overset{\smile}{=} T_1^\ell(\chi)\epsilon_A{}^B\mathcal{D}_B\boldsymbol{Y}^{\ell m}
$$

$$^{(1)}\chi_{AB}^{TT(-)\ell m} :\overset{\vee}{=} T_2^\ell(\chi)\epsilon^C{}_{(A}\mathcal{D}_{B)}\mathcal{D}_C Y^{\ell m}$$

$$T_0^\ell(\chi) = \mathcal{P}_2^\ell(\chi), \quad \mathcal{P}_n^\ell(\chi) := \frac{1}{\sqrt{\sinh\chi}}\boldsymbol{P}_{n-\frac{1}{2}}^{\ell+\frac{1}{2}}(\cosh\chi), \quad T_1^\ell(\chi)\overset{\vee}{=}\mathcal{P}_0^\ell(\chi)$$

$$T_2^\ell(\chi)\overset{\vee}{=}\frac{\sinh^2\chi}{(\ell-1)(\ell+2)}\left[\partial_\chi + 2\frac{\cosh\chi}{\sinh\chi}\right]\mathcal{P}_0^\ell(\chi) \quad \text{for } \ell\neq 1,-2.$$

DEFINITION:

$^{(1)}\hat{g}_{ab}$ given by eq.(6) is called the *first order asymptotic structure* of an AFTI-1 spacetime, where $^{(1)}\boldsymbol{F}$, $^{(1)}\boldsymbol{\beta}_a$, $^{(1)}\psi$ and $^{(1)}\chi_{ab}$ takes the form eq.(8) on $\breve{\imath}^+$, in the Poisson gauge.

In such a spacetime, it can be calculated that

$$^{(1)}\boldsymbol{E}_{ab}\overset{\vee}{=}\frac{1}{2}(\boldsymbol{D}_a\boldsymbol{D}_b - \boldsymbol{h}_{ab})^{(1)}\boldsymbol{F} \qquad \text{and} \qquad ^{(1)}\boldsymbol{B}_{ab}\overset{\vee}{=}\frac{1}{2}\epsilon_{ra}{}^s\boldsymbol{D}^{r(1)}\chi_{bs}. \tag{9}$$

Next we introduce the notion of *asymptotic stationarity* of ATI-n spacetimes, and prove that an ATI-1 spacetime that is asymptotically stationary to order 2 must be an asymptotically Schwarzschild spacetime. The plan of the proof is: 1) we derive, in the Poisson gauge, the reduced first asymptotic structure of an ATI-1 spacetime which is asymptotically stationary to order 2 in the lemma; 2) we transform the gauge as to show explicitly that such an asymptotic structure gives the Schwarzschild spacetime in the theorem.

A Killing vector field $\hat{\xi}^a$ is a vector field with respect to which the Lie derivative of the metric vanishes, $\mathcal{L}_{\hat{\xi}}\hat{g}_{ab} = 0$. This fact motivates us to define an asymptotic Killing vector field and its order as follows.

DEFINITION: An ATI-m spacetime $(\hat{\mathcal{M}},\hat{g}_{ab})$ is said to admit an *asymptotic Killing field* $\hat{\xi}^a$ to order n if

$$\lim_{-\breve{\imath}^+}\Omega^{-n}(\mathcal{L}_{\hat{\xi}}\hat{g})_{\hat{\mu}\hat{\nu}} = 0 \tag{10}$$

where $(\mathcal{L}_{\hat{\xi}}\hat{g})_{\hat{\mu}\hat{\nu}} := \Psi^*((\hat{e}_\mu)^a(\hat{e}_\nu)^b\mathcal{L}_{\hat{\xi}}\hat{g}_{ab})$.

Next, we consider how to define asymptotic stationarity, using this notion of an asymptotic Killing vector field. A spacetime is said to be stationary if it admits a timelike Killing vector field. If such a vector field $\hat{\xi}^a$ exists, it can be always normalized as to satisfy $\hat{g}_{ab}\hat{\xi}^a\hat{\xi}^b = -1$. Hence, we define an asymptotic stationary Killing vector field as follows.

DEFINITION: A vector field ξ^a is said to be an asymptotic stationary Killing vector field to order n of an ATI-m spacetime if $\hat{\xi}^a$ is admitted as an asymptotic Killing vector field to order n in the ATI-m spacetime and satisfies

$$\hat{g}_{ab}\hat{\xi}^a\hat{\xi}^b\overset{\vee}{=}-1 \tag{11}$$

on $\check{\imath}^+$.

Now we are ready to prove the following lemma. The proof is given in Ref. [1]

LEMMA: An ATI-1 spacetime is asymptotically stationary to order 2 if and only if $\boldsymbol{a}_{\ell m} \overset{\vee}{=} 0$ for $\ell \neq 0$ and $\boldsymbol{b}_{\ell m} \overset{\vee}{=} 0$.

The fact that $\boldsymbol{a}_{\ell m} \overset{\vee}{=} 0$ for $\ell \neq 0$ and $\boldsymbol{b}_{\ell m} \overset{\vee}{=} 0$ means that in such an ATI-1 spacetime, the first asymptotic structure takes the simple form

$$^{(1)}\boldsymbol{F} \overset{\vee}{=} \boldsymbol{a}_{00} \, ^{(1)}\boldsymbol{F}^{00}, \qquad ^{(1)}\psi \overset{\vee}{=} -\frac{1}{2} \boldsymbol{a}_{00} \, ^{(1)}\boldsymbol{F}^{00}, \qquad ^{(1)}\beta \overset{\vee}{=} 0 \qquad \text{and} \qquad ^{(1)}\chi_{ab} \overset{\vee}{=} 0. \qquad (12)$$

Before we show that such an ATI-1 spacetime is an asymptotically Schwarzschild spacetime, we remark an important fact relating to the definition of the angular-momentum of an asymptotically flat spacetime.

THEOREM: An ATI-1 spacetime which is asymptotically stationary to order 2 is an asymptotically Schwarzschild spacetime with mass a_{00} in the sense that the metric takes the form

$$\hat{g}_{ab} = \hat{g}_{ab}^{\text{SCH}} + O(r^{-2}) + O(t^{-2}), \qquad (13)$$

where

$$\hat{g}_{ab}^{\text{SCH}} := -(1 - \frac{2a_{00}}{r})(dt)_a(dt)_b + (1 - \frac{2a_{00}}{r})^{-1}(dr)_a(dr)_b + r^2(d\sigma)_{ab} \qquad (14)$$

in which $t > r$.

Let us summarize. In this presentation, we have proved that the asymptotically flat spacetime as defined in the definition of Sec. II is an asymptotically Schwarzschild spacetime in the sense of eq.(13), if the energy-momentum of the spacetime falls off at the rate faster than $O(\Omega^3)$ and the spacetime is asymptotically stationary to order 2 in the sense that $(\mathcal{L}_{\hat{\xi}}\hat{g})_{\hat{\mu}\hat{\nu}}$ falls off at the rate faster than $O(\Omega^2)$ for the asymptotically timelike vector $\hat{\xi}^a$, $\hat{g}_{ab}\hat{\xi}^a\hat{\xi}^b \overset{\vee}{=} -1$.

REFERENCES

1. U. Gen and T. Shiromizu, J. Math. Phys. **40**, 2021(1999)
2. A. Ashtekar and J.D. Romano, Class. Quantum Grav. **9** 1069(1992).
3. U. Gen and T. Shiromizu, J. Math. Phys. **39**, 6573(1998)

The Vaidya Metric, the Bach Tensor and Radiated Power

E.N. Glass

Physics Department, University of Windsor, Ontario N9B 3P4

Abstract.
A conformal map between a special case of the Vaidya metric and a 2-fluid Schwarzschild atmosphere is given. The Bach tensor and the conformal map allow the definition of a new conserved current. This current provides a bound to radiated power.

I INTRODUCTION

A spherical astrophysical system (star, star cluster, galaxy) described by the Vaidya metric has short-wavelength photons streaming outward in the region beyond the source. A conformal map has been constructed between a special case of the Vaidya metric and a 2-fluid Schwarzschild atmosphere [1], revealing a homothetic scale symmetry of the Vaidya solution.

The generator ξ^β of the conformal map allows the construction of a conserved current in the same sense that a Killing vector k^β allows the current j^α to be formed from the Einstein tensor: $j^\alpha = G^\alpha{}_\beta k^\beta$. The current j^α is conserved since the Einstein tensor is divergence-free and k^β satisfies Killing's equation. In the same way we define

$$J^\alpha := B^\alpha{}_\beta \xi^\beta \tag{1}$$

where $B_{\alpha\beta}$ is the Bach tensor, a symmetric, divergence-free, trace-free, conformally covariant tensor [2], [3], [4]

$$B_{\alpha\beta} = \nabla^\mu \nabla^\nu C_{\alpha\mu\beta\nu} + \frac{1}{2} R^{\mu\nu} C_{\alpha\mu\beta\nu}. \tag{2}$$

$B_{\alpha\beta} = 0$ for spacetimes which are conformally flat, or vacuum, or conformal to an Einstein space.

J^α is conserved since the Bach tensor is divergence-free and the symmetrized covariant derivative of the conformal generator ξ^β yields a 2-fluid metric multiplied

CP493, *General Relativity and Relativistic Astrophysics*, edited by C. P. Burgess and R. C. Myers
© 1999 American Institute of Physics 1-56396-905-X/99/$15.00

by a time-dependent function. Since the Bach tensor is trace-free and conformally covariant it follows that $B_{\alpha\beta}\xi^{(\alpha;\beta)}$ vanishes.

When J^{α} is computed for the homothetic case of the Vaidya metric and integrated over a $u = const$ three-volume, we obtain the radiated power as a function of r along the null surface. For r at, or beyond, the homothetic horizon the radiated power is bounded.

II   THE VAIDYA METRIC

The Vaidya metric in outgoing null coordinates, with arbitrary mass function $m(u)$ and $A := 1 - 2m(u)/r$, is given by

$$ds^2_{Vaidya} = Adu^2 + 2dudr - r^2(d\theta^2 + \sin^2\theta \; d\varphi^2). \tag{3}$$

Calculations are most easily done using a Newman-Penrose null tetrad:

$$l^{\alpha}\partial_{\alpha} = \partial_r, \qquad\qquad l_{\alpha}dx^{\alpha} = du, \tag{4a}$$

$$n^{\alpha}\partial_{\alpha} = \partial_u - \frac{1}{2}A\partial_r, \qquad n_{\alpha}dx^{\alpha} = dr + \frac{1}{2}Adu, \tag{4b}$$

$$m^{\alpha}\partial_{\alpha} = \frac{1}{\sqrt{2}r}(\partial_\vartheta + \frac{i}{\sin\theta}\partial_\varphi), \quad m_{\alpha}dx^{\alpha} = -\frac{r}{\sqrt{2}}(d\vartheta + i\sin\theta \; d\varphi). \tag{4c}$$

The Riemann curvature tensor is composed of the Weyl and Ricci tensors, and the only non-zero Weyl tensor component is $\Psi_2 = -m(u)/r^3$ where

$$C^{\alpha\mu\beta\nu} = \frac{m(u)}{r^3}(U^{\alpha\mu}V^{\beta\nu} + V^{\alpha\mu}U^{\beta\nu} + M^{\alpha\mu}M^{\beta\nu} + c.c.). \tag{5}$$

The Weyl tensor has been given in terms of a set of anti self-dual bivectors

$$U^{\alpha\beta} = 2\bar{m}^{[\alpha}n^{\beta]}, \tag{6a}$$

$$M^{\alpha\beta} = 2l^{[\alpha}n^{\beta]} - 2m^{[\alpha}\bar{m}^{\beta]}, \tag{6b}$$

$$V^{\alpha\beta} = 2l^{[\alpha}m^{\beta]}. \tag{6c}$$

The Ricci tensor is

$$R_{\mu\nu} = \frac{2\dot{m}}{r^2}l_{\mu}l_{\nu}. \tag{7}$$

III   THE VAIDYA MAP

A homothetic map of the Vaidya metric is generated by

$$\xi^{\alpha}\partial_{\alpha} = (u_0 + u)\partial_u + r\partial_r \tag{8}$$

where

$$\mathcal{L}_\xi g_{\alpha\beta}^{Vaidya} = 2g_{\alpha\beta}^{Vaidya} \tag{9}$$

subject to the condition $(u_0 + u)\dot{A} + rA' = 0$. For $A = 1 - 2m(u)/r$, this condition has solution $m(u) = m_0 u + m_1$ with $m_1 = m_0 u_0$. The Vaidya metric under the homothetic map is restricted to

$$ds_{homV}^2 = (1 - 2m_0 u/r - 2m_1/r)du^2 + 2dudr - r^2 d\Omega^2. \tag{10}$$

We define a new radial coordinate $y \mapsto r = yu$ and rewrite (10) as

$$ds_{homV}^2 = (1 - 2m_0/y - 2m_1/yu + 2y)du^2 + 2udual dy - y^2 u^2 d\Omega^2.$$

Now we factor out u^2 and introduce a new time coordinate $w \mapsto u = e^w$ to obtain

$$ds_{homV}^2 = u^2 \left[(1 - 2m_0/y - 2m_1/yu + 2y)dw^2 + 2dwdy - y^2 d\Omega^2 \right].$$

With $M(w, y) = m_0 + m_1 e^{-w} - y^2$ we have

$$ds_{homV}^2 = e^{2w} \left[(1 - 2M/y)dw^2 + 2dwdy - y^2 d\Omega^2 \right]. \tag{11}$$

We see from Eq.(9) that generator $\xi^\alpha \partial_\alpha = (u_0 + u)\partial_u + r\partial_r$ conformally relates the homothetic restriction of the Vaidya metric to a 2-fluid Schwarzschild atmosphere [1] i.e.

$$\mathcal{L}_\xi g_{\alpha\beta}^{homV} = 2e^{2w} g_{\alpha\beta}^{2fluid}. \tag{12}$$

The 2-fluid atmosphere consists of the Vaidya null radiation fluid and a spherical string fluid.

IV   THE CONSERVED CURRENT

The Bach tensor, given above in Eq.(2), is written in terms of the conformal and Ricci tensors. The Bianchi identities allow an alternate expression which is easier to evaluate

$$B_{\alpha\beta} = \nabla^\mu \nabla_{[\alpha} P_{\mu]\beta} + \frac{1}{2} P^{\mu\nu} C_{\alpha\mu\beta\nu}. \tag{13}$$

Here $P_{\mu\nu} = R_{\mu\nu} - (R/6)g_{\mu\nu}$. Substitution of the Weyl tensor Eq.(5) and the Ricci tensor Eq.(7) yields

$$B_{\alpha\beta}^{Vaidya} = \left(\frac{2\dddot{m}}{r^3}\right)l_\alpha l_\beta + \left(\frac{2\dot{m}}{r^4}\right)(l_\alpha n_\beta + n_\alpha l_\beta + m_\alpha \bar{m}_\beta + \bar{m}_\alpha m_\beta). \tag{14}$$

We write the conformal map generator ξ^β of Eq.(8) using the null tetrad vectors (4a) and (4b):

$$\xi^\beta = [(u_0 + u)\frac{A}{2} + r)]l^\beta + (u_0 + u)n^\beta. \tag{15}$$

The conserved four-current J^α, for the homothetic metric, is thus

$$J^\alpha = B^\alpha{}_\beta \xi^\beta \tag{16}$$

$$= l^\alpha \left[(u_0 + u)A\frac{m_0}{r^4} + \frac{2m_0}{r^3}\right] + n^\alpha \left[(u_0 + u)\frac{2m_0}{r^4}\right].$$

The interpretation of J^α as a power current is due to Lindquist, Schwartz, and Misner [5].

V   RADIATED POWER

On any $u = const$ null surface $\mathcal{N}$ the integrated power current will be

$$\dot{E} : = \int_{\mathcal{N}} J^\alpha l_\alpha \sqrt{-g}\, dr d\vartheta d\varphi \tag{17}$$

$$= \int_{\mathcal{N}} 8\pi \frac{m_0(u_0 + u)}{r^2} dr.$$

We integrate from r beyond the horizon r_h to future null infinity $\mathcal{I}^+$:

$$\dot{E} = \left[-8\pi \frac{m_0(u_0 + u)}{r}\right]_r^\infty \tag{18}$$

$$= 8\pi\, m_0(u_0 + u)/r.$$

For $r \geq r_h = 2m$ there is an upper bound to the radiated power

$$\dot{E} \leq 4\pi\, m_0(u_0 + u)/m$$
$$\dot{E} \leq 4\pi. \tag{19}$$

This bound is an artifact of the homothetic map which requires $m(u) = m_0(u_0 + u)$. The bound should exist for more general maps.

CONVENTIONS

Greek indices range over $(0,1,2,3) = (u, r, \vartheta, \varphi)$. Sign conventions are $2A_{\nu;[\alpha\beta]} = A_\mu R^\mu{}_{\nu\alpha\beta}$, and $R_{\alpha\beta} = R^\nu{}_{\alpha\beta\nu}$. Overdots abbreviate $\partial/\partial u$ and primes denote $\partial/\partial r$. The metric signature is $(+,-,-,-)$

REFERENCES

1. E.N. Glass and J.P. Krisch, Class. Quantum Grav. **16**, 1175 (1999).
2. P. Szekeres, Proc. Roy. Soc. A **304** 113 (1968).
3. C.N. Kozameh, E.T. Newman, and P. Tod, Gen. Rel. Grav. **17** 343 (1985).
4. R. Penrose and W. Rindler, *Spinors and Space-Time* (Cambridge Univ. Press, Cambridge, 1986) Vol 2, p129
5. R.W. Lindquist, R.A. Schwartz. and C.W. Misner, Phys. Rev. B **137**, 1364 (1965).

II. BLACK HOLES

On the Nature of Black Hole Entropy

Ted Jacobson

Institute for Theoretical Physics, University of California, Santa Barbara, CA 93106
Department of Physics, University of Maryland, College Park, MD 20742

Abstract. I argue that black hole entropy counts only those states of a black hole that can influence the outside, and attempt (with only partial success) to defend this claim against various objections, all but one coming from string theory. Implications for the nature of the Bekenstein bound are discussed, and in particular the case for a holographic principle is challenged. Finally, a generalization of black hole thermodynamics to "partial event horizons" in general spacetimes without black holes is proposed.

BLACK HOLE ENTROPY AND INTERNAL STATES

Let me begin by giving several reasons why we should not think that the Bekenstein-Hawking entropy $S_{\text{BH}} = A/4\hbar G$ of a black hole counts the number of internal states of the black hole. (By "the entropy" of a black hole I will always mean $A/4\hbar G$ in this article.) These reasons have been enunciated in a thoughtful article by Rafael Sorkin [1], which I will borrow from here.

1. The spatial region inside a black hole horizon can have arbitrarily large volume, with room for an arbitrarily large number of states. For example a Friedmann universe of any size can be joined to the interior of a Schwarzschild black hole. Thus the number of possible internal states of a black hole is unbounded.

2. A black hole is not in "internal equilibrium", so why should its thermodynamic entropy refer to its interior states?

3. Conditions inside the horizon are causally disconnected from the outside, so how can the states inside be thermodynamically relevant to the outside?

4. According to local quantum field theory the evaporation of a black hole is unitary, at least until the final stages, and the Hawking radiation is correlated to field degrees of freedom inside the black hole. The number of internal states of the black hole must therefore remain large enough to store all the correlations maintaining the purity of the total state. As a black hole evaporates, however, its area and therefore its entropy decreases. Thus the entropy must not be counting the number of internal states.

CP493, *General Relativity and Relativistic Astrophysics*, edited by C. P. Burgess and R. C. Myers
© 1999 American Institute of Physics 1-56396-905-X/99/$15.00

Regarding point 1, it should be mentioned that the example given will have a white hole horizon and singularity in its past (assuming the weak energy condition holds) so it is not a configuration that would evolve from an ordinary collapse process [2]. It is nevertheless a possible state of the black hole.

There is by now a "standard" argument against points 3 and 4, namely, that local quantum field theory may be inapplicable. This argument is suggested by (but not restricted to) string theory, in which local quantum field theory is only an approximation valid under certain conditions. It has been argued both on general principles [3] and in string theory [4] that there are no truly local observables in quantum gravity and that for this reason the decomposition of the Hilbert space into sectors inside and outside the black hole is invalid from the beginning. While this may indeed be true at some fundamental level, the relevant question here is whether local quantum field theory holds to a sufficiently good approximation for points 3 and 4 to be valid. Since the black hole can be macroscopic and the curvature can be very small compared with the string length or Planck length, it is hard for me to see why the local field theory approximation should fail in this regard. To postulate such a mysterious failure, when simpler scenarios exist, seems to me uncalled for radicalism, although it is a hypothesis favored by many physicists today.

BLACK HOLE ENTROPY AND SURFACE STATES

The previous arguments point to the conclusion that black hole entropy is a measure of only those states that can influence the outside of the black hole.[1] These states must be associated with the presence of the horizon, otherwise they would simply be counted as ordinary states of the exterior itself.

One interpretation of this "surface entropy" is that it measures the information in the entanglement of the vacuum across the horizon ("entanglement entropy") [7]. For fields on a fixed background this is equivalent [8] to the entropy of the thermal state ("thermal atmosphere") that results when the state is restricted to the outside [9]. This entropy diverges, but gives something of the correct order of magnitude if a Planck scale cutoff is imposed.

It is insufficient to consider fields on a fixed background however. For one thing, although the contributions of quantum fields can be thought of as "loop corrections" to the black hole entropy, there is also a classical contribution coming from the gravitational action itself. On can imagine an induced gravity scenario [10–13], in which the entire gravitational action is induced by matter, however there is still

---

[1] The case for a surface interpretation of black hole entropy has been made by various authors. In particular, an article by Banks [5] (written before the age of D-branes) makes the case with many of the same arguments as used here, and the argument that the universality of black hole entropy (in spite of the *non*-universal history of the black hole) arises from the universality of the near-horizon geometry was made in a paper by Parentani and Piran [6].

another problem: for non-minimally coupled scalar fields or gauge fields, the entanglement entropy is not equal to the corresponding contribution to the entropy computed from the induced gravitational action. It seems that the difference between these two entropies can be understood as a consequence of the fact that the background itself varies when the temperature is varied [14,15]. Physically, this means that to understand the entropy one must count states in the coupled matter-gravity vacuum.

The large and universal number of states per unit of surface area seems to be explained by the infinite redshift at the horizon: many states at short distances near the horizon have the same, low, energy. In fact, the number would appear to be *infinite* from perturbative counting, but the final count requires knowledge of only the low energy effective gravitational action and the associated low energy Newton constant, as long as the spacetime curvature is small compared with Planck curvature [16,11,17]. Although we are unable to compute the renormalized Newton constant from quantum gravity, its (finite) value can be measured and used in the entropy formula.

Entanglement entropy and the generalized second law

Sorkin proposed a derivation of the generalized second law based on the entanglement interpretation of black hole entropy [1,18]. His idea was that the total entropy $S_{outside} = S_{horizon} + S_{rest}$ of the reduced density matrix outside the horizon receives a large universal contribution $S_{horizon}$ from the vicinity of the horizon and the rest S_{rest} is primarily just the ordinary entropy of a mixed state outside. Invoking the dynamical autonomy of the evolution outside the horizon, Sorkin argues that $S_{outside}$ cannot decrease, which amounts to the usual generalized second law provided $S_{horizon}$ can be identified with the black hole entropy. This explanation of the generalized second law seems so natural that it is hard to believe there is not some truth in it. Unfortunately, as mentioned above, the entanglement interpretation of black hole entropy does not seem to work, but perhaps this conclusion is premature. Perhaps the black hole entropy could yet be understood in terms of entanglement entropy if, as proposed in [19], the division of the system into inside and outside is referred to an intrinsic feature of the fluctuating geometry such as the minimal throat area on some preferred spacelike slice.

OBJECTIONS

Objections can be raised to the assertion that black holes have many more states than are counted by the black hole entropy. I believe that all of these objections are wrong, but it is challenging and instructive to try to point to exactly where they are wrong. I will try to do so here with regard to several objections, all but the first coming from string theory.

Black hole pair creation amplitudes

Semiclassical calculations of black hole pair creation rates display a factor $\exp S_{\mathrm{BH}}$ which admits the natural interpretation as a density of states factor [24]. This seems to lend solid support to the interpretation of $\exp S_{\mathrm{BH}}$ as the number of states of the black hole. If the black hole had more states, would they not contribute to the pair creation rate? This question has been discussed in the past, with conflicting conclusions [20,21,5,22,23], and it deserves to be discussed further. Here I will only state the reason for my belief that the answer is no[2]: Pair creation is an exponentially suppressed tunneling process, and any "unnecessary" decoration of the black holes would, it seems, be even more suppressed. All the extra internal states are unnecessary decoration, and are therefore essentially irrelevant to the pair creation rates.

String theory

Calculations of black hole entropy in string theory and its descendents have been carried out in several contexts yielding agreement with the Bekenstein-Hawking entropy. In all cases it appears that one is indeed counting all of the states of the object identified with a black hole. Can this be compatible with the claim that black hole entropy does *not* count all of the states? I will attempt to argue that it can, pointing to where one might find the other states. My attempts are only partly successful, and are particularly weak in the context of the AdS/CFT duality.

D-branes

The entropy of certain near-extremal configurations of D-branes has been found to agree with the semiclassical entropy of the black hole configurations with the same set of charges (see for example [25,26]). In the extremal case, for the supersymmetric BPS states, this is understood as a consequence of the fact that the D-brane configuration evolves into the black hole as the string coupling is increased from weak to strong, all the while maintaining the supersymmetry. The enumeration of BPS states is independent of the coupling, hence the agreement in the count of states. In the D-brane picture there is nothing corresponding to the inside of the black hole where extra states can reside so, given the agreement with the black hole entropy, how could a black hole have any more states? For the BPS states the answer is simple: the black hole also does not have any interior in the sense that on a spacelike slice orthogonal to the timelike Killing field the horizon is infinitely far away and has no other side.

The D-brane and black hole entropies also agree for near-extremal states however. In these cases, one can not give such a simple answer. Imagine for instance a

---

[2] I was asked this question during my talk and had no quick answer. After the talk Renaud Parentani suggested the following answer.

configuration that has been maintained at fixed energy above extremality for a long time with the help of an influx of energy equal in magnitude to the Hawking flux. In the black hole picture there is an arbitrarily large amount of information stored in the correlations between the inside and outside of the black hole, so there must be a correspondingly large number of states for the interior. In the D-brane description however there is nothing that corresponds to the interior. How could there be such a drastic mismatch between the total number of states in the two descriptions and still be such agreement on not only the entropy but also the rate of Hawking emission (i.e. the "greybody factors")?

I can give no really satisfying answer to this question. Surely one has less control over the correspondence between strong and weak coupling away from the BPS sector. It is conceivable that the initial rates for Hawking radiation agree but the details about the correlations that develop over time do not match. In this scenario, there would simply be more non-BPS states at strong coupling than there are at weak coupling. This is not so hard to imagine, since in going from weak to strong coupling the causal structure of the background spacetime is distorted into that of a black hole. An analogy that may be useful is the coupling constant dependence of the state space of electrons in an atom. At sufficiently strong electric coupling the ground state becomes unstable and electrons can be absorbed into the nucleus, at which point the nuclear Hilbert space comes into play in resolving the physics. A strength of this analogy is that in the black hole case the ergoregion inside the (non-extremal) horizon also manifests a kind of instability of the ground state.

AdS/CFT duality

The near-horizon limit of the D-brane physics led to the celebrated Maldacena conjecture, according to which supergravity/string theory in an asymptotically Anti-deSitter spacetime is equivalent to a superconformal field theory on the conformal boundary of that spacetime [27]. An example of this is the duality between superstring theory on $AdS_5 \times S^5$ and a $U(N)$ super-Yang-Mills theory on $S^3 \times R$, where N is related to the string coupling g_s the string length ℓ_s and the AdS radius R by $R^4 = 4\pi g_s N \ell_s^4$, and the Yang-Mills and string couplings are related by $g_{YM}^2 = 4\pi g_s$. There is much remarkable evidence in favor of the AdS/CFT duality, and no evidence against it to date. Hence, for the sake of argument, let us suppose it is valid and ask about the consequences for black holes.

In the $AdS_5 \times S^5$ example it has been shown that the entropy of a black hole which is large compared to the AdS radius (and is hence stable) is 3/4 of the entropy of a thermal state in the Yang-Mills theory at weak 't Hooft coupling ($g_{YM}^2 N \ll 1$) at the corresponding Hawking temperature. Moreover, there is reason to believe that the entropy would only change by a factor of order unity if the calculation could be done at strong 't Hooft coupling (which is what is required by for the case of large AdS radius).

We thus have a puzzle similar to that in the case of the D-brane state counting, but now far from extremality. In the Yang-Mills theory it seems there can be no missing states corresponding to the degrees of freedom inside of the black hole. The entropy of the thermal state simply counts all states so, if the Maldacena conjecture is really true, one infers that there can be no independent degrees of freedom inside the black hole. Can this conclusion be evaded?

A simple evasion is to suppose that the equivalence conjectured by Maldacena actually relates the supergravity observables only *outside* the horizon to the Yang-Mills observables in the boundary theory (see Fig. 1(a)). This would be consistent

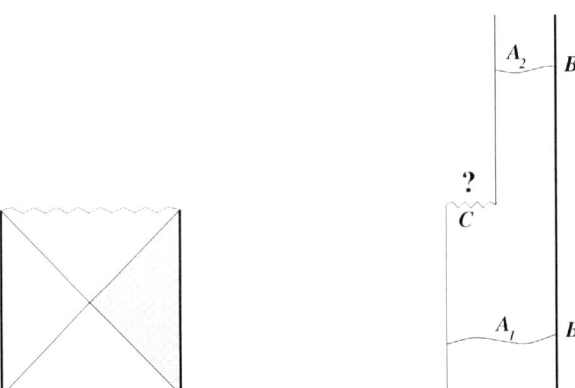

FIGURE 1. (a) Schwarzschild-AdS spacetime. The conformal boundary is the pair of thick vertical lines. A single copy of the Yang-Mills theory on the boundary may just correspond to the gravity theory in the shaded wedge. (b) Black hole formation and evaporation in Anti-de Sitter spacetime. The algebra of observables $\mathcal{A}_2$ may be a proper subalgebra of $\mathcal{A}_2$.

with causality and would certainly explain why all the states inside the black hole are not seen in the Yang-Mills theory. In fact, something like this seems almost necessary in view of the fact that the full Schwarzschild-AdS spacetime has a boundary with two disconnected pieces, the dark vertical lines in Fig. 1(a). What would be the role for the states in the Yang-Mills theory on the left if the one on the right already covered all the states inside the black hole?

A different evasion is required if we consider not an eternal black hole but rather a black hole, small compared with the AdS radius, that forms from collapse and then evaporates (see Fig. 1(b)). In this case the AdS/CFT duality presumably states that *all* observables in the spacetime have correspondents in the CFT on the single boundary component. In particular, observables in the algebra C localized behind the horizon (which is of course only defined relative to a particular state $|\psi\rangle$ of the CFT which corresponds to matter collapsing to form a black hole) must be contained within the full algebra B of observables in the CFT. This in itself is not mysterious, since the field equations allow us to express any observable in C as an observable in the algebra A_1 localized at a spacelike slice before the black hole ever formed.

The question of whether there are independent states of the black hole interior is perhaps most sharply formulated here as the question whether the black hole evaporation is unitary from the viewpoint of the exterior [28]. Since the CFT itself is unitary, the question amounts to whether the algebra A_2 of observables on a spacelike slice after the black hole has evaporated completely is equal to A_1 or is rather a proper subalgebra of A_1. In the latter case one would need also the observables in the algebra C behind the horizon to fill out the complete algebra. Moreover, causality would suggest that C and A_2 would commute relative to the state $|\psi\rangle$, that is, expectation values of the ideal generated by the commutator algebra $[C, A_2]$ would vanish.

Many practitioners of duality have argued that the equality $A_2 = A_1$ is assured because the situations before collapse and after evaporation are similar: Anti-de Sitter spacetime with some matter. In particular, the "initial" configuration could have been the result of a prior black hole formation and evaporation, or of many cycles of formation and evaporation. If each black hole has internal observables not captured on the outside after the black hole is gone, then one seems to be requiring that the CFT contains within it (relative for an appropriate state) commuting subalgebras of observables corresponding to an infinite number of such black hole interiors, all of which commute with an algebra of outside observables such as A_2. This requirement seems difficult to reconcile with reasonable expectations about the number of states in the CFT at a given energy.

Can one really arrange a sequence of black hole formations and evaporations where each black hole is made from the Hawking radiation into which the previous black hole evaporated? If not, then a state which produces many black holes must contain to begin with energy corresponding to each black hole. In this case there there are perhaps more states so let us suppose, to be difficult, that one can indeed repeatedly refocus the Hawking radiation to form an endless cycle of black holes with a finite amount of energy. In this case either the collection of commuting interior subalgebras exists, or one must deny the independence of the interior observables. Most string theorists support the second alternative. I prefer the first since it requires only nonintuitive behavior of the unfamiliar strongly coupled, (astronomically) large N gauge theory, rather than gross violations of locality where they would not otherwise be expected.

Matrix Theory

Matrix theory (a candidate for a nonperturbative formulation of string theory) can purportedly describe formation and evaporation of black holes, and the theory is manifestly unitary. There seems to be no room in matrix theory for any states corresponding to the interior of a black hole, left over after all particles in the Hawking radiation have dissipated [29]. I have not yet learned enough about matrix theory to think carefully about whether or not there is any loophole through which this conclusion can be evaded.

NATURE OF THE BEKENSTEIN BOUND

The "Bekenstein bound" [30] on the entropy that can be associated with a closed 2-surface Σ is

$$S_\Sigma \leq \frac{A_\Sigma}{4\hbar G}. \tag{1}$$

This is (presently) a heuristic notion motivated by the generalized second law of thermodynamics as follows. Suppose that by tossing in a suitable arrangement of matter the surface Σ could be made to coincide with a slice of the horizon of a black hole. Then the entropy of that black hole would be $A_\Sigma/4\hbar G$, which would violate the second law unless the entropy S_Σ associated with Σ if the extra matter is *not* tossed in is *less*, i.e. unless the bound (1) holds.

In describing the Bekenstein bound I was careful to refer to S_Σ as the entropy *associated with* Σ, rather than the entropy *contained within* Σ, since the meaning of the bound (1) inferred by the black hole formation argument depends on the interpretation of the black hole entropy. If the black hole entropy is the logarithm of the number of states of the the black hole including the interior states, then we infer a "volume bound" on the entropy contained within Σ. If however, as argued above, the black hole entropy reflects only those states that can influence the exterior, then we infer only a "surface bound" on the surface states of Σ.

I do not consider the volume bound interpretation to be viable. Not only can it not be inferred from the second law with the surface interpretation of black hole entropy, but it seems contradicted by the example used in the first section of this paper: since the volume of the region interior to the surface could be arbitrarily large it could contain an arbitrarily large amount of entropy. It also suffers from a species problem, that is, the entropy inside could be arbitrarily large if the number of independent fields in nature is arbitrarily large (but see [33] for another point of view). (On the other hand, if the number of species is sufficient for an order unity violation of the bound, then a black hole would be unstable to explosive evaporation on a timescale of ordet the light crossing time, and so the original rationale for the bound would be lost [34].)

As an important side remark, note that the black hole formation argument suggesting the bound (1) does not apply to every closed 2-surface, since not every such surface can be made to coincide with a slice of the horizon of a black hole. Consider for instance an outer trapped surface inside a black hole. The future pointing null congruences orthogonal to this surface are converging on both sides, whereas the horizon generators are always non-converging according to the area theorem. For another example, consider the intersection of the past light cones of two spacelike related points p and q. The future pointing null congruences orthogonal to this intersection surface are converging (to p and q) on both sides. (This surface is not compact, but one can build a compact 2-surface out of pieces like this.) The restriction on surfaces is certainly necessary for the volume interpretation of the bound (although as discussed above I do not consider this interpretation to be viable in any case), since otherwise it is easy to find surfaces with arbitrarily little area enclosing a large volume. For example, a trapped surface near the singularity of a Schwarzschild black hole can have arbitrarily small area and still bound a finite volume. For another example, one can make a spacelike surface of arbitrarily small area enclose any volume by wiggling the surface "up and down" in the timelike direction.

An interpretation of the bound (1) that is neither a volume nor a surface interpretation has been proposed by Bousso [35]. In this interpretation, S_Σ is the entropy crossing any segment of a null hypersurface, meeting Σ orthogonally, that is expanding towards Σ. The validity of this bound in a variety of contexts has been argued for in Ref. [35].

The volume bound interpretation of (1) suggests the "holographic principle" [31,36] according to which all the physics in the volume should be describable by a theory on the bounding surface Σ. The surface bound interpretation on the other hand does not have any holographic connotation. Bousso suggests that his bound motivates a holographic priciple which refers to the null surface segments, but these segments do not in general span the volume. It thus seems to me that the holographic principle, while it may be a property of quantum gravity and/or of the AdS/CFT duality, is *not* logically suggested by the Bekenstein bound.

BLACK HOLE ENTROPY WITHOUT BLACK HOLES

I have argued above that black hole entropy is not determined by the number of internal states of the black hole, but rather by the number of states, associated with the presence of the horizon, that can influence the outside world. This suggests that the notion of black hole entropy should apply not just to black holes but to any causal horizon.

In fact, some approaches to computing the entropy associated with horizons do yield the result 1/4 per Planck area of a Rindler horizon or a deSitter horizon, both of which are observer dependent horizons. For example, in a recent paper Carlip [37] finds this result from the representation theory of a conformal subgroup of the diffeomorpism group associated with any (non-degenerate) Killing horizon, and he points out that the Euclidean path integral approach also yields an entropy for deSitter horizons [38]. Also, the black hole pair creation probability is weighted by $\exp(\Delta A_{\mathrm{accel}}/4)$ where $\Delta A_{\mathrm{accel}}$ is the associated increase of the area of an acceleration horizon [41,40]. This suggests a state-counting role for the entropy of acceleration horizons (although Hawking and Horowitz [41] argue that one should nevertheless *not* attribute an entropy to the acceleration horizon because of its observer-dependent nature).

As a more direct way to establish the validity of horizon entropy without black holes, I will will now argue that there are general laws of horizon thermodynamics, strictly analogous to those for black holes, for a class of causal horizons which

I will call "partial event horizons". Recall that the global event horizon of an asymptotically flat spacetime is the boundary of the past of future null infinity $\mathcal{I}^+$. I define similarly a partial event horizon (PEH) as the boundary $\partial I^-[p]$ of the past of a single point[3] $p \in \mathcal{I}^+$. In flat spacetime a PEH is just a Rindler (acceleration) horizon, and in an asymptotically flat spacetime a PEH asymptotically approaches a Rindler horizon.

Although a PEH has cross sections with infinite area, it satisfies Hawking's classical area theorem in the local sense that the expansion of its null generators is nowhere negative. The proof is similar to but slightly simpler than that for the event horizon since the assumption of cosmic censorship can be applied directly to rule out the possibility that a null generator leaves the PEH before reaching $\mathcal{I}^+$. Thus *changes* in the area are nonnegative, so a PEH satisfies a classical "second law of horizon mechanics".

A quasistationary region of a PEH also satisfies a "first law of horizon mechanics" that is strictly analogous to the first law of black hole mechanics $dM = (\kappa/8\pi)dA$. This law for black holes can be understood in a quasi-local fashion, called the "physical process version" in Ref. [42], which applies to variations away from a quasi-stationary configuration with approximate horizon generating Killing field χ^a. In this setting dM is interpreted as the flux $\int T_{ab}\chi^a d\Sigma^b$ of "boost energy" across the horizon or a part thereof. A generic PEH will possess many quasistationary regions, to which the physical process version of first law will apply for the same reason as for black hole horizons. (The normalization ambiguity of the boost Killing field scales both dM and κ in the same way, so the first law is independent of this ambiguity [43].

Finally, as for the generalized second law, note that Sorkin's proposal for the origin of the generalized second law described above applies to any causal horizon, and in particular it applies to a PEH. Moreover, it seems that all gedanken experiments supporting the generalized second law for quasistationary processes involving black hole horizons would apply as well to quasistationary regions of PEH's.

CONCLUDING REMARKS

What distinguishes a black hole horizon from a more general causal horizon is that it is universally defined with reference only to the global causal structure of the spacetime. The absence of reference to particular observers or classes of observers is thus its key distinguishing feature. In practice, however, this universality is irrelevant. For example, the universe may be spatially compact, and yet we have no reservations in applying the laws of black hole thermodynamics to approximately isolated "black holes". It is always we who divide the system into the "outside" and the "inside". It thus seems entirely natural that the notion of black hole entropy extends to general causal horizons. This generalized notion of horizon entropy preserves the the formula $S = A/4\hbar G$, whose universality is understood

---

[3] One could of course consider the boundary of the past of any subset of $\mathcal{I}^+$.

as arising from the ultraviolet dominance of the "density of surface states", much as the universal form of the short distance limit of quantum field correlations is understood.

ACKNOWLEDGEMENTS

I am grateful to numerous colleagues for even more numerous discussions on the topics discussed here. This work was supported in part by the National Science Foundation under grants No. PHY98-00967 at the University of Maryland and PHY94-07194 at the Institute for Theoretical Physics.

REFERENCES

1. R.D. Sorkin, "The statistical mechanics of black hole thermodynamics," in *Black Holes and Relativistic Stars*, Chicago: The University of Chicago Press, 1998, ch. 9, pp. 177-194; gr-qc/9705006.
2. R.D. Sorkin, R.M. Wald, and Z.Z. Jiu, "Entropy of self-gravitating radiation", *Gen. Rel. Grav.* **13**, 1127 (1981).
3. G. 't Hooft, "The Scattering matrix approach for the quantum black hole: An Overview," *Int. J. Mod. Phys.* **A11**, 4623 (1996); gr-qc/9607022.
4. D.A. Lowe, J. Polchinski, L. Susskind, L. Thorlacius and J. Uglum, "Black hole complementarity versus locality," *Phys. Rev.* **D52**, 6997 (1995); hep-th/9506138.
5. T. Banks, "Lectures on black holes and information loss," *Nucl. Phys. Proc. Suppl.* **41**, 21 (1995); hep-th/9412131.
6. R. Parentani and T. Piran, "The Internal geometry of an evaporating black hole," *Phys. Rev. Lett.* **73**, 2805 (1994); hep-th/9405007.
7. R.D. Sorkin, "On the Entropy of the vacuum outside a horizon," in B. Bertotti, F. de Felice, and A. Pascolini (eds.), *Tenth International Conference on General Relativity and Gravitation, Contributed Papaers*, Vol. 2 (Consiglio Nazionale Delle Ricerche, 1983), 734; L. Bombelli, R.K. Koul, J. Lee and R.D. Sorkin, "A Quantum source of entropy for black holes," *Phys. Rev.* **D34**, 373 (1986).
8. See, for example, D. Kabat and M.J. Strassler, "A Comment on entropy and area," *Phys. Lett.* **B329**, 46 (1994); hep-th/9401125.
9. G. 't Hooft, "On the Quantum structure of a black hole," *Nucl. Phys.* **B256**, 727 (1985).
10. A.D. Sakharov, "Vacuum quantum fluctuations in curved space and the theory of gravitation," *Sov. Phys. Dokl.* **12**, 1040 (1968).
11. T. Jacobson, "Black hole entropy and induced gravity," gr-qc/9404039.
12. V.P. Frolov, D.V. Fursaev and A.I. Zelnikov, "Statistical origin of black hole entropy in induced gravity," *Nucl. Phys.* **B486**, 339 (1997); hep-th/9607104.
13. V.P. Frolov and D.V. Fursaev, "Mechanism of generation of black hole entropy in Sakharov's induced gravity," *Phys. Rev.* **D56**, 2212 (1997); hep-th/9703178.
14. V.P. Frolov, D.V. Fursaev and A.I. Zelnikov, "Black hole entropy: Off-shell vs on-shell," *Phys. Rev.* **D54**, 2711 (1996); hep-th/9512184.

15. V.P. Frolov and D.V. Fursaev, "Thermal fields, entropy, and black holes," *Class. Quant. Grav.* **15**, 2041 (1998); hep-th/9802010.

16. L. Susskind and J. Uglum, "Black hole entropy in canonical quantum gravity and superstring theory," Phys. Rev. **D50**, 2700 (1994); hep-th/9401070.

17. F. Larsen and F. Wilczek, "Renormalization of black hole entropy and of the gravitational coupling constant," *Nucl. Phys.* **B458**, 249 (1996); hep-th/9506066.

18. R.D. Sorkin, "Toward an explanation of entropy increase in the presence of quantum black holes," *Phys. Rev. Lett.* **56**, 1885 (1986).

19. A.O. Barvinsky, V.P. Frolov and A.I. Zelnikov, "Wave function of a black hole and the dynamical origin of entropy," *Phys. Rev.* **D51**, 1741 (1995); gr-qc/9404036.

20. T. Banks and M. O'Loughlin, "Classical and quantum production of cornucopions at energies below 10^{18} GeV," *Phys. Rev.* **D47**, 540 (1993); hep-th/9206055.

21. T. Banks, M. O'Loughlin and A. Strominger, "Black hole remnants and the information puzzle," *Phys. Rev.* **D47**, 4476 (1993); hep-th/9211030.

22. S.B. Giddings, "Comments on information loss and remnants," *Phys. Rev.* **D49**, 4078 (1994); hep-th/9310101.

23. S.B. Giddings, "Why aren't black holes infinitely produced?," *Phys. Rev.* **D51**, 6860 (1995); hep-th/9412159.

24. D. Garfinkle, S.B. Giddings and A. Strominger, "Entropy in black hole pair production," *Phys. Rev.* **D49**, 958 (1994); gr-qc/9306023.

25. J.M. Maldacena, "Black holes in string theory," hep-th/9607235.

26. A.W. Peet, "The Bekenstein formula and string theory (N-brane theory)," *Class. Quant. Grav.* **15**, 3291 (1998); hep-th/9712253.

27. O. Aharony, S.S. Gubser, J. Maldacena, H. Ooguri and Y. Oz, "Large N field theories, string theory and gravity," hep-th/9905111.

28. See for example D.A. Lowe and L. Thorlacius, "AdS/CFT and the information paradox," hep-th/9903237.

29. L. Susskind, private communication.

30. In the form given here the first reference I know of is Ref. [31]. The name "Bekenstein bound" is actually somewhat of a misnomer, since the bound usually discussed by Bekenstein [32] ($S < 2\pi ER$) is actually a bound on the entropy given an "energy" E in a region of "size" R. See however [33] for a discussion of the bound (1).

31. G. 't Hooft, "Dimensional reduction in quantum gravity," in *Salamfestschrift*, edited by A. Ali, J. Ellis, S. Randjbar-Daemi, Singapore: World Scientific, 1994; gr-qc/9310026.

32. J.D. Bekenstein, "A Universal upper bound on the entropy to energy ratio for bounded systems," *Phys. Rev.* **D23**, 287 (1981).

33. J.D. Bekenstein, "Do we understand black hole entropy?," *The Seventh Marcel Grossmann Meeting on Recent Developments in Theoretical and Experimental General Relativity, Gravitation, and Relativistic Field Theories: Proceedings*, edited by R.T. Jantzen, G. Mac Keiser, and Remo Ruffini (eds.), Singapore: World Scientific, 1996; gr-qc/9409015.

34. T. Jacobson, to be published.

35. R. Bousso, "A Covariant entropy conjecture," *JHEP* **07**, 004 (1999); hep-th/9905177; "Holography in general space-times," *JHEP* **06**, 028 (1999); hep-th/9906022.

36. L. Susskind, "The World as a hologram," *J. Math. Phys.* **36**, 6377 (1995); hep-th/9409089.

37. Carlip, S., "Entropy from conformal field theory at Killing horizons," gr-qc/9906126.

38. G.W. Gibbons and S.W. Hawking, "Cosmological event horizons, thermodynamics, and particle creation," *Phys. Rev.* **D15**, 2738 (1977).

39. S.W. Hawking, G.T. Horowitz and S.F. Ross, "Entropy, area, and black hole pairs," *Phys. Rev.* **D51**, 4302 (1995); gr-qc/9409013.

40. S. Massar and R. Parentani, "Gravitational instanton for black hole radiation," *Phys. Rev. Lett.* **78**, 3810 (1997); gr-qc/9701015.

41. S.W. Hawking and G.T. Horowitz, "The Gravitational Hamiltonian, action, entropy and surface terms," *Class. Quant. Grav.* **13**, 1487 (1996); gr-qc/9501014.

42. Wald, R.M., *Quantum Field Theory in Curved Spacetime and Black Hole Thermodynamics*, Chicago: The University of Chicago Press, 1994.

43. T. Jacobson, "Thermodynamics of space-time: the Einstein equation of state," *Phys. Rev. Lett.* **75**, 1260 (1995); gr-qc/9504004.

Holography in the Flat Space Limit

Leonard Susskind

Department of Physics, Stanford University
Stanford, CA 94305-4060

Abstract. Matrix theory and the AdS/CFT correspondence provide nonperturbative holographic formulations of string theory. In both cases the finite N theories can be thought of as infrared regulated versions of flat space string theory in which removing the cutoff is equivalent to letting N go to infinity.

In this paper we consider the nature of this limit. In both cases the holographic mapping becomes completely nonlocal. In matrix theory this corresponds to the growth of D0-brane bound states with N. For the AdS/CFT correspondence there is a similar delocalization of the holographic image of a system as N increases. In this case the limiting theory seems to require a number of degrees of freedom comparable to large N matrix quantum mechanics.

I   INTRODUCTION

According to the holographic principle, a physical system of dimensionality D which includes gravity, should be described by a quantum system which lives in fewer dimensions. We have seen a good deal of evidence for the holographic principle from both matrix theory and the AdS/CFT correspondence but very little real understanding of how it works, in other words, how a general configuration of a D dimensional system is coded by lower dimensional degrees of freedom. My main purpose in this paper is to provoke discussion about the mechanism of holography [1], [2]. Most of the things that I will discuss I do not understand very well. In trying to formulate them precisely I have mainly encountered frustration. Nevertheless I think they are important and deserve to be discussed.

One of the characteristic features of a real hologram is that it codes information in a highly nonlocal way. For example by casually looking at a hologram of several distinct objects it is impossible to tell how many objects it describes or their size and shape. These details are completely delocalized on the hologram. The point of this paper is to argue that quantum gravity is holographic in exactly this sense.

Two concrete realizations of holographic theories now exit, namely matrix theory [3] and the AdS/CFT correspondence [4], [5], [6], [7]. In both theories the hologram is a large N super Yang Mills (SYM) theory. Furthermore in both cases N serves as a kind of infrared regulator. In the limit $N \to \infty$ keeping the Yang Mills

CP493, *General Relativity and Relativistic Astrophysics*, edited by C. P. Burgess and R. C. Myers
© 1999 American Institute of Physics 1-56396-905-X/99/$15.00

coupling fixed both theories describe physics in infinite flat space. Furthermore, as we shall see, as N grows, the mapping between the hologram and the system it describes becomes more and more nonlocal. In this respect the mapping is like a real hologram. In this paper I will raise some unanswered questions about the nature of the holographic mapping, especially in the limit of infinite flat spacetime. As we shall see, the large N limit involved in going to flat space is quite different than the usual 't Hooft limit in which the coupling shrinks to zero as N increases. The flat space limits in Matrix and AdS/CFT theories both involve letting N go to infinity with fixed gauge coupling. Thus the 't Hooft coupling parameter $g_{ym}^2 N$ tends to infinity and the fixed point becomes infinitely strongly coupled.

Imagine a system composed of point sources of light (particles). Assume that the light from the different sources is coherent as long as they are within a coherence length L_c. All of this takes place in the 3-dimensional half space $z > 0$. At $z = 0$ in the x, y plane there is a photographic film which records the light from the particles. As long as the particles are separated by distance greater than L_c they form two separate blobs of light on the film. If we made a movie from such photos we could follow the individual particles' motion from these blobs. However as soon as they approached within L_c the individual identities would disappear. However the details would not be lost. At this point the details such as the number and position of point sources would become encoded holographically, that is nonlocally distributed over the coherence length L_c. As the coherence length increases the information becomes completely delocalized over the entire hologram. For an ordinary hologram the information is in the interference patterns created by the coherent light sources. For matrix theory and the AdS/CFT correspondence the coding is more obscure but in both cases it involves the $N \times N$ matrix degrees of freedom of Super Yang Mills theories. In both these theories we will see the same kind of delocalization with a coherence length that increases like $N^{1/3}$ in matrix theory and $N^{1/4}$ in the AdS/CFT correspondence.

II HOLOGRAPHY AND MATRIX THEORY

Let us begin with matrix theory. For the present purposes we are interested in uncompactified matrix theory described by $0 + 1$ dimensional SYM theory. For a review of matrix theory and notations we refer the reader to [8].

Matrix theory can be thought of as the Discrete Light Cone Quantization (DLCQ) [9] of M-Theory in which the spacetime is compactified on an almost light like circle X^-. The discrete conjugate momentum is related to the gauge group rank N by $P_- R = N$. Thus we see that if we fix the momentum P_-, removing the IR cutoff (letting $R \to \infty$) is tantamount to letting $N \to \infty$.

What has not been sufficiently realized is that N also plays the role of an infrared cutoff in the transverse dimensions. To see why, let us first consider the 10 dimensional metric and dilaton describing a collection of N coincident D0-branes in the near horizon limit [10].

$$ds^2 = f^{-1/2}dt^2 + f^{1/2}dx^i dx^i$$
$$\exp(2\phi) = f^{3/2}$$
$$f = \frac{Nl_{11}^9}{R^2 r^7} \tag{II.1}$$

where l_{11} is the 11 dimensional Planck scale.

Now consider the limits of validity of (2.1). At small r the ten dimensional supergravity description breaks down because the effective string coupling gets large. In 11 dimensional terms, the local value of the radius of the 11th direction becomes bigger than l_{11}. This happens at $r \sim N^{1/7}l_{11}$

While we have to give up the duality between 10D supergravity and D0-brane physics at this point we can replace it with a duality between D0-branes and 11D supergravity. This is the basis for matrix theory. Thus there is no limit on the matrix theory/Supergravity duality at $r \sim N^{1/7}l_{11}$

At the large distance end another limitation is reached. The scalar curvature $\mathcal{R}$ of the 10D metric satisfies

$$\mathcal{R} \sim \frac{r^{3/2}R}{N^{1/2}l_{11}^{9/2}}$$

It is monotonically increasing with r and exceeds the string scale at $r \sim N^{1/3}l_{11}$. At this point the supergravity description completely breaks down. The region $r > N^{1/3}l_{11}$ is the region where the D0-brane quantum mechanics can be treated perturbatively. From the supergravity point of view $r = N^{1/3}l_{11}$ represents an infrared cutoff beyond which classical supergravity is no longer applicable. This means that when two colliding objects in matrix theory approach each other from infinity, semiclassical gravity will not generally describe their interactions correctly until $r < N^{1/3}l_{11}$. Although it is true that matrix theory with 16 supersymmetries sometimes agrees with DLCQ tree graph supergravity to asymptotic distances this probably has more to do with the tight constraints of maximal supersymmetry than with any general reason for agreement. A more typical example is matrix theory on a blown up orbifold where for finite N the supergravity and one loop matrix theory disagree [11]. If this case is typical, we would expect agreement only when $N > (\frac{r}{l_{11}})^3$.

These considerations suggest although the compactification radius R is allowed to be vanishingly small, the D0-branes create a bubble of space whose transverse size grows as $N^{1/3}$ so that the limit $N \to \infty$ is effectively decompactified. It is therefore interesting to ask if we can see the scale $N^{1/3}l_{11}$ occurring in matrix theory. In the original matrix theory conjecture [3] it was speculated that the threshold bound state describing a supergraviton would grow with N. One estimate was based on the well known v^4/r^7 velocity dependent effective interaction between D0-brane clusters and suggested that the bound state radius grows like $N^{1/9}l_{11}$. A second estimate based on a perturbative large N argument gave the even more rapid growth $N^{1/3}l_{11}$. Recently Polchinski has given a rigorous proof [12] that the growth is at least as fast as $N^{1/3}l_{11}$. Polchinski's argument is based on the virial theorem. The argument I will give here is a less rigorous paraphrase of Polchinski's but gives some intuition about the nature of the bound state.

Let us use the gauge freedom of matrix theory to work in a basis in which one of the 9 X-coordinates, say X_1, is diagonal. The eigenvalues can be thought of as the locations of the constituent D0-branes along the X_1 axis. Let us suppose that they are smoothly spread over a region of size L. Now consider the quantity $\langle Tr(X_1)^2 \rangle$. This obviously satisfies

$$\langle Tr(X_1)^2 \rangle \sim N L^2 \tag{II.2}$$

Consider the quantity $\langle TrY^2 \rangle$ where Y is any of the other 8 X's. The off diagonal elements of the matrix Y are described by harmonic oscillators in the background of X with frequency of order

$$\omega \sim \frac{LR}{l_{11}^3} \tag{II.3}$$

and fluctuation $(\Delta Y)^2 \sim \frac{l_{11}^3}{L}$. Since there are of $\sim N^2$ such elements we find

$$\langle TrY^2 \rangle \sim \frac{N^2 l_{11}^3}{L} \tag{II.4}$$

But now we can use rotational symmetry to equate $\langle Tr(X_1)^2 \rangle$ and $\langle TrY^2 \rangle$ giving

$$L \sim N^{1/3} l_{11} \tag{II.5}$$

The typical conjugate momentum of a matrix element is also easily estimated and is given by

$$\Delta P_{ij} = N^{1/6}/l_{11} \tag{II.6}$$

Thus we see that the bound state grows large with N, extending to the boundaries of the region of validity of 10D supergravity. As seen from eq(2.3) the matrices X have very high frequency oscillations reminiscent of the high frequency zero point oscillations of free strings which also lead to a growth of the wave function but in this case only a logarithmic growth [13], [14]. Finally, the kinetic energy of the D0-branes is estimated as follows. The total kinetic energy is $\frac{R}{2} Tr P_\perp^2$. Using (2.6) and the fact that there are N^2 matrix elements we find the total kinetic energy to be of order $\frac{RN^{7/3}}{l_{11}^2}$. This is to be compared with the typical energy scale in DLCQ M-theory $\frac{R}{NL_s^2}$. Evidently, on the scale of the energies of physical processes the kinetic energies are huge. The kinetic energy per D0-brane is

$$E/N = \frac{RN^{4/3}}{l_{11}^2} \tag{II.7}$$

This enormous energy is cancelled by the quartic and fermionic terms in the hamiltonian but this estimate gives an idea of the energy scales involved.

In light of the above, let us consider a collision between two gravitons. Most of the literature on scattering in matrix theory makes the implicit assumption that the "wave function effects" are not important. What this means is that the scattering objects are described by little clusters of D0-branes which are much smaller than the distance separating them. As we shall see this is completely incorrect.

For simplicity take the gravitons to have equal light cone momenta and therefore equal values of N. In the transverse center of mass frame they have equal and opposite transverse momenta $P_\perp$ and $-P_\perp$. The light cone energy is

$$E_{lc} = \frac{P_\perp^2}{P_-} = R\frac{P_\perp^2}{N} \tag{II.8}$$

and the Mandelstam invariant center of mass energy is

$$S = 2P_\perp^2 \tag{II.9}$$

Suppose $P_\perp$ is fixed and of order $1/l_{11}$. If Matrix theory is consistent then the scattering amplitude must tend to a finite limit in 11D Planck units as N increases. But as we have seen the size of the bound state wave functions grow as $N^{1/3}$. Each particle is huge blob of eigenvalues and the blobs begin to overlap long before the particles come close in the usual sense. During the period of overlap the constituents of each blob lose their identity. This is obvious because of the very large energy scales involved in the D0-brane dynamics eq(2.7). The puny available energy in eq (2.8) is not enough to significantly modify the correlations in the ground state. Therefore the state of the system should more closely resemble the ground state of the $2N \times 2N$ matrix theory than two overlapping but distinguishable subsystems.

Thus the history of the scattering process has two very different but equivalent descriptions. In the usual space time supergravity description two small particles come in from infinity and remain essentially noninteracting until they come within a distance of order l_{11}. They interact for a short time and then separate into final particles which cease to interact as soon as they are separated by l_{11}. In light cone units the interaction lasts for a time $\frac{l_{11}N}{P_\perp R}$.

The holographic matrix description also begins with asymptotically distant non-interacting objects. In this description the constituents begin to merge and interact when their separation is of order $N^{1/3}l_{11}$. As they approach, the many body wave function begins to more and more resemble the ground state. The system remains in this entangled state for a light cone time of order $\frac{l_{11}N^{4/3}}{P_\perp R}$ and then separate into noninteracting final clusters. The situation is particularly perplexing if the energy is not very large and the impact parameter is much larger than l_{11}. In this case the gravity description the particles miss each other and just continue without significant deflection. Exactly how this miracle happens from the SYM description is still a mystery. We will see exactly the same puzzles in the AdS/CFT correspondence.

III HOLOGRAPHY AND THE ADS/CFT DUALITY

String theory in $AdS_5 \times S_5$ is dual to SYM theory on the boundary of the space [4], [5]. As pointed out by Witten, this is another example of a holographic connection [6]. For our purposes AdS is best thought of as a finite cavity with reflecting walls. The metric is given by

$$ds^2 = R^2 dS^2 \qquad \text{(III.1)}$$

where R is the radius of curvature of the AdS and dS^2 is the metric of a "unit" AdS. The unit AdS metric is

$$dS^2 = \frac{(1 + r^2)^2}{(1 - r^2)^2} dt^2 - \frac{4}{(1 - r^2)^2}(dr^2 + r^2 d\Omega) \qquad \text{(III.2)}$$

with $d\Omega$ being the unit 3-sphere.

The full geometry is $AdS_5 \times S_5$. The S_5 factor is a 5-sphere of radius R. Although the boundary, $r = 1$ is an infinite proper distance from any point in the interior of the ball $r < 1$ the time for a light signal to reflect off the boundary is finite. A light signal originating at $r = 0$ (with vanishing S_5 momentum) will return after a coordinate time π. Thus as far as light signals are concerned the space behaves like a finite radiation cavity with reflecting walls. The effect of the inhomogeneous metric is to slow the light velocity at the center to half its value at the boundary. In other words the bulk sphere has a varying dielectric constant. A very simple example of the equivalence of bulk physics with the boundary theory is given by the restriction of causality. Consider a signal originating at a point on the boundary. At a later time it will reappear at the antipodal point on the boundary. In the SYM description it travels with the speed of light on the boundary taking a time π to get to the antipode. In the dual bulk theory the signal travels through the center of the ball, $r = 0$, along a light like geodesic. A simple calculation shows that it again arrives after coordinate time π.

Massive particle trajectories (timelike geodesics) are all periodic in time with period 2π. These trajectories never reach the boundary. The cavity walls repel massive particles with a force which diverges near $r = 1$. The force is proportional to the mass of the particle as is always the case in gravity. From the point of view of the AdS_5 the particles carrying momentum along the 5-sphere are massive. A massive particle which starts at $r = 0$ with velocity v will move outward on a radial trajectory for a time $\pi/2$ at which point it reaches a maximum radial coordinate satisfying

$$v^2 = \frac{4r_{max}^2}{(1 + r_{max}^2)^2} \qquad \text{(III.3)}$$

In describing the SYM theory we will use the dimensionless metric dS^2. This means that all SYM quantities will be treated as dimensionless. The corresponding quantities in the bulk theory carry their usual dimensions. To go from one to the other the conversion factor is R. For example an SYM energy of order 1 corresponds to an energy of order $1/R$ in the bulk theory. A coordinate time interval t is an interval Rt in bulk units.

The dimensionless parameters of the bulk theory are the 10 dimensional string coupling constant g_s and the ratio of the radius of curvature to the string length scale R/l_s. The parameters of the dual SYM theory are the SYM coupling g_{ym} and the rank of the gauge group N. The connection between these parameters was given by Maldacena,

$$g_s = g_{ym}^2$$
$$R/l_s = (Ng_s)^{1/4} \tag{III.4}$$

The fact that by increasing N the radius of curvature in eq(3.11) can be made to increase while keeping the string coupling fixed leads to a conjecture for a new nonperturbative definition of IIB string theory in terms of SYM theory. The $AdS \times S_5$ geometry can be thought of as an infrared regulator for type IIB string theory. As $R \to \infty$ the space becomes locally flat 10 dimensional Minkowski space. To formulate this precisely let us begin with Euclidean SYM theory in the Euclidean version of the metric (3.2).

$$dS^2 = \frac{-(1+r^2)^2}{(1-r^2)^2}d\tau^2 - \frac{4}{(1-r^2)^2}(dr^2 + r^2 d\Omega) \tag{III.5}$$

I will refer to these coordinates and their Minkowski counterparts as "cavity coordinates" where τ is Euclidean time. It is very convenient to transform to "1/2-plane" coordinates with metric

$$ds^2 = -R^2 \frac{(dx^i dx^i + dy^2)}{y^2} \tag{III.6}$$

The 4 noncompact coordinates x^i are parallel to the boundary and can also be used as coordinates for the SYM theory. The coordinate y runs perpendicular to the boundary and varies from zero to infinity.

The transformation from 1/2-plane to cavity coordinates is given as follows. First transform (x^i, y) to 5 dimensional polar coordinates $\rho, \theta, \alpha, \beta, \gamma,$.

$$y = \rho \cos \theta$$
$$x^1 = \rho \sin \theta \cos \alpha$$
$$x^2 = \rho \sin \theta \sin \alpha \cos \beta$$
$$x^3 = \rho \sin \theta \sin \alpha \sin \beta \cos \gamma$$
$$x^4 = \rho \sin \theta \sin \alpha \sin \beta \sin \gamma \tag{III.7}$$

Now set $\rho = e^\tau$ and $\cos \theta = \frac{1-r^2}{1+r^2}$. The three angles α, β, γ are the coordinates of the unit sphere Ω.

We will be interested in correlation functions of various fields in the supcrcon-formally invariant SYM theory. Thus consider a set of points x_a on the boundary of the 1/2-plane coordinates. For each pair of points a, b define $x_{ab} \equiv |x_a - x_b|^2$. In terms of Euclidean cavity coordinates x_{ab} is given by

$$x_{ab} = e^{(\tau_a + \tau_b)}(\cosh \tau_{ab} - \cos\phi_{ab}) \tag{III.8}$$

where $\tau_{ab} \equiv \tau_a - \tau_b$ and ϕ_{ab} is the angular separation between the points in Ω. It is also convenient to define $Z_{ab} = (\cosh \tau_{ab} - \cos\phi_{ab})$.

Euclidean correlation functions of the SYM theory are typically homogeneous functions of the x_{ab} of degree determined by the dimensions of the operators. To express the corresponding correlators in cavity coordinates just replace each x_{ab} by Z_{ab}. The various of e^τ cancel the Jacobian factors in the transformation of fields with nonvanishing dimensions. Thus the correlators are homogeneous functions of the Z_{ab}. As an example, the correlation function of two scalar fields Φ of dimension 4 is of the form Z_{ab}^{-4}

It is now a simple matter to pass to Minkowski signature by replacing τ by it. Thus the correlator becomes

$$\langle \Phi(x_a)\Phi(x_b) \rangle = (\cos t_{ab} - \cos\phi_{ab})^{-4} \tag{III.9}$$

The singularity when $\cos t_{ab} = \cos\phi_{ab}$ is the usual light cone singularity.

Strictly speaking there is no true S matrix in AdS space. As I have emphasized, AdS is for all practical purposes a finite cavity with reflecting walls. Asymptotic states can not be defined in such a geometry. The strategy that we follow is to introduce sources on the walls of the cavity which act as particle sources and de-tectors. This will allow us to define a finite time version of the S matrix. When the size of the box is allowed to increase, keeping fixed the energies, impact param-eters and other physical quantities the finite time S matrix should tend to a true asymptotic scattering amplitude.

Before discussing the boundary sources further we need to determine what quan-tities should be kept fixed as $R \to \infty$ in order to recover flat space string theory. First of all we must keep the microscopic parameters of string theory fixed. This means letting $R/l_s \to \infty$ with g_s fixed. In terms of Yang Mills quantities

$$g_{ym} = fixed$$
$$N \to \infty \tag{III.10}$$

In addition the energy scale of physical processes should be fixed in string units. In terms of the dimensionless energy of the SYM theory E

$$E \sim (g_{ym}^2 N)^{1/4} \tag{III.11}$$

Thus we see that the flat space limit involves the high energy limit of large N SYM theory. We will also require restrictions on the angular momenta of particles.

We will define a spacetime region called the "lab". The lab is centered at $r = t = 0$. Its linear dimensions L in both space and time are fixed in string units but are much larger than l_s. At the end we may take L/l_s as big as we like. As $N \to \infty$ the entire region of the lab becomes accurately described by flat spacetime. The sources will be constructed in such a way as to insure that the entire collision process takes place within the lab.

A particle can carry momentum components in both the AdS_5 directions and in the S_5 directions. We will call these p and k respectively. For the moment we will ignore k. Consider a massless particle that is inside the lab with momentum p. Its angular momentum l is necessarily less than Lp. Since the cavity is spherically symmetric, The angular momentum of a freely moving incoming particle is conserved. Therefore if the particle is to arrive in the lab it must be emitted from the boundary with $l < Lp$. This restriction guarantees that the "beam" is focused to pass through the lab.

As an example we will consider scattering amplitudes for dilatons carrying vanishing momentum in the S_5 directions. The dilatons will be emitted in such a way that they propagate freely toward the region $r = 0$ where they meet and interact within the lab. Since the wavelength of the particles is vanishingly small by comparison with the radius of the AdS space, the propagation of the wave packets toward the lab can be treated by geometrical optics. The time it takes for a wave packet to travel from the boundary to the lab is $\pi/2$, just half the time for a light signal to cross the AdS space. Therefore the initial sources must act at $t = -(\frac{\pi}{2} \pm \frac{L}{R})$. Similarly the final detector-sources must act at $t = +(\frac{\pi}{2} \pm \frac{L}{R})$.

The appropriate SYM operators for emitting all massless 10 dimensional particles are known. In particular the operator that creates a dilaton at the boundary is the dimension 4 operator $Tr F_{\mu\nu} F^{\mu\nu} \equiv FF$. Let us consider the emission operator for a zero angular momentum dilaton of bulk energy p. The obvious choice is

$$A^{in}(p) \sim \int dt d\Omega e^{ipRt} FF \qquad \text{(III.12)}$$

However in order to build wave packets which arrive at the lab at $t = 0 \pm \frac{L}{R}$ we need modify the definition of A. This can be done by replacing the factor e^{ipRt} by a wave packet of finite extent. Let $f_{in}\left[(t - \frac{\pi}{2})\frac{R}{L}\right]$ be a smooth function (such as a gaussian) which is peaked at $t = \pi/2$. The definition of A is

$$A^{in}(p) \sim \int dt d\Omega f_{in}\left[(t - \frac{\pi}{2})\frac{R}{L}\right] e^{ipRt} FF \qquad \text{(III.13)}$$

A similar expression defines the operators representing the final particles.

$$A^{out}(p) \sim \int dt d\Omega f_{out}\left[(t + \frac{\pi}{2})\frac{R}{L}\right] e^{-ipRt} FF \qquad \text{(III.14)}$$

To create particles of arbitrary angular momentum the integral over Ω should contain the relevant $O(4)$ spherical harmonic.

The recipe for computing bulk S matrix elements from SYM quantities is straight-forward.

$$S = \langle 0| \prod_{out} Z_{out} A^{out} \prod_{in} Z_{in} A^{in} |0\rangle \qquad (III.15)$$

The factors Z are inverse boundary-bulk propagators which are needed to amputate the external AdS propagators.

The above prescription for recovering flat space amplitudes can be generalized to include nonvanishing momenta along the 5-sphere. The operators which create particles with nonvanishing $O(6)$ angular momentum n are schematically of the form $TrFFXXXX...$ where F represents components of the Yang Mills Field strength and $XXXX..$ is a polynomial of order n in the scalar fields which transform as vectors under $O(6)$. These are operators of mass dimension $4 + n$. We must also integrate these operators with functions of time and Ω in order to project out definite energy and $O(4)$ angular momentum. Again the frequencies should be of order $g_{ym}^2 N^{1/4}$ in order to keep the physical momentum of the bulk particles of order unity in string units.

Thus we see that passing to the flat space limit generally involves operators in the SYM theory which are high frequency components of high dimension operators.

To actually compute scattering amplitudes from conformal field theory data, a useful strategy might be to use the operator product expansion for the operators $A^{in,out}$. Consider for example a two particle scattering process in which the incoming (outgoing) particles are emitted (absorbed) at time $t_{in,out} = \pm\pi/2$. The angular positions of the incoming particles are $\Omega_{1,2}$ and the outgoing particles $\Omega_{3,4}$. The 4 points $(1,2,3,4)$ are far from each other in spacetime and it is not obvious why the operator product expansion is useful. However, consider the case where there is a small momentum transfer $(p_1 - p_3) << p$. Then the locations of 1 and 3 will be almost light-like with respect to each other. In the rules for continuation from Euclidean to Minkowski signature in AdS space the almost light like separation between 1 and 3 maps to an almost vanishing Euclidean separation so that the OPE should provide an expansion for small angle scattering. Obviously, the operators of low dimensionality in the operator product of $A(1)A(3)$ correspond to massless exchange. In addition we also expect contributions corresponding to massive string exchange with masses of order l_s^{-1}. From the point of view of the operator product expansion this means operators of dimensionality $\sim g_s N^{1/4}$. We will leave it to a future publication, hopefully by someone else, to work out the detailed rules for computing on shell scattering amplitudes from CFT data in the flat limit.

IV THE INFRARED ULTRAVIOLET CONNECTION

The connection between the boundary SYM theory and the ideas of holography rely on an important connection between the ultraviolet behavior of the SYM theory and the infrared behaviour of the bulk supergravity [7]. We will begin by reviewing the argument for counting the number of degrees of freedom of the system. By now it is well known that an ultraviolet cutoff at wavelength δ in the SYM is equivalent to a cutoff in the radial coordinate r at $r = 1 - \delta$. The cutoff SYM describes the bulk supergravity in the interior of the ball $r < 1 - \delta$. Now the number of cutoff cells of coordinate size δ on the boundary of this ball is of order $1/\delta^3$. Assuming that each independent SYM field has one degree of freedom per cells, the total number of degrees of freedom is

$$N_{dof} \sim \frac{N^2}{\delta^3} \qquad (IV.1)$$

If we use

$$R = l_s(Ng_s)^{1/4}$$
$$Area = \frac{R^8}{\delta^3}$$
$$G = g_s^2 l_s^8 \qquad (IV.2)$$

where $Area$ is the area of the cutoff 3-sphere times S_5 and G is the 10D gravitational coupling constant we find the typical holographic behaviour

$$N_{dof} \sim Area/G \qquad (IV.3)$$

Let us push this reasoning to the extreme and take the cutoff δ such that the proper volume of the 4 dimensional ball $r < \delta$ is R^4 (dimensionless volume ~ 1). The area of the cutoff boundary sphere is then $\sim R^3$ and the number of degrees of freedom is just N^2. In other words it takes N^2 degrees of freedom to describe all the states of the bulk theory which are supported within a sphere of proper size R. This means that the physics, within a neighborhood small enough so that curvature can be ignored, is coded by the N^2 matrix degrees of freedom and that the inhomogeneous spatial modes of the SYM are unexcited. This suggests the possibility that the states supported within such a neighborhood might be described by an $N \times N$ matrix quantum mechanics.

As an illustration of the IR-UV connection consider a graviton carrying momentum k along the S_5 and momentum $p >> k$ in the radial direction along r. Its total energy is $E = \sqrt{p^2 + k^2}$. It is created by applying the operator

$$A^{in}(p,k) = \int dt d\Omega f_{in} \left[(t - \frac{\pi}{2}) \frac{R}{L} \right] e^{iERt} Tr FFXXXXX... \qquad (IV.4)$$

where there are $n = kR$ factors inside the trace. This means that the total energy is divided among n SYM quanta and the energy ν of each SYM quantum is

$$\nu = \sqrt{\frac{(p^2 + k^2)}{k^2}} \approx \frac{p}{k} \qquad (IV.5)$$

This corresponds to a cutoff in the SYM theory $\delta \approx 1/\nu \approx \frac{k}{p}$. The implication is that in the bulk theory the particle created by A appears not at the boundary but at $1 - r \approx \frac{k}{p}$. This makes good sense for the following reason. From the point of view of AdS_5 a particle with S_5 momentum k is a massive particle with mass k. The classical trajectory of such a massive particle with (bulk) energy E has a turning point (vanishing velocity) given by (3.3). In terms momentum components the turning point is at $1 - r \approx \frac{k}{p}$. Hence the particle starts out at the outermost point on its trajectory.

The fact that massive particles originate in the interior of the AdS space does not require a modification of the rules for constructing the A operators. Although they start closer to the interaction point at $r = 0$, the time that it takes to arrive at $r = 0$ is independent of the mass.

From the above discussion it seems that the cutoff theory with a given value of δ can describe the sector of the theory containing particles with $\frac{k}{p} \geq \delta$. Suppose for example, all the particles in a given reaction have $\frac{k}{p} \sim 1$. In this case only the lowest modes of the SYM theory are excited corresponding to configurations which are spatially homogeneous (in the boundary theory). In other words physical processes involving such particles always appear completely smeared and nonlocal in the holographic SYM description. The situation is very similar to the matrix case.

At this point it is interesting to consider just what problems we in principle would know how to set up and solve if we could completely master SYM theory and find all its correlators, and energy levels. First of all we could apply the recipe described in section(3) to compute any scattering amplitude involving 10 dimensional massless particles. Since we do not expect any other stable particles in the theory, this exhausts all IIB flat space scattering amplitudes.

In addition we could compute the thermodynamics of the theory and discover the existence of a phase transition at (dimensionless) temperature ~ 1. This corresponds to the formation of a large black hole at bulk temperature $\sim R^{-1}$. This object has no significance in the flat space limit since it has a size of order the radius of curvature.

However most of ordinary flat space physics would remain out of reach even though it implicitly must be described by the SYM theory. As an example consider the problem of describing an ordinary 10 dimensional Schwarzschild black hole of finite mass and entropy as $N \to \infty$. For simplicity the black hole could be located near the center of the AdS at some point on the 5-sphere. If the proper distance of the black hole from the center is kept fixed as $R \to \infty$ then its coordinates will tend to the origin at $r = 0$. The image will become completely symmetric on the 3-sphere. What features of the SYM state contain the information of the exact position or even the fact that it is a black hole or any other object of the same mass and angular momentum is not known. In fact it is not even clear how to distinguish this configuration from a pair of distant black holes or other objects of the same total mass if their separation is much smaller than the radius of curvature

R. In principle the AdS/CFT correspondence requires the SYM theory to contain these objects. Recognizing them from their SYM description requires deciphering the holographic code.

V DECODING THE HOLOGRAM

Exactly how information is holographically stored in either matrix theory or the AdS/CFT correspondence is a mystery. I will try to give some thoughts about it. Lets begin with matrix theory.

The $N^{1/3}$ increase in the size of the low energy wave functions of D0-brane wave functions is caused by the ground state oscillations of an increasing number ($\sim N^2$) of high frequency modes. The situation is parallel to that in free string theory where as the number of modes is increased the ground state expands [13]. In the free string case the expansion is only logarithmic but for any finite coupling it will eventually grow like a power [2]. The increase in the sizes of images eventually blurs the details of a system. For example if the system consists of two distinct objects separated by transverse distance Δ then when $N^{1/3} > \frac{\Delta}{l_{11}}$ the holographic images become entangled. Given a state of the matrix theory at very large N it would be very difficult to decipher its meaning.

The trick in decoding the hologram is to get rid of the high frequency oscillations. This can be done by averaging over time but the right thing has to be averaged. For example we could define a density of D0-branes along the X^1 axis in terms of the distribution of its eigenvalues. This however is a very slowly varying quantity which does not have high frequency oscillations. The right thing to average is the Heisenberg operators representing the matrix elements $X_{a,b}$. For example, we may average X over a time δt. If we work in the eigenbasis of the Hamiltonian, the averaging is equivalent to throwing away all (quantum) matrix elements $\langle E_1 | X_{ab} | E_2 \rangle$ with $|E_1 - E_2| > 1/\delta t$. The resulting quantum operators will have a modified distribution of eigenvalues. Since modes of frequency $> 1/\delta t$ are now absent the distribution should have a smaller spread. Therefore the holographic image of several objects should become clear. It is evident that all of this is a manifestation of the stringy space, time uncertainty relation [17].

In order to be a little more quantitative I will make an assumption that is motivated by a particular view of the large N limit. According to this view, the large N limit is a fixed point of a kind of renormalization group associated with integrating out rows and columns of the $N \times N$ matrix degrees of freedom to produce a theory with smaller matrices. What I will assume is that time averaging over δt, or equivalently, integrating out high frequency modes is equivalent to replacing the original $N \times N$ matrix system by another with smaller $n \times n$ matrices. The maximum relevant frequency for the original system is given by eq(2.3) with $L = N^{1/3} l_{11}$. We will call this the characteristic frequency ω_N.

$$\omega_N = \frac{N^{1/3} R}{l_{11}^2} \qquad \qquad (\text{V.1})$$

110

If we identify the characteristic frequency of the $n \times n$ model to be $\omega_n = (\delta t)^{-1}$ then

$$\frac{n}{N} = \frac{\omega_n^3}{\omega_N^3} = \frac{l_{11}^6}{(R\delta t)^3 N} \tag{V.2}$$

Furthermore since the size of the eigenvalue distributions of X scale like $N^{1/3}$ we should find the spread diminished by the factor $\frac{\omega_n}{\omega_N}$. According to this estimate, by averaging over $\delta t = l_{11}^2/R$ resolution of order the Planck length should be restored for a pair of gravitons.

For the AdS/CFT correspondence decoding the hologram seems to be very different. In the flat space limit the SYM dimensionless energy of a given system increases like $(g_s N)^{1/4}$. On the other hand the information is coded in the longest wavelength modes on the unit sphere. These modes have frequency ~ 1 in dimensionless units which corresponds to very long bulk time scales of order R. In other words, information seems to be coded in extremely slow degrees of freedom. At the moment I have no idea how this works.

VI ACKNOWLEDGEMENTS

Much of what is written in this paper was stimulated during discussion with Juan Maldacena and Andy Strominger while I was visiting Harvard in September of this year. This of course does not mean that they are responsible for the confusion and fuzzyness of my ideas. I also benefited from many conversations with Steve Shenker over the last couple of years about these issues.

REFERENCES

1. G. 't Hooft, Dimensional Reduction in Quantum Gravity, gr-qc/9310026
2. L. Susskind, The World as a Hologram, hep-th/9409089
3. T. Banks, W. Fischler, S.H. Shenker, L. Susskind, M Theory As A matrixModel: A Conjecture. hep-th/9610043
4. Juan M. Maldacena, The Large N Limit of Superconformal Field Theories and Supergravity. hep-th/9711200
5. S.S. Gubser, I.R. Klebanov, A.M. Polyakov, Gauge Theory Correlators from Non-Critical String Theory. hep-th/9802109
6. Edward Witten, Anti De Sitter Space And Holography. hep-th/9802
7. L. Susskind, Edward Witten, The Holographic Bound in Anti-de Sitter Space hep-th/9805114
8. Daniela Bigatti, Leonard Susskind, Review of Matrix Theory, hep-th/9712072
9. Leonard Susskind, Another Conjecture about M(atrix) Theory hep-th/9704080
10. G. Horowitz and A. Strominger, Nucl. Phys. B360 (1990)197.
11. Michael R. Douglas, Hirosi Ooguri, Stephen H. Shenker, Issues in M(atrix) Theory Compactification, hep-th/9702203

12. J. Polchinski, unpublished

13. M. Karliner, I. Klebanov and L. Susskind, Size And Shape Of Strings, Int.J.Mod.Phys. A3 (1988) 1981

14. Leonard Susskind, Particle Growth and BPS Saturated States, hep-th/9511116

15. Massimo Bianchi, Michael B. Green, Stefano Kovacs, Giancarlo Rossi Instantons in supersymmetric Yang-Mills and D-instantons in IIB superstring theory hep-th/9807033 [abs, src, ps, other]

16. M.B. Green , Configurations of two D-instantons, Phys.Lett. B398 (1997) 69

17. Miao LI, Tamiaki Yoneya, D-Particle Dynamics and The Space-Time Uncertainty Relation hep-th/9611072, Journal-ref: Phys.Rev.Lett. 78 (1997) 1219-1222

Black Holes, Strings and Polymers

Ramzi R. Khuri

Baruch College, CUNY, 17 Lexington Ave., New York, NY 10010[1]
Graduate School and University Center, CUNY, 365 5th ave., New York, NY 10036
Center for Advanced Mathematical Sciences, American University of Beirut, Beirut, Lebanon

Abstract. Quantum aspects of black holes represent an important testing ground for a theory of quantum gravity. The recent success of string theory in reproducing the Bekenstein-Hawking black hole entropy formula provides a link between general relativity and quantum mechanics via thermodynamics and statistical mechanics. Here we speculate on the existence of new and unexpected links between black holes and polymers and other soft-matter systems.

The standard model of elementary particle physics has been successful in describing three of the four fundamental forces of nature. In the most optimistic scenario, the standard model can be generalized to take the form of a grand unified theory, in which quantum chromodynamics, describing the strong force, and the electroweak theory, unifying the weak interaction with electromagnetism, are synthesized into a single theory in which all three forces have a common origin. The underlying framework of particle physics is quantum mechanics, in which the natural length scale associated with a particle of mass m (such as an elementary particle) is given by the Compton wavelength $\lambda = \hbar/mc$, where $\hbar$ is Planck's constant divided by 2π and c is the speed of light. Scales less than λ are therefore unobservable within the context of the quantum mechanics of this particle.

Quantum mechanics, however, has so far proven unsuccessful in describing the fourth fundamental force, gravitation. The successful theory in this case is that of general relativity, which, however, does not lend itself to a straightforward attempt at quantization. The main problem in such an endeavour is that the divergences associated with trying to quantize gravity cannot be circumvented (or "renormalized") as they are for the strong, weak and electromagnetic forces.

Among the most interesting objects predicted by general relativity are black holes, which represent the endpoint of gravitational collapse. According to relativity, an object of mass m under the influence of only the gravitational force (*ie* neutral with respect to the other three forces) will collapse into a region of spacetime bounded by a surface, the event horizon, beyond which signals cannot be trans-

---

[1] Supported by NSF Grant 9900773 and by PSC-CUNY Award 669663.

CP493, *General Relativity and Relativistic Astrophysics*, edited by C. P. Burgess and R. C. Myers
© 1999 American Institute of Physics 1-56396-905-X/99/$15.00

mitted to an outside observer. The event horizon for the simplest case of a static, spherically symmetric black hole of mass m is located at a radius $R = 2Gm/c^2$, the Schwarzschild radius, from the collapsed matter at the center of the sphere, where G is Newton's constant.

In trying to reconcile general relativity and quantum mechanics, a natural question to ask is whether they have a common domain. This would arise when an elementary particle exhibits features associated with gravitation, such as an event horizon. This may occur provided $\lambda \lesssim R$, which implies that, even within the framework of quantum mechanics, an event horizon for an elementary particle may be observable. Such a condition is equivalent to $m \gtrsim m_P = \sqrt{\hbar c/G} \sim 10^{19} GeV$, the Planck mass, or $\lambda \lesssim l_P = \sqrt{\hbar G/c^3}$, the Planck scale. It is in this domain that one may study a theory that combines quantum mechanics and gravity, the so-called *quantum gravity* (henceforth we use units in which $\hbar = c = 1$).

A problem, however, arises in this comparison, because most black holes are thermal objects, and hence cannot reasonably be identified with pure quantum states such as elementary particles. In fact, in accordance with the laws of *black hole thermodynamics* [1], black holes radiate with a (Hawking) temperature constant over the event horizon and proportional to the surface gravity: $T_H \sim \kappa$. Furthermore, black holes possess an entropy $S = A/4G$, where A is the area of the horizon (the area law), and $\delta A \geq 0$ in black hole processes. So only a black hole with zero area can correspond to a pure state with $S = 0$ such as an elementary particle, while a black hole with nonzero area, and therefore nonzero entropy, corresponds to an *ensemble* of states. A question, then, that can be posed of a theory of quantum gravity is the following: since the basis of ordinary thermodynamics is (quantum) statistical mechanics, can one recover the laws of black hole thermodynamics by the counting of microscopic states? In particular, can one recover the area law from a quantum mechanical entropy arising as the logarithm of the degeneracy of quantum states?

At the present time, string theory, the theory of one-dimensional extended objects, is the only known reasonable candidate theory of quantum gravity. The divergences inherent in trying to quantize point-like gravity seem not to arise in string theory. Furthermore, string theory has the potential to unify all four fundamental forces within a common framework. At an intuitive level, one can see how point-like divergences may be avoided in string theory by considering scattering amplitudes in string theory [2]. Unlike those of field theory, the four-point amplitudes in string theory do not have well-defined vertices at which the interaction can be said to take place, hence no corresponding divergences associated with the zero size of a particle. A simpler way of saying this is that the finite size of the string smooths out the divergence of the point particle.

For the purpose of understanding black hole thermodynamics, an important feature of string theory is that classical solutions [3] may be easily constructed as composites of single-charged fundamental constituents. Identifying these constituents with states in string theory, one can compare the Bekenstein-Hawking entropy obtained from the area of the classical solution to the quantum-mechanical micro-canonical counting of ensembles of states [4]. For example, the extremal Reissner-Nordström charged black hole solution of Einstein-Maxwell theory arises in string theory as the composite of four charges, N_1, N_2, N_3 and N_4, normalized to correspond to number operators in string theory. The area law then yields a Bekenstein-Hawking entropy $S_{BH} = 2\pi\sqrt{N_1 N_2 N_3 N_4}$. The counting of the degeneracy of the states forming this black hole leads to the same quantity $S_{QM} = \ln d(N_i) = S_{BH}$. Even in the black hole picture, this result can be seen to arise from the number of ways in which the various constituents combine. It is straightforward to show [5] that one can write four-centered solutions each with charge N_i of a given species. A black hole with nonzero area is formed when all charges are brought together to the same point. The precise partition function [6] yielding the correct degeneracy $d(N_i) = exp(S_{BH})$ is obtained provided both bosonic and fermionic excitations of a supersymmetric string-like object along various dimensions are taken into account.

The recovery of the area law in a wide variety of contexts in string theory suggests that we have accounted for the microscopic degrees of freedom of the black hole. However, the ensemble of string states on the one hand and the black hole on the other represent two very different objects, so we must try to understand the correspondence between them [7]. For simplicity, let us consider the case of a long, self-gravitating string in $D = 4$ dimensions [8]. At level N, a free string has mass $M \sim \sqrt{N}/l_s$, size $L \sim N^{1/4} l_s$ and entropy $S \sim \sqrt{N}$, where l_s is the string scale. This picture is valid provided the string coupling $g << 1$, where g is related to Newton's constant G via $G \sim g^2 l_s^2$. This picture represents a random walk [9] with $n = \sqrt{N}$ steps, each a single string "bit" of length l_s [10].

Let us now slowly increase the coupling g. As shown by Horowitz and Polchin-ski [8], gravitational effects start becoming strong at $g_0 \sim N^{-3/8} = n^{-3/4}$, after which the string collapses until it reachers the size of the string scale l_s. At the critical coupling $g_c \sim N^{-1/4} = n^{-1/2}$, the Schwarzschild radius $R = 2GM$ of a black hole with the same mass becomes of the order of the string scale, and one can sensibly start thinking of the string as a black hole. At this point, too, the entropies match: $S_{BH} \sim R^2/G = 1/g_c^2 = \sqrt{N} = n$. For $g > g_c$, the black hole picture prevails. In the intermediate range $g_0 < g < g_c$, the size of the string state was shown using a thermal scalar field theory to be [8]

$$L \sim \frac{l_s}{g^2 N^{1/2}} = \frac{l_s}{g^2 n},$$ (1)

which smoothly interpolates between the random walk size and the string scale. Note that for n large, the coupling is small throughout the ranges we are considering. This is an interesting result with a specific prediction for the coupling dependence of the size of the string as it collapses into a black hole. A natural question to ask is whether this sort of result also arises in analogous physical systems already considered. Since random walks with interactions arise in polymer physics [11,10], the relation (1) should also hold for a self-attracting polymer chain.

We start with a random walk with n steps each of size a, so that the size of the polymer is initially given by $L_0 = \sqrt{n}a$. Suppose we place the polymer in a medium of scatterers of number density ρ and (dimensionless) potential strength u. Then the size of the polymer was shown to be [12]

$$L^2 = x^{-2} \left(1 - \exp(-nx^2a^2)\right), \tag{2}$$

where $x = u\rho a^2$ can be thought of as an effective scattering cross section.

To compare with a self-gravitating string with $a = l_s$, the scatterers are taken to coincide with the positions of the string bits themselves. For large n and in a mean-field approximation, the number density of n bits in a volume L_0^3 is given by

$$\rho = \frac{n}{\left(n^{3/2}l_s^3\right)} = n^{-1/2}l_s^{-3}. \tag{3}$$

For g small, the leading order interaction potential is given by

$$\frac{u}{l_s} \sim \sum_{i,j} \frac{g^2}{|\vec{r}_i - \vec{r}_j|} \sim \frac{g^2 n^2}{L_0} = \frac{g^2 n^{3/2}}{l_s}, \tag{4}$$

where $\vec{r}_i$ is the position of the ith link. It follows that $x \sim ng^2/l_s$, so that

$$L^2 = l_s^2 n^{-2} g^{-4} \left(1 - \exp(-n^3 g^4)\right). \tag{5}$$

For $g < g_0 = n^{-3/4}$, $L^2 \simeq n l_s^2$ which is the random walk, corresponding to the free string. As in the string case, a transition occurs at $g \sim g_0$. As g is increased past g_0, the size quickly shrinks to $L^2 \simeq l_s^2/n^2 g^4 = l_s^2/g^2 N$, as in (1). This kind of relation holds[2] until $g \sim g_c \sim n^{-1/2}$, when $L \sim R$, the Schwarzschild radius of the polymer, and the black hole picture dominates.

This connection between black holes, strings and polymers is very interesting and merits further investigation. Similar links with other soft-matter systems have also been noted by Callaway [13], where the area law was recovered for the case of a liquid field theory and where it was argued that the area law contributions to the free energy are primarily responsible for liquid surface tension. The speculation was also made that the area law arises in the context of protein folding.

Connections between physical and biological systems are always exciting. The cases discussed above are especially so since quantum gravity is generally considered too remote to have relevance to other areas of physics, much less other fields of science. In particular, the fascinating possibility arises that mathematical techniques used to study black holes can be useful in understanding biological questions, such as protein dynamics, while methods of polymers physics can potentially shed light on quantum gravity.

---

[2] Once the self-interaction of the polymer becomes strong, the simple result (5) is no longer exact and a more precise computation is required. Nevertheless, it is clear that one obtains a smooth transition from the random walk to the Schwarzschild radius via a nonperturbative coupling dependence, so that even if (1) is not exactly recovered, it remains a good approximation for the collapse of the polymer.

REFERENCES

1. J. Bekenstein, Lett. Nuov. Cimento 4 (1972) 737; Phys. Rev. **D7** (1973) 2333; Phys. Rev. **D9** (1974) 3292; S. W. Hawking, Nature **248** (1974) 30; Comm. Math. Phys. **43** (1975) 199.

2. M. B. Green, J. H. Schwarz and E. Witten, *Superstring Theory*, Cambridge University Press, Cambridge (1987).

3. See M. J. Duff, R. R. Khuri and J. X. Lu, Phys. Rep. **B259** (1995) 213, M. Cvetic and D. Youm, Phys.Rev. **D54** (1996) 2612, M. Cvetic and A. A. Tseytlin, Nucl. Phys. **B477** (1996) 499 and references therein.

4. A. Strominger and C. Vafa, Phys. Lett. **B379** (1996) 99; J. Maldacena, hep-th/9607235 and references therein; K. Sfetsos and K. Skenderis, hep-th/9711138; R. Arguiro. F. Englert and L. Houart, hep-th/9801053.

5. J. Rahmfeld, Phys. Lett. **B372** (1996) 198.

6. T. M. Apostol, *Introduction to Analytic Number Theory*, Springer Verlag (1976).

7. L. Susskind, hep-th/9309145; G. T. Horowitz and J. Polchinski, Phys. Rev. **D55** (1997) 6189.

8. G. T. Horowitz and J. Polchinski, Phys. Rev. **D57** (1998) 2557. See also S. Kalyana Rama, Phys. Lett. **B424** (1998) 39.

9. P. Salomonson and B. S. Skagerstam, Nucl. Phys. **B268** (1986) 349; Physica **A158** (1989) 499; D. Mitchell and N. Turok, Phys. Rev. Lett. **58** (1987) 1577; Nucl. Phys. **B294** (1987) 1138.

10. See C. B. Thorn, hep-th/9607204 and references therein; see also O. Bergman and C. B. Thorn, Nucl. Phys. **B502** (1997) 309.

11. See M. Doi and S. F. Edwards, *The Theory of Polymer Dynamics*, Clarendon Press, Oxford (1986) and references therein.

12. S. F. Edwards and M. Muthukumar, J. Chem. Phys. **89** (1988) 2435; S. F. Edwards and Y. Chen, J. Phys. **A21** (1988) 2963.

13. D. J. E. Callaway, Phys. Rev. **E53** (1996) 3738.

Entanglement/Brick-Wall Entropy's Correspondence

Shinji Mukohyama

Department of Physics and Astronomy, University of Victoria, Victoria, BC, Canada V8W 3P6
Yukawa Institute for Theoretical Physics, Kyoto University, Kyoto 606-8502, Japan

Abstract. There have been many attempts to understand the statistical origin of black-hole entropy. Among them, entanglement entropy and the brick wall model are strong candidates. In this paper we show a relation between entanglement entropy and the brick wall model: the brick wall model seeks the maximal value of the entanglement entropy. In other words, the entanglement approach reduces to the brick wall model when we seek the maximal entanglement entropy .

I INTRODUCTION

Black hole entropy is given by a mysterious formula called the Bekenstein-Hawking formula [1,2]: $S_{BH} = A/4l_{pl}^2$, where A is area of the horizon. There have been many attempts to understand the statistical origin of the black-hole entropy.

Entanglement entropy [3,4] is one of the strongest candidates of the origin of black hole entropy. It is originated from a direct-sum structure of a Hilbert space of a quantum system: for an element $|\psi\rangle$ of the Hilbert space $\mathcal{F}$ of the form

$$\mathcal{F} = \mathcal{F}_I \otimes \mathcal{F}_{II}, \qquad (1)$$

the entanglement entropy S_{ent} is defined by

$$S_{ent} = -\mathbf{Tr}_I[\rho_I \ln \rho_I], \quad \rho_I = \mathbf{Tr}_{II}|\psi\rangle\langle\psi|. \qquad (2)$$

Here $\otimes$ denotes a tensor product followed by a suitable completion and $\mathbf{Tr}_{I,II}$ denotes a partial trace over $\mathcal{F}_{I,II}$, respectively.

On the other hand, there is another strong candidate for the origin of black hole entropy: the brick wall model introduced by 'tHooft [5]. In this model, thermal atmosphere in equilibrium with a black hole is considered. In this situation, we encounter with two kinds of divergences in physical quantities. The first is due to infinite volume of the system and the second is due to infinite blue shift near the horizon. We are not interested in the first since it represents contribution from

CP493, *General Relativity and Relativistic Astrophysics*, edited by C. P. Burgess and R. C. Myers
© 1999 American Institute of Physics 1-56396-905-X/99/$15.00

matter in the far distance. Hence we introduce an outer boundary in order to make our system finite. It is the second divergence that we would like to associate with black hole entropy. Namely, it can be shown by introducing a Planck scale cutoff that entropy of the thermal atmosphere near the horizon is proportional to the area of the horizon in Planck units.

In this paper, we show that the brick wall model seeks the maximal value of the entanglement entropy.

II MODEL DESCRIPTION

For simplicity, we consider a minimally coupled, real scalar field described by the action

$$S = -\frac{1}{2} \int d^4x \sqrt{-g} \left[g^{\mu\nu} \partial_\mu \phi \partial_\nu \phi + m_\phi^2 \phi^2 \right], \tag{3}$$

in the spherically symmetric, static black-hole spacetime

$$ds^2 = -f(r)dt^2 + \frac{dr^2}{f(r)} + r^2 d\Omega^2. \tag{4}$$

We denote the area radius of the horizon by r_0 and the surface gravity by κ_0 ($\neq 0$):

$$f(r_0) = 0, \quad \kappa_0 = \frac{1}{2} f'(r_0). \tag{5}$$

We quantize the system of the scalar field with respect to the Killing time t in a Kruskal-like extension of the black hole spacetime. The corresponding ground state is called the Boulware state and its energy density is known to diverge near the horizon. Although we shall only consider states with bounded energy density, it is convenient to express these states as excited states above the Boulware ground state for technical reasons. Hence, we would like to introduce an ultraviolet cutoff α with dimension of length to control the divergence. The cutoff parameter α is implemented so that we only consider two regions satisfying $r > r_1$ (shaded regions I and II in *Figure* 1), where r_1 ($> r_0$) is determined by

$$\alpha = \int_{r_0}^{r_1} \frac{dr}{\sqrt{f(r)}}. \tag{6}$$

[Evidently, the limit $\alpha \to 0$ corresponds to the limit $r_1 \to r_0$. Thus, in this limit, the whole region in which $\partial/\partial t$ is timelike is considered.] Strictly speaking, we also have to introduce outer boundaries, say at $r = L$ ($\gg r_0$), to control the infinite volume of the constant-t surface. However, even if there are outer boundaries, the following arguments still hold.

In this situation, there is a natural choice for division of the system of the scalar field: let $\mathcal{H}_I$ be the space of mode functions with supports in the region I and $\mathcal{H}_{II}$

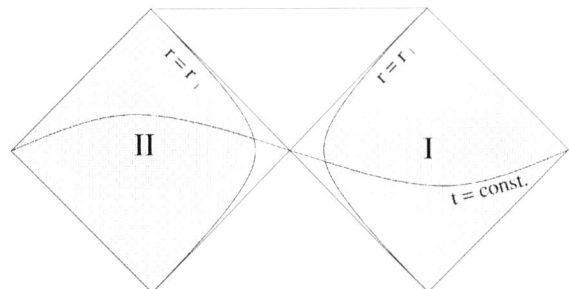

FIGURE 1. The Kruskal-like extension of the static, spherically symmetric black-hole space-time. We consider only the regions satisfying $r > r_1$ (the shaded regions I and II).

be the space of mode functions with supports in the region II. Thence, the space $\mathcal{F}$ of all states are of the form (1), where $\mathcal{F}_I$ and $\mathcal{F}_{II}$ are defined as symmetric Fock spaces constructed from $\mathcal{H}_I$ and $\mathcal{H}_{II}$, respectively:

$$\mathcal{F}_{I,II} \equiv C + \mathcal{H}_{I,II} + (\mathcal{H}_{I,II} \cdot \mathcal{H}_{I,II})_{sym} + \cdots. \tag{7}$$

Here $(\cdots)_{sym}$ denotes the symmetrization.

III SMALL BACKREACTION CONDITION

Let us investigate what kind of condition should be imposed for our arguments to be self-consistent. A clear condition is that the backreaction of the scalar field to the background geometry should be finite. For the brick wall model this condition is satisfied. Namely, in Ref. [6], it was shown that the total mass of the thermal atmosphere of quantum fields is actually bounded. Thus, also for our system, we would like to impose the condition that the contribution ΔM of the subsystem $\mathcal{F}_I$ to the mass of the background geometry should be bounded in the limit $\alpha \to 0$.

It is easily shown that ΔM is given by

$$\Delta M \equiv - \int_{x \in I} T^t_t 4\pi r^2 dr = H_I, \tag{8}$$

where H_I is the Hamiltonian of the subsystem $\mathcal{F}_I$ with respect to the Killing time t. Hence, the expectation value of ΔM with respect to a state $|\psi\rangle$ of the scalar field is decomposed into the contribution of excitations and the contribution from the zero-point energy:

$$\langle \psi | \Delta M | \psi \rangle = E_{ent} + \Delta M_B, \tag{9}$$

where E_{ent} is entanglement energy defined by

$$E_{ent} \equiv \langle \psi | : H_I : | \psi \rangle, \tag{10}$$

and ΔM_B is the zero-point energy of the Boulware state. Here, the colons denote the usual normal ordering. [This definition of entanglement energy corresponds to $E_{ent}^{(I')}$ in Ref. [7] and $\langle : H_2 : \rangle$ in Ref. [8].]

Since the Boulware energy ΔM_B diverges as $\Delta M_B \sim -AT_H\alpha^{-2}$ in the limit $\alpha \to 0$ [6], we should impose the condition

$$E_{ent} \simeq |\Delta M_B|, \tag{11}$$

where $A = 4\pi r_0^2$ is the area of the horizon, $T_H = \kappa_0/2\pi$ is the Hawking temperature. We would like to call this condition *the small backreaction condition (SBC)*. Note that the right hand side of SBC (11) is independent of the state $|\psi\rangle$.

IV MAXIMAL ENTANGLEMENT ENTROPY

Now, we shall show that the Hartle-Hawking state is a maximum of the entanglement entropy in the space of quantum states satisfying SBC. For this purpose, we would like to state a more general statement for a quantum system with a state-space of the form (1): *a state of the form*

$$|\psi\rangle = \mathcal{N} \sum_n e^{-E_n/2T} |n\rangle_I \otimes |n\rangle_{II} \tag{12}$$

is a maximum of the entanglement entropy in the space of states with fixed expectation value of the operator E_I defined by

$$E_I = \left(\sum_n E_n |n\rangle_I \cdot {}_I\langle n| \right) \otimes \left(\sum_m |m\rangle_{II} \cdot {}_{II}\langle m| \right), \tag{13}$$

provided that the real constant T is determined so that the expectation value of E_I is actually the fixed value. Here, $\{|n\rangle_I\}$ and $\{|n\rangle_{II}\}$ ($n = 1, 2, \cdots$) are bases of the subspaces $\mathcal{F}_I$ and $\mathcal{F}_{II}$, respectively, and E_n are assumed to be real and non-negative. (For a proof of this statement, see Ref. [13].) We would not like to give a proof of this statement here, but just would like to mention that this statement is almost the same as the following statement in statistical mechanics: a canonical state is a maximum of statistical entropy in the space of states with fixed energy, provided that the temperature of the canonical state is determined so that the energy is actually the fixed value.

Note that the expectation value of E_I is equal to the entanglement energy (10), providing that $|n\rangle_I$ and E_n are an eigenstate and an eigenvalue of the normal-ordered Hamiltonian $: H_I :$ of the subsystem $\mathcal{F}_I$. Hence, for the system of the scalar field, the above general statement insists that the state (12) is a maximum of the entanglement entropy in the space of states satisfying SBC, which corresponds to fixing the entanglement entropy. Off course, in this case, the constant T should be determined so that SBC (11) is satisfied. The value of T is easily determined as $T = T_H$ by using the well-known fact that the negative divergence in the Boulware energy density can be canceled by thermal excitations if and only if temperature with respect to the time t is equal to the Hawking temperature.

Finally, we obtain the statement that the Hartle-Hawking state [10] is a maximum of entanglement entropy in the space of quantum states satisfying SBC since the Hartle-Hawking state is actually of the form (12) with $T = T_H$ [11]. [Strictly speaking, in order to obtain the Hartle-Hawking state, we have to take the limit $\alpha \to 0$ (and $L \to \infty$). However, the following arguments still hold for a finite value of α (and L).] The corresponding reduced density matrix is the thermal state with temperature equal to the Hawking temperature. Therefore, the maximal entanglement entropy is equal to the thermal entropy with the Hawking temperature, which is sought in the brick wall model.

In summary the brick wall model seeks the maximal value of entanglement entropy. In other words, the entanglement approach reduces to the brick wall model when we seek the maximal entanglement entropy.

Our arguments suggests strong connection among three kinds of thermodynamics: black hole thermodynamics, statistical mechanics, and entanglement thermodynamics [9,7,8,12]. It will be interesting to investigate close relations among them in detail.

ACKNOWLEDGMENTS

The author would like to thank Professors W. Israel and H. Kodama for their continuing encouragement. This work was supported partially by the Grant-in-Aid for Scientific Research Fund (No. 9809228).

REFERENCES

1. J. D. Bekenstein, Phys. Rev. **D7**, 949 (19973).
2. S. W. Hawking, Commun. Math. Phys. **43**, 199 (1975).
3. L. Bombelli, R. K. Koul, J. Lee and R. D. Sorkin, Phys. Rev. **D34**, 373 (1986).
4. M. Srednicki, Phys. Rev. Lett. **71**, 666 (1993).
5. G. 'tHooft, Nucl. Phys. **B256**, 727 (1985).
6. S. Mukohyama and W. Israel, Phys. Rev. **D58**, 104005 (1998).
7. S. Mukohyama, M. Seriu and H. Kodama, Phys. Rev. **D58**, 064001 (1998).
8. S. Mukohyama, *"The origin of black hole entropy"*, Doctoral thesis, gr-qc/9812079.
9. S. Mukohyama, Phys. Rev. **D58**, 104023 (1998).
10. J. B. Hartle and S. W. Hawking, Phys. Rev. **D13**, 2188 (1976).
11. W. Israel, Phys. Lett. **57A**, 107 (1976).
12. S. Mukohyama, M. Seriu and H. Kodama, Phys. Rev. **D55**, 7666 (1997).
13. S. Mukohyama, *"Hartle-Hawking state is a maximum of entanglement entropy"*, gr-qc/9904005.

Some Brane-Theoretic No-Hair Results (and Their Field-Theory Duals)

Donald Marolf

Institute for Theoretical Physics, University of California, Santa Barbara, CA 93106
Physics Department, Syracuse University, Syracuse, NY 13244

Abstract. This contribution to the proceedings of the 1999 Canadian Conference on General Relativity and Relativistic Astrophysics is a brief exposition of earlier work, with Sumati Surya (hep-th/9805121) Amanda Peet (hep-th/9903213), addressing certain results in higher dimensional supergravity that are related to black hole no-hair theorems. Its purpose is to describe, in language appropriate for an audience of relativists, how these results can be related to the Maldacena conjecture (aka, the AdS/CFT correspondence). The end product may be taken as a new kind of quantitative evidence in support of the Maldacena conjecture.

I INTRODUCTION

The Maldacena conjecture [1] (aka the AdS/CFT correspondence) has been a topic of much interest and discussion in most communities with interests in quantum gravity. Here we provide some commentary on a few recent results which, in the end, provide a new type of quantitative check on this conjecture and its relatives [2], beyond those previously known (see [3] for a review). These new results relate to the effective 'delocalization' of charge near a black hole horizon, a phenomenon associated with black hole no-hair theorems. After discussing this phenomenon in the familiar context of 3+1 Einstein-Maxwell theory, we describe a related feature of ten-dimensional supergravity, which can then be related to the Maldacena conjecture. We will find that a corresponding phenomenon occurs in the so-called dual field theory, and that the supergravity and field theory results match at both the qualitative and quantitative levels.

Due to a shortage of space, both citations of the literature and inclusion of technical details will be minimal. In particular, as the intention is to make this paper accessible to newcomers to string theory and the Maldacena conjecture, stringy (and super Yang-Mills) details will be particularly sparse, and true string/field theorists are encouraged to go directly to the original works [4,5]. The goal of this

CP493, *General Relativity and Relativistic Astrophysics,* edited by C. P. Burgess and R. C. Myers

paper is merely to provide a rough feel for the results and, perhaps, to motivate the reader to examine the original works.

II ON NO-HAIR RESULTS

Let us begin with a brief reminder of certain results associated with black hole no-hair theorems. For definiteness, consider Einstein-Maxwell theory (in 3+1 dimensions) in the presence of charged dust. Then we know that all stationary black hole solutions are parameterized by their mass M, their charge Q, and their angular momentum J. In particular, if $J = 0$ then the solution is spherically symmetric.

Now suppose that we take a bit of charged dust and drop it into a black hole. The result will be a new black hole which will eventually settle down to a stationary state. In particular, if both the black hole and the bit of dust had $J = 0$, the result would be spherically symmetric. Thus, even though the bit of dust approaches from one side of the black hole, the electric field becomes spherically symmetric in the far future. As the charge approaches the black hole horizon, the electric field begins to become spherically symmetric due to the fact that the spacetime curvature will bend the electric field lines around the black hole. The result is that the electric charge appears (when viewed from far away) to be 'spread out' over the horizon of the black hole.

In fact, this effect is not sensitive to whether the electric charge actually falls through the horizon. Let us suppose that, at some point, the charge is attached to a powerful rocket that keeps it from falling further into the black hole, perhaps moving instead along one of the worldlines shown in the conformal diagram on the left below. The curvature of spacetime bends the electric field lines and tends to make the electric field spherically symmetric. A quasi-artistic impression of this process is shown in the diagram on the right below. There, the small dot denotes the bit of charge and the lines describe the electric field produced by the charge. The extent to which the electric field appears spherical is determined by just how close to the horizon the charge actually sits. In the limit in which the charge approaches the horizon, the electric field becomes spherically symmetric.

As a final conceptual jump, let us dispense with the rockets used above by passing to the extremal case, in which both the charged dust and the black hole satisfy $Q = +M$. Now, the electrostatic repulsion alone is sufficient to keep the charge from falling into the black hole. We could consider a family, labeled by a parameter Δ as shown on the left below, of such extremal charges which do not fall toward the black hole at all. Instead, they sit close to an extremal hole following the integral curves of a timelike killing vector field. Once again, in the limit where the worldline of the charge approaches the horizon, the electric field becomes spherically symmetric and the charge appears to have spread out over the horizon of the black hole. We will refer to this effect as "delocalization" of the charge. Note that the diagram on the right below shows only the field lines produced by the additional charge, and not the field lines produced by the black hole itself.

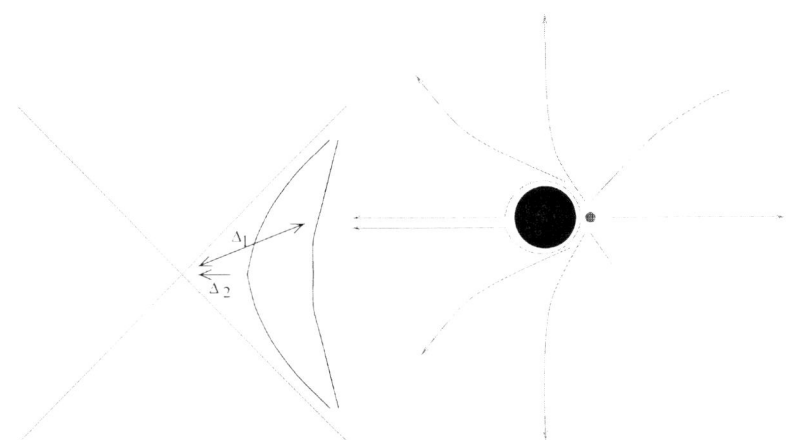

We will shortly be interested in analogous phenomena in 10-d supergravity. What we will do is to use some qualitative features of this effect, along with certain aspects of the supergravity/field theory correspondence, to suggest an analogue in the field theory. Assuming this guess to be correct, the field theory makes both qualitative predictions about when effects of this sort should occur and also *quantitative* predictions[1] as to how fast the charge should delocalize as we adjust the parameter Δ. Thus, it is worth thinking for a moment about how to build a quantitative measure of the delocalization in a black hole solution. A natural choice is to decompose the electric field using spherical harmonics, to measure the dipole, quadrapole, and higher moments of the electric field and to declare that the charge has delocalized on an angular scale $\theta \sim 1/l$ when the charge is close enough to the black hole that the spherical harmonics of order l have become small. However, to do so, one must introduce a foliation of spacetime by spheres and an action of the rotation group on those spheres. Since, at finite Δ, the spacetime is not spherically symmetric, there is some ambiguity here. A convenient choice is to use the spheres and SO(3) action defined by isotropic coordinates centered on the black hole. A calculation of the delocalization rate is then straightforward, as our spacetimes are described by the Majumdar-Papapetrou solutions [6], for which the fields take a simple form in isotropic coordinates. The detailed results are not important here, though they will be for the 10-d supergravity analogues discussed below.

---

[1] The term 'prediction' should perhaps be taken with a grain of salt. Even assuming both that the Maldacena conjecture is correct and that the field theory analogue of our effect has been correctly guessed, it is not a priori obvious that naive calculations of the type that we will discuss must be quantitatively correct. Nonetheless, that such naive results will in fact precisely agree with the supergravity seems rather impressive.

III THE SUPERGRAVITY VERSION

The main focus of this exposition is a version of charge delocalization in $9+1$ dimensional supergravity theories. Recall that, in addition to black holes, higher dimensional supergravity has what are known as black *brane* solutions. A black brane is like a black hole except that its horizon does not have the topology of a sphere, and is often not compact. For example, in an n dimensional spacetime, an event horizon with topology $S^{n-2} \times \mathbf{R}$, where the $\mathbf{R}$ factor is associated with a null direction, would be referred to as a black hole, while an event horizon with topology $S^{n-3} \times \mathbf{R} \times \mathbf{R}$ is a black string, and $S^{n-2-p} \times \mathbf{R}^p \times \mathbf{R}$ is a black p-brane.

We will be particularly interested in the class of branes known as "D-branes" in the context of $9+1$ supergravity theories. Here, the label D stands for 'Dirichlet' and D in no way denotes the dimension of the brane. Thus, we will often refer to Dp-branes; e.g., a D1-brane is a string of the "Dirichlet" type. For a discussion of what the term "Dirichlet" means in this context, see [7]. Below, we focus on the extremal limits of these solutions[2]. The associated supergravity solutions containing only a one brane have a single length scale, $r_p \propto Q_p^{1/(7-p)}$, where Q_p is the charge of the brane.

A relevant fact about Dp-branes is that branes of different dimensions (different values of p) carry different kinds of charge. Thus, even in the extremal limit, one cannot in general construct static solutions containing Dp-branes with different values of p, as the gravitational attraction is not balanced by the electrostatic repulsion. However, there is also a dilaton field in the supergravity theories of interest, and this produces a repulsive force between two branes even when their dimensions differ. For the right combinations of extremal branes, it can be arranged for these dilatonic forces to hold the branes apart, so that exactly static solutions can in fact be constructed. A class of examples on which we will focus here consists of static spacetimes containing Dp-branes and D$(p-4)$-branes. Roughly speaking, static solutions exist whenever the D$(p-4)$-branes are 'parallel' to the Dp-branes. That is, the final solutions have a $(p-4)$ dimensional set of commuting spacelike Killing vector fields[3]. We will focus on such systems from now on and, as a result, we can refer to the Dp-brane simply as 'the big brane' and the D$(p-4)$-brane simply as 'the little brane' or 'the smaller brane.' As before, we may consider a family of such solutions labeled by a parameter Δ which describes the separation of the two branes.

Let us now think of the larger (p-dimensional) D-brane as the analogue of the black hole in section II, and let us think of the smaller ($(p-4)$-dimensional) brane as the analogue of the charge. We may again ask if the charge of the smaller brane 'spreads out over the larger brane' as the separation Δ goes to zero. If it does not,

---

[2] Unfortunately, the extremal limits. Instead, they are singular static solutions, with the norm of the Killing field vanishing at the singularity. The singularity is null for $p \leq 5$ but timelike for $p \geq 6$.

[3] In fact, the symmetry group contains the Poincare' group of $(p-4)+1$ dimensional Minkowski space.

then by taking the limit $\Delta \rightarrow 0$, one could form 'hairy' brane solutions which could not be completely characterized by the amount of big- and little-brane charge they carried. Instead, they would also require the specification of the *distribution* of the little-brane charge over the big brane.

Now, it is not a priori clear what our 3+1 Einstein-Maxwell intuition should tell us about the delocalization of little-brane charge in the current context. We now consider to a different gravitating theory and, while the objects being described are similar to black holes, they are singular. It turns out that it is again possible to study delocalization by direct calculation as the analogues of the Majumdar-Papapetrou metrics can be found for an arbitrary separation Δ between the branes. Such solutions were studied in [4] for the case of $p = 5$, and more generally in [5]. The final result is that the little-brane delocalizes for the cases $p = 4$ and $p = 5$, but not for $p = 6$. That is, for $p = 4, 5$ the limiting soliton as $\Delta \rightarrow 0$ in fact has a p-dimensional group of translation symmetries in addition to the rotational symmetries that one would expect of a p-brane. However, for $p = 6$, the only new symmetries of the limiting solution are the rotational ones. Corresponding localized solutions were in fact constructed explicitly in [8] in what is known as the "near-core limit." One might say that, for $p = 6$, the 2-brane charge spreads out in the angular directions around the 6-brane, but not in the translational directions along the 6-brane. It is probably not a coincidence that D4-branes and D5-branes have (naked) null singularities, while the D6-brane in fact has a naked timelike singularity. Following a common practice, we will not discuss cases with $p \geq 7$. The reason for this is that such solutions have a more complicated structure at infinity. Since our branes live in a (9+1)-dimensional spacetime, $p = 7$ branes behave like point particles in 2+1 dimensions, producing conical deficit angles at infinity. The $p = 8$ branes produce fields which do not fall off at infinity, and the $p = 9$ branes extend to infinity in all directions.

One can measure the rate at which charge delocalizes in a manner similar to that discussed in section II. Here, we are most interested in the delocalization in the directions along the brane. Thus, we can simplify our lives by considering not just a single (fully localized) little brane, but a shell of such branes placed in a spherically symmetric manner about the bigger-brane. Thus, even at finite Δ, we may consider solutions that have the symmetries of S^{8-p} in addition to a $(p - 4)$-dimensional translational symmetry.

These spacetimes may be equipped with a nice set of coordinates [5] analogous to the isotropic coordinates of the Majumdar-Papapetrou solutions. This introduces a radial function r and an action of the four-dimensional translation group that moves the little brane along the big-brane. To measure the extent to which little-brane charge is delocalized, consider some surface $r = r_0$ outside the shell of little branes and Fourier transform the fields on this surface with respect to the four-dimensional translation group. For a shell at $r = \Delta < r_0$, we may say that the solution has delocalized on a length scale λ when the corresponding Fourier component falls below, say, e^{-10} times the value it has when the shell is at $r = r_0$. In fact, we may take a limit $r_0 \rightarrow \infty$ to make this definition independent of r_0. We display the results here so that the reader may properly appreciate the comparison with the field theoretic calculations in section IV. In the limit where the separation Δ between the branes is small, the distance scale over which the solution is delocalized behaves like $\lambda \sim r_4^{3/2}/\Delta^{1/2}$ for the case $p = 4$, and it behaves like $\lambda \sim r_5\sqrt{\ln(r_5/\Delta)}$ for $p = 5$, while the $p = 6$ case does not delocalize.

After reading this section, the traditional relativist may feel caught up in a swirl of D's, p's, Q's and various other letters of the alphabet. Such a reader should take a moment to collect their thoughts, as this final section will not be more familiar. In regard to many statements below, the traditional relativist may feel unqualified to just what is 'reasonable' or 'natural.' This is, of course, to be expected with a new subject, and I urge such a reader not to worry overly much. The main goal should be to take away a broad overview of the argument and some appreciation for the results.

IV FIELD THEORY DUALS

Now, describing charge delocalization in supergravity is all well and good, but the current excitement in string theory concerns the so-called Maldacena conjecture. This conjecture states that aspects of supergravity are described by certain quantum field theories, even though those theories do not include gravity when viewed in the usual way. For details, see the original papers [1,2] or the recent review [3]. Here we content ourselves with an extremely rough statement of the conjecture, which is that supergravity physics near the 'horizon' (the locus where the norm of the static Killing field vanishes) of a Dp-brane is in fact completely described by a (non-gravitating) quantum field theory. The idea is that simple supergravity quantities supergravity may in principle be quite complicated when written in terms of the gauge theory, but that nevertheless such a 'dictionary' that tells us how to translate supergravity physics into field theory physics can in fact be constructed. Furthermore, this dictionary has the property that classical gravitational effects on the supergravity side of this correspondence are mapped to strongly quantum mechanical effects on the field theory side. For a discussion of how our current context of multiple separated branes fits into the Maldacena conjecture, see [5].

For now, we simply quote a few results that will paint a backdrop for the connections we wish to make between supergravity and field theory. The fact that we are interested in the near-horizon physics of the Dp-brane means that we will be considering $SU(N)$ super Yang-Mills theory in $p + 1$ dimensions. Here, the value of N is related to the charge of the branes, the string length l_s, and the string coupling g_s through $r_p = l_s(g_sN)^{1/(7-p)}$. Now, it is well known that Yang-Mills theory in 4 Euclidean dimensions contains instantons. As a result, Yang-Mills theory in $p+1$ dimensions (for $p \geq 4$) contains solitons which are just the lift of the instanton solutions to $p+1$ dimensions. The result is a $(p-4)$-brane shaped soliton, and such solitons in the Yang-Mills theory are to be associated in the Maldacena conjecture with the D$(p-4)$-branes of the supergravity theory.

The next part of our story will be to guess what aspect of the Yang-Mills theory should correspond to the spreading of the charge in the supergravity solution. This is certainly a guess, as it is not something which can be derived from the known aspects of the correspondence. However, we will see that there is an extremely natural candidate. Now, a soliton can be viewed as a coherent 'lump' of classical field that holds itself together through the non-linear dynamics of the field. It turns out that, in super Yang-Mills theory, the size ρ of this lump has no preferred value. Taking any instanton solution to the equations of motion and scaling it by a constant factor again gives a static solution. Thus, there are static soliton solutions in our $p + 1$ Yang-Mills theory with any value of the scale size ρ.

Furthermore, one may allow ρ to vary over the $(p - 4) + 1$ dimensional world-volume of the soliton. Such solutions are not static, but if the distortions are small and of long wavelength (much larger than the string length) it turns out that $\rho(x)$ behaves just like a massless $(p - 4) + 1$ dimensional field. To be a bit more precise, because the soliton can point in roughly N directions in gauge space, ρ can be thought of as a roughly N-dimensional vector of massless scalars. Thus, ρ^2 acts like a sum $\sum_{i=1}^{N} \phi_i^2$ over massless scalar fields. Normalizing the scalar fields ϕ_i canonically, one finds $\rho^2 = l_s^{(p-3)} g_s \sum_{i=1}^{N} \phi_i^2$.

The soliton in the Yang-Mills theory gives a nice picture of a D$(p - 4)$-brane sitting inside a Dp-brane, but one might ask how to encode a separation of the two branes into the Yang-Mills theory. This is done through another set of fields and, for our purposes, the important property is that, when these fields take some nonzero value Δ, a mass scale $m(\Delta)$ is generated which interacts with the field ρ. If the D$(p-4)$-brane shell is located at $r = \Delta$, then this mass is known to be $m = l_s^{-2}\Delta$, in units where $\hbar = 1$. This means that our description of ρ as a free field is really only valid for wavelengths shorter than some infra-red cutoff $\Lambda_{IR}(\Delta) = l_s^2/\Delta$ and for wavelengths longer than the string scale due to the short-distance cutoff described above.

So, since ρ tells us how spread out the instanton is in the appropriate directions, it is natural to expect that it corresponds, via the Maldacena conjecture, to the spread of the D$(p - 4)$-brane charge in supergravity, at least when the D$(p - 4)$ brane is close to the horizon of the Dp-brane. Now, in order to match the energy of the extremal supergravity solution, the energy in the field theory must be that of a ground state. Let us therefore ask how large ρ should be in a typical low-energy quantum state of the gauge theory. Classically, ρ can be set to any size we desire without changing the energy. Quantum mechanically, an attempt to confine ρ to a small range of values, minimizing the uncertainty in ρ, requires a large momentum conjugate to ρ and thus a large energy. Thus, quantum fluctuations will effectively cause the value of ρ to be non-zero in the ground state. We can estimate a scale for the effective size of ρ by computing, for example, $\langle \rho^2 \rangle$ for the various cases. To do so, we use the fact that, between the appropriate infra-red and ultra-violet cutoff scales, ρ acts much like a free field. Outside this range of wavelengths, it is reasonable to assume that the fluctuations of ρ are small. Thus, we can estimate $\langle \rho^2 \rangle$ using the properties of massless scalar fields in various dimensions. The result is

$$\begin{aligned}
&\text{For p} = 4, \quad \langle \rho^2 \rangle = (l_s g_s) N (\Lambda_{IR} - \Lambda_{UV}), \\
&\text{For p} = 5, \quad \langle \rho^2 \rangle = (l_s^2 g_s) N \ln(\Lambda_{IR}/\Lambda_{UV}), \\
&\text{For p} = 6, \quad \langle \rho^2 \rangle = (l_s^3 g_s) N (\Lambda_{UV}^{-1} - \Lambda_{IR}^{-1}).
\end{aligned} \tag{1}$$

Converting to the supergravity parameters and using the near-horizon limit, these results become

$$\begin{aligned}
&\text{For p} = 4, \quad \sqrt{\langle \rho^2 \rangle} \sim r_4^{3/2} \Delta^{-1/2}, \\
&\text{For p} = 5, \quad \sqrt{\langle \rho^2 \rangle} \sim r_5 \sqrt{\ln(\Delta/r_5)}, \\
&\text{For p} = 6, \quad \sqrt{\langle \rho^2 \rangle} \sim r_6^{1/2} l_s^{1/2}.
\end{aligned} \tag{2}$$

Note that for $p = 4$ and $p = 5$ these agree precisely with the delocalization rates of the supergravity solutions, while for $p = 6$ the field theory predicts that the delocalization is bounded as $\Delta \to 0$ by a value proportional to a positive power of the string scale; i.e., that it is small compared to the scales of classical supergravity. As explained in [5], similar arguments in the field theory tell us that certain other kinds of charge delocalization should also occur in supergravity. Due to their more complicated nature, these predictions are harder to check directly. Nevertheless, some preliminary investigations [9] seem to support these predictions. Thus, the phenomenon of charge delocalization provides a new kind of evidence in support of the Maldacena conjecture. We hope that it may also provide new insight into the nature of the supergravity/field theory correspondence.

Acknowledgments

The author would like to thank Andrés Gomberoff, David Kastor, Jennie Traschen, Sumati Surya, and especially Amanda Peet for the pleasure of collaborating with them on the work which let to this commentary. Thanks also to Jorma Louko and Amanda Peet for comments on an earlier draft. This work was supported in part by NSF grants PHY94-07194 and PHY97-22362 and by funds provided by Syracuse University.

REFERENCES

1. J. Maldacena, *Adv. Theor. Math. Phys.* 2 (1998) 231, hep-th/9711200.
2. N. Itzhaki, J. M. Maldacena, J. Sonnenschein, and S. Yankielowicz, Phys. Rev. D58 (1998) 046004, hep-th/9802042.
3. O. Aharony, S.S. Gubser, J. Maldacena, H. Ooguri, and Y. Oz, hep-th/9905111.
4. S. Surya and D. Marolf, *Phys. Rev. D* 58 (1998) 124013, hep-th/9805121.
5. D. Marolf and A.W. Peet, hep-th/9903213.
6. S. Majumdar, *Phys. Rev.* **72** (1947) 930; A. Papapetrou, *Proc. Roy. Irish Acad.* **A51** (1947) 191.
7. J. Polchinski, "String Theory," (Cambridge, New York, 1998).

8. N. Itzhaki, A. Tseytlin, and S. Yankielowicz, *Phys. Lett.* B432 (1998) 298, hep-th/9803103.
9. A. Gomberoff, D. Kastor, D. Marolf, and J. Traschen, hep-th/9905094.

Topological Censorship and Black Hole Topologies

Kristin Schleich and Donald M. Witt

Department of Physics and Astronomy
University of British Columbia
Vancouver, BC Canada V6T 1Z1

Abstract. In contrast to asymptotically flat black hole solutions for which all horizons settle down rapidly to have the topology of a sphere, certain asymptotically anti-de Sitter black hole solutions have horizons with the topology of a higher genus surface. The fact that the nontrivial topology of these horizons is not hidden might appear to indicate that generalizations of topological censorship do not hold for such spacetimes. However, this is not the case: topological censorship holds in anti-de Sitter spacetimes. Specifically, we prove every causal curve with endpoints on scri can be deformed to a curve at scri. From this it follows that the genus of the black hole horizon is set by the large scale topology of the spacetime. Our results clarify the connection between topological censorship and the non-spherical horizon topologies of locally anti-de Sitter black holes. Furthermore, the proof making this connection is valid for any spacetime that satisfies the principle of topological censorship.

An enduring idea in popular culture has been the use of topological structures to produce almost magical effects. Two dimensional right handed beings in flatland journey around a mobius strip and return as left handed. Spaceships traverse wormholes to travel between distant points almost instantaneously. Appealing though such fictional travels may be, in actual fact they cannot take place in asymptotically flat spacetimes. In particular, the topological censorship theorem of Friedman, Schleich and Witt proved that isolated topological structures in physically reasonable asymptotically flat spacetimes are shrouded [1]. Intuitively, this result means that small scale topology collapses to a black hole faster than it can be traversed by any causal curve. Futhermore, as pointed out by Galloway [2], this result implies that the spacetime exterior to any black holes must be simply connected. Additionally, Jacobson and Venkataramani [3] used this theorem to show that black hole horizons in asymptotically flat spacetimes have spherical topology, generalizing a result of Hawking [4]. Thus topological censorship also severely restricts the topology of black hole horizons and that of the spacetime exterior to them.

Recently, there have appeared examples of black holes in asymptotically anti-de Sitter spacetimes (adS spacetimes) which have horizons with the topology of

CP493, *General Relativity and Relativistic Astrophysics,* edited by C. P. Burgess and R. C. Myers
© 1999 American Institute of Physics 1-56396-905-X/99/$15.00

higher genus surfaces (See for example [5,6] and other references in [7]). Indeed, these examples have led some to suggest that anti-de Sitter spacetimes do not obey topological censorship. However, this is not the case; in this talk we will summarize a recent proof that asymptotically adS spacetimes satisfy the **Principle of Topological Censorship** (PTC) : *Every causal curve whose initial and final endpoints belong to the boundary at infinity ($\mathcal{I}$) is fixed endpoint homotopic to a curve on $\mathcal{I}$* [7]. Furthermore, using a restatement of the theorem in terms of loops on the spacetime exterior to the black hole horizons, we show that $\sum_{i=1}^{n} g_i \leq g_0$ where g_i are the genera of a good spacelike cut of the black hole horizons and g_0 the genera of a spacelike cut of $\mathcal{I}$. Finally the homology of the spacetime exterior to the black hole horizons is completely determined by its betti numbers. Thus the topology of $\mathcal{I}$ dictates that of the black hole horizons.

Precisely, we will consider a spacetime $\mathcal{M}$, with metric g_{ab}, which can be conformally included into a spacetime-with-boundary $\mathcal{M}' = \mathcal{M} \cup \mathcal{I}$, with metric g'_{ab}, such that $\partial \mathcal{M}' = \mathcal{I}$ is timelike (*i.e.*, is a Lorentzian hypersurface in the induced metric) and $\mathcal{M} = \mathcal{M}' \setminus \mathcal{I}$. Note that $\mathcal{I}$ can have multiple components. The conditions on the conformal factor $\Omega \in C'^1(\mathcal{M}')$, are standard: (a) $\Omega > 0$ and $g'_{ab} = \Omega^2 g_{ab}$ on $\mathcal{M}$, and (b) $\Omega = 0$ and $d\Omega \neq 0$ pointwise on $\mathcal{I}$. If $\partial \mathcal{M}'$ is not connected, let $\mathcal{I}_0$ denote a single component of $\partial \mathcal{M}'$. Let $\mathcal{D} = I^+(\mathcal{I}_0) \cap I^-(\mathcal{I}_0)$ be the domain of outer communications (DOC) of $\mathcal{M}$ with respect to $\mathcal{I}_0$. Assume that $\mathcal{D}$ does not meet any other components of $\partial \mathcal{M}'$. Using the fact that $\mathcal{I}_0$ is timelike, it follows that $\mathcal{D}$ is connected and the closure of $\mathcal{D}$ in $\mathcal{M}'$ contains $\mathcal{I}_0$. Then $\mathcal{D}' := \mathcal{D} \cup \mathcal{I}_0$ is a connected spacetime-with-boundary, with $\partial \mathcal{D}' = \mathcal{I}_0$ and $\mathcal{D} = \mathcal{D}' \setminus \mathcal{I}_0$.

Theorem 1. *Let $\mathcal{D}$ be the domain of outer communications with respect to $\mathcal{I}_0$ as described above, and assume the following conditions hold: (i) $\mathcal{D}' = \mathcal{D} \cup \mathcal{I}_0$ is globally hyperbolic* [1]. *(ii) $\mathcal{I}_0$ admits a compact spacelike cut. (iii) For each point p in $\mathcal{M}$ near $\mathcal{I}_0$ and any future complete null geodesic $s \to \eta(s)$ in $\mathcal{D}$ starting at p, $\int_0^\infty \mathrm{Ric}(\eta', \eta') \, ds \geq 0$. Then the PTC holds on $\mathcal{D}$.*

As detailed in [7], this theorem is proven using a lemma stating that $\mathcal{I}_0$ cannot communicate with any other component of $\mathcal{I}$ by a causal curve in $\mathcal{M}$. A key factor in proving this lemma is that asymptotically adS spacetimes satisfy a modified form of the Averaged Null Energy Condition (ANEC) as stated in (*iii*). If one assumes that the Einstein equations with cosmological constant $R_{ab} - \frac{1}{2} R g_{ab} + \Lambda g_{ab} = 8\pi T_{ab}$ hold, then for any null vector X, $\mathrm{Ric}(X, X) = R_{ab} X^a X^b = 8\pi T_{ab} X^a X^b$. Thus the integrand $\mathrm{Ric}(\eta', \eta')$ in (*iii*) could be replaced by $T(\eta', \eta')$. Clearly, the presence and sign of the cosmological constant is irrelevant to whether or not a spacetime satisfying the Einstein equations will satisfy condition (*iii*). One then uses this lemma to prove the theorem by applying a covering space argument. Namely any causal curve not fixed endpoint homotopic to a causal curve on $\mathcal{I}_0$ will begin or

---

[1] As in the case of spacetimes without boundary, a spacetime-with-boundary $\mathcal{M}'$ is defined to be *globally hyperbolic* if $\mathcal{M}'$ is strongly causal and the sets $J^+(p, \mathcal{M}') \cap J^-(q, \mathcal{M}')$ are compact for all $p, q \in \mathcal{M}'$. Observe that $\mathcal{M}$ is not globally hyperbolic but $\mathcal{M}'$ is in many examples of interest such as asymptotically anti-de Sitter spacetimes.

end on a different component of $\mathcal{I}$. However, such a curve cannot exist by the aforementioned lemma. Hence the result.

As stated, Theorem 1 applies to many examples of local adS spacetimes. However, a more general form of the theorem also holds in which one removes the compactness condition (ii) and replaces the energy condition (iii) by the the generic condition and the ANEC: $\int_{-\infty}^{\infty} \mathrm{Ric}(\eta', \eta')\, ds \geq 0$ along any complete null geodesic $s \to \eta(s)$ in $\mathcal{D}$. Note also that the global hyperbolicity assumption Theorem 1 can be weakened in a manner similar to that done in the AF case in [8].

Next the PTC can be conveniently restated in terms of the fundamental group of the DOC. Observe that that the inclusion map $i : \mathcal{I} \to \mathcal{D}'$ induces a homomorphism of fundamental groups $i_* : \Pi_1(\mathcal{I}) \to \Pi_1(\mathcal{D}')$. Then

Theorem 2. *If the PTC holds for $\mathcal{D}'$, then the group homomorphism $i_* : \Pi_1(\mathcal{I}) \to \Pi_1(\mathcal{D}')$ induced by inclusion is surjective.*

Theorem 2 says roughly that every loop in $\mathcal{D}$ is deformable to a loop in $\mathcal{I}$. Moreover, it implies that $\Pi_1(\mathcal{D})$ is isomorphic to the factor group $\Pi_1(\mathcal{I})/\ker i_*$. In particular, if $\mathcal{I}$ is simply connected then so is $\mathcal{D}$, thus generalizing the result of [2].

The event horizon is defined as boundary of the region of spacetime visible to observers at $\mathcal{I}$. This horizon is a set of one or more null surfaces, also called black hole horizons, generated by null geodesics that have no future endpoints but possibly have past endpoints. The topology of these black hole horizons is constrained in spacetimes obeying the PTC because the topology of the DOC is constrained by the PTC—intuitively, causal curves that can communicate with observers at $\mathcal{I}$ cannot link with these horizons in a non-trivial way. Rather they only carry information about the non-triviality of curves on $\mathcal{I}$.

One can obtain further information about these horizons if one considers the topology of the intersections of certain spacelike hypersurfaces with the horizons; those for which this intersection is a set of closed spacelike 2-manifolds (good cuts of the horizons). Precisely, when the domain of outer communications is globally hyperbolic to the future of a cut of $\mathcal{I}$ and if the PTC holds on this sub-domain, then the PTC will constrain the topology of this sub-domain, of its Cauchy surface and ultimately of good cuts of the horizons. Thus, though the PTC does not determine the topology of arbitrary embedded hypersurfaces or the cuts they make on the horizons, it does do so for hypersurfaces homeomorphic to Cauchy surfaces for these sub-domains that make good cuts of the horizons.

Precisely, let $\mathcal{M}$ and $\mathcal{I}$ be as described in Theorem 1. Let K be a cut of $\mathcal{I}$, and let $\mathcal{I}_K$ be the portion of $\mathcal{I}$ to the future of K, $\mathcal{I}_K = \mathcal{I} \cap I^+(K)$. Let $\mathcal{D}_K$ be the domain of outer of communications with respect to $\mathcal{I}_K$, $\mathcal{D}_K = I^+(\mathcal{I}_K) \cap I^-(\mathcal{I}_K) = I^+(K) \cap I^-(\mathcal{I})$. Theorems 1 applies equally as well to $\mathcal{D}_K$. This procedure, first discussed by Jacobson and Venkataramani [3], allows one to study the topology of cuts on the future event horizon $\mathcal{H} = \partial I^-(\mathcal{I})$ of the form $\partial I^+(\mathcal{I}_K) \cap \mathcal{H}$. Then for asymptotically adS spacetimes one has

Theorem 3. *Let $\mathcal{D}_K$ be the domain of outer communications to the future of the cut K on $\mathcal{I}$ as described above. Assume $\mathcal{D}_K'$ is globally hyperbolic and satisfies*

the PTC. Suppose V' is a Cauchy surface for $\mathcal{D}'_\lambda$ such that its closure $V = \overline{V'}$ in $\mathcal{M}'$ is a compact topological 3-manifold-with-boundary whose boundary ∂V (corresponding to the edge of V' in M') consists of a disjoint union of compact 2-surfaces, $\partial V = \sqcup_{i=0}^{k} \Sigma_i$ where Σ_0 is on $\mathcal{I}$ and the Σ_i, $i = 1, \ldots, k$, are on the event horizon. Then

$$\sum_{i=1}^{k} g_i \leq g_0 \quad , \tag{1}$$

where $g_j = $ the genus of Σ_j, $j = 0, 1, \ldots, k$. In particular, if Σ_0 is a 2-sphere then so is each Σ_i, $i = 1, \ldots, k$.

The proof if this result is given in [7]. Observe that this theorem is readily generalized to any space satisfying PTC, though for technical precision, its statement is given for asymptotically adS spacetimes only.

Finally further application of techniques of algebraic topology results in

Theorem 4. If $i_* : \Pi_1(\Sigma_0) \to \Pi_1(V)$ is onto, then the integral homology $H_*(V, Z)$ is torsion free, and hence is completely determined by the Betti numbers. Furthermore, (a) $b_1 = \sum_{i=0}^{k} g_i$, and (b) $b_2 = k$.

We wish to emphasize that the results concerning black hole topology in no way contradict the numerical findings concerning the existence in principle of temporarily toroidal black holes in asymptotically flat spacetimes. The consistency of topological censorship with asymptotically flat models containing temporarily toroidal black hole horizons has been clearly elucidated in [9]. The results of Theorem 3 on the topology of cuts of black hole horizons use Cauchy slices of a sub-region of the DOC to the future of a cut on $\mathcal{I}$. Slices exhibiting temporarily toroidal black hole horizons are not such slices. Moreover, the method requires such a slice to have non-empty edge which meets the horizons in C^0 compact surfaces. This requirement is also not always satisfied. As an example, consider a 3 dimensional spacetime formed from a line segment L in the $t = 0$ plane by removing all points of $J^+(L)$ of that lie above a hyperboloid intersecting it to the future. The horizon $\partial I^-(\mathcal{I}^+)$ of the remaining globally hyperbolic black hole spacetime is generated by null geodesics that all begin on L.

The Cauchy surface for $\mathcal{D}$ will have topology $\mathbb{R}^2$ and does not cross the horizon. The topology of the horizon, however can be studied by considering spacetimes corresponding to the future of a cut of $\mathcal{I}$. However, not every cut of $\mathcal{I}$ will produce a spacetime with a Cauchy slicing with the correct properties. Such a bad cut of $\mathcal{I}$, V_K is illustrated in Figure 1. Additionally not all spacelike surfaces need be Cauchy surfaces of the spacetime to the future of a cut of $\mathcal{I}$. The hyperboloid T in Figure 1 is an example of such a surface. As recognized in [9] in a similar model, a family of such surfaces exhibits formation of the analog of a temporarily toroidal black hole horizon; these surfaces intersect the horizon in a pair of topological circles. The circles increase in size and eventually meet, whence the horizon topology changes. After this point, these surfaces meet the horizon at a circle. In contrast, with respect to constant-t surfaces, the horizon forms completely at the $t = 0$ instant.

135

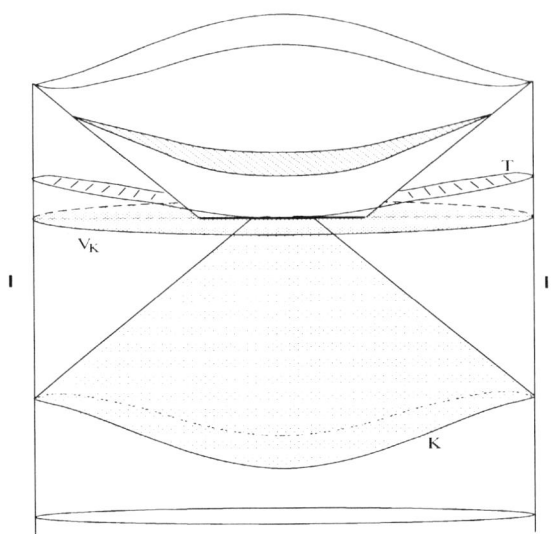

FIGURE 1. A bad cut of $\mathcal{I}$. The boundary of the causal future of this cut intersects the horizon at a segment I of L. The topology of the Cauchy slice V_K for its DOC is $R^2 \setminus I$. Its closure intersects the horizon at I; thus the closure of this slice has no inner boundary, being in this case R^2. The slice T is not a Cauchy slice for the spacetime to the future of a cut of $\mathcal{I}$.

The apparent change of horizon topology is an effect entirely dependent on the choice of hypersurface. The only unambiguous description of this black hole is that no causal curve is able to link with the horizon; *i.e.*, that the PTC is not violated.

REFERENCES

1. Friedman, J., Schleich, K. and Witt, D., *Phys. Rev. Lett.* **71** (1993), 1486.
2. Galloway, G., *Class. Quantum Gravit.* **12**, L99 (1995).
3. Jacobson, T. and Venkataramani, S., *Class. Quantum Gravit.* **12**, 1055 (1995).
4. Hawking, S.,*Commun. Math. Phys.* **25**, 152 (1972).
5. Lemos, J.,*Phys. Lett.* B **352**, 46 (1995).
6. Åminneborg, S., Bengtsson, I., Holst, S., and Peldán, P., *Class. Quantum Gravit.* **13**, 2707 (1996).
7. Galloway, G., Schleich, K., Witt, D. and Woolgar, E., preprint gr-qc/9902061 to be published in *Phys. Rev. D*.
8. Galloway, G. and Woolgar, E., *Class. Quantum Gravit.* **14**, L1 (1997).
9. Shapiro, S., Teukolsky, S. and Winicour, J., *Phys. Rev.* D **52**,6982 (1995), Husa, S. and Winicour, J., preprint gr-qc/9905039.

Higher Genus Horizons as Least Area Surfaces

E. Woolgar

Dept. of Mathematical Sciences and Theoretical Physics Institute
University of Alberta, Edmonton AB, Canada T6G 2G1
ewoolgar@math.ualberta.ca

Abstract. By a trivial modification of Hawking's well-known topology theorem for black holes, one can obtain a lower bound of $\frac{4\pi(g-1)}{-\Lambda}$ for the area of an apparent horizon of genus g in a vacuum spacetime with cosmological constant $\Lambda < 0$. We discuss a derivation in the special case of an apparent horizon at a moment of time symmetry, so that the horizon is a minimal surface.

There are now two approaches to the problem of determining the topology of black hole horizons. One such method, based on topological censorship techniques, is described elsewhere in this volume ([1], see also [2] and references therein), so we say no more about it here. Instead, we consider the second method, the variational technique introduced by Hawking [3,4]. By an argument that drew on both the Hodge decomposition of differential forms (in this case, zero-forms, *i.e.*, functions) and the fact that under certain global conditions trapped surfaces do not lie outside the horizon, Hawking was able to conclude that smooth stationary event horizons must be spherical (somewhat more precisely, they must have spherical cross-sections). Later, citing a suggestion by Gibbons, Hawking was able to apply his methods to show that smooth apparent horizons must be spherical [5].

More recently, black hole solutions with toroidal and higher-genus horizons have been found. They fall into two categories: the "temporarily toroidal" horizons found using numerical collapse codes in asymptotically flat spacetimes ([6] and references therein), and the stationary black holes embedded in locally anti-de Sitter backgrounds ([7] and references therein). The temporarily toroidal event horizons occur during black hole formation, when new generators are entering the horizon at "cross-over points" (where two or more generators begin). At such points, Hawking's smoothness assumption is violated. In the case of the black holes in locally anti-de Sitter backgrounds, another of Hawking's assumptions, the dominant energy condition, is violated. In these circumstances, Hawking's arguments do not

CP493, *General Relativity and Relativistic Astrophysics,* edited by C. P. Burgess and R. C. Myers
© 1999 American Institute of Physics 1-56396-905-X/99/$15.00

constrain the horizon topology (though in the latter case, topological censorship arguments still result in certain constraints [1,2]).

Until recently, it had gone unnoticed that Hawking's variational technique gives useful information even when the dominant energy condition is violated. To see this, we reconsider the technique in a special case, that of an apparent horizon in a negatively curved Einstein spacetime with Ricci curvature $\mathcal{R}_{ab}$:

$$R_{ab} = \Lambda g_{ab} \quad , \quad \Lambda < 0 \quad . \tag{1}$$

Moreover, we will further simplify to the case where the apparent horizon lies in a time-symmetric slice H, a special case recently treated by Gibbons [9]; see [10] for an argument that holds more generally. For time-symmetric slices, there is a local time coordinate t that is zero on H and $t \to -t$ is an isometry in some neighbourhood of H. From these two conditions and the Gauss equation, we can compute the scalar curvature R of the induced metric h_{ab} on H. It is simply

$$R = 2\Lambda \quad . \tag{2}$$

Now recall that at every point of a smooth apparent horizon Γ, the expansion θ of the outbound null vectors orthogonal to the horizon is zero. As well, for any smooth compact orientable surface lying outside Γ, θ must be positive somewhere on the surface, since otherwise the surface would be outer trapped, a contradiction. Moreover, it is well-known that if this horizon lies in a moment of time symmetry, then these conditions translate to conditions on the mean curvature Π of Γ. Essentially, at a moment of time symmetry H, any expansion in the null vectors orthogonal to Γ must be due to expansion in the spacelike normals to Γ in H, and this is measured by the mean curvature. The condition that θ be positive somewhere on every surface outside Γ becomes the condition that no mean concave surface lies outside Γ. We remark that a surface with vanishing mean curvature is called a *minimal surface*. Now consider a variation of Γ defined by some vector field $e^y n^a$ where n^a is the outward unit normal field to Γ and $y : \Gamma \to R$ is some function. This vector field is initial data for a geodesic congruence in H. By moving each point of Γ a parameter distance t along the geodesic through that point, we define a family of surfaces $\Gamma_s \subset H$, where Γ is the $s = 0$ surface ($\Gamma = \Gamma_0$). In passing, we note that, to first order in s, the variation in area $A(\Gamma_s) - A(\Gamma_0)$ is given by the classical variation of area formula

$$\delta A(\Gamma) = \int_{\Gamma_0} e^y \Pi d^2 \Gamma_0 \quad . \tag{3}$$

Thus, as in Galloway [8], we may describe an apparent horizon embedded in a moment of time symmetry H as (i) a surface of least area which (ii) cannot have a mean concave surface lying outside it (since such a surface would be a trapped surface). We now consider condition (ii).

What is of interest is the variation of mean curvature, which is

$$\frac{d\Pi}{ds} = -\Delta y - \nabla y \cdot \nabla y - \frac{1}{2}\Pi^{ab}\Pi_{ab} - \frac{1}{2}\Pi^2 + K - \frac{1}{2}R \quad , \tag{4}$$

where ∇ is the covariant derivative induced on Γ_s, Δ is the Laplacian on Γ_s, Π_{ab} is the extrinsic curvature of Γ_s, and K is the Gauss curvature of Γ_s. Hawking noticed that since Γ_s is a Riemann surface, then by Hodge theory one can always solve

$$\Delta y + \frac{1}{2}\Pi^{ab}\Pi_{ab} + \frac{1}{2}\Pi^2 - K + \frac{1}{2}R = c \quad , \tag{5}$$

iff the constant c (which can vary with s but not within Γ_s) is chosen such that

$$cA(\Gamma_s) = \int_{\Gamma_r} \left[\frac{1}{2}\Pi^{ab}\Pi_{ab} + \frac{1}{2}\Pi^2 - K(t) + \frac{1}{2}R \right] d\Gamma_s \quad , \tag{6}$$

in which case (4) simplifies to

$$\frac{d\Pi}{ds} = -c - \nabla y \cdot \nabla y \quad . \tag{7}$$

Now the argument is simply that the constant c must be ≤ 0 for $r = 0$, for otherwise from (7) we could construct a mean concave surface outside Γ_0. Then this condition and (6) yield the inequality

$$0 \geq \int_{\Gamma_0} \left[\frac{1}{2}R - K \right] d\Gamma_0 = \Lambda A(\Gamma_0) - 2\pi\chi(\Gamma_0) \quad , \tag{8}$$

where to get the last part of this equation we use (2) and the Gauss-Bonnet theorem. The Euler characteristic is given by $\chi(\Gamma_0) = 2(1 - g)$ where g is the genus of Γ_0. If $\Lambda = 0$, then $g = 0, 1$, which was essentially Hawking's topology theorem argument, but instead we will note that for $g > 1$ then Λ must be negative and, moreover, we can then deduce that

$$A(\Gamma_0) \geq \frac{4\pi(g - 1)}{-\Lambda} \quad . \tag{9}$$

This is the area bound for higher genus horizons in the special case of a time-symmetric slice in a vacuum spacetime.

Gibbons has also obtained the area bound for horizons in a time-symmetric slice. As above, he constructed a one-parameter variation of the apparent horizon, but he considered the *second* variation of area, and sought *stable* minimal surfaces. These are true minima of the area functional, for which the first variation of area vanishes and the second variation is positive. An apparent horizon in a time-symmetric slice of an extreme black hole is a stable minimal surface.

A general argument that does not assume H to be a moment of time symmetry and that allows for a non-zero energy momentum tensor can be found in [10]. That bound also applies to event horizons.

139

There are $\Lambda < 0$ Einstein-Maxwell solutions [11]

$$ds^2 = V(r)dt^2 - \frac{1}{V(r)}dr^2 - r^2\left(d\theta^2 + \sinh^2\theta d\phi^2\right) \tag{10a}$$

$$V(r) = -1 - \frac{\Lambda}{3}r^2 - \frac{2m}{r} + \frac{Z^2}{r^2} \tag{10b}$$

that have higher genus horizons. As discussed in [10], the the "extreme" case of this form of solution, wherein the horizon radius a is a double root of $V(r)$, saturates the area bound. That root is given by

$$a^2 = \frac{1 + \sqrt{1 - 4\Lambda Z^2}}{-2\Lambda} \quad , \tag{11}$$

for Z^2 the sum of the squares of the electric and magnetic charges.

Acknowledgements. I thank Greg Galloway for discussions.

REFERENCES

1. D.M. Witt, this volume.
2. G.J. Galloway, K. Schleich, D.M. Witt, and E. Woolgar, *Phys. Rev.* D, to appear (1999).
3. S.W. Hawking, *Commun. Math. Phys.* **25**, 152–166 (1972).
4. S.W. Hawking and G.F.R. Ellis, *The Large Scale Structure of Space-Time*, Cambridge: Cambridge University Press, 1973.
5. S.W. Hawking, in *Black Holes*, eds. C. DeWitt and B. DeWitt, New York: Gordon and Breach, 1973, pp. 1–55.
6. S.L. Shapiro, S.A. Teukolsky, and J. Winicour, *Phys. Rev.* **D52**, 6982 (1995); J. Winicour, this volume.
7. R.B. Mann, in *Internal Structure of Black Holes and Spacetime Singularities*, eds. L. Burko and A. Ori, *Ann. Israeli Phys. Soc.* **13**, 311 (1998).
8. G.J. Galloway, *Contemp. Math.* **170**, 113 (1994).
9. G.W. Gibbons, *Class. Quantum Gravit.* **16**, 1677 (1999).
10. E. Woolgar, *Class. Quantum Gravit.*, to appear (1999).
11. R.B. Mann, *Class. Quantum Gravit.* **14**, 2927 (1997).

Critical Collapse of a Massless Scalar Field: Perturbative Approach

Andrei V. Frolov[1]

Physics Department, University of Alberta
Edmonton, Alberta, Canada, T6G 2J1

Abstract. This talk discusses general perturbations of the continuously self-similar critical solution of the gravitational collapse of a massless scalar field (the Roberts solution). The exact analysis of the perturbation equations reveals that there are no growing non-spherical perturbation modes. Formation of discretely self-similar structure by growing spherically symmetric modes is considered.

I INTRODUCTION

Critical phenomena in the gravitational collapse have been relatively recent and interesting development in the established field of general relativity. Following numerical work of Choptuik on the spherically symmetric collapse of the minimally coupled massless scalar field [1], critical behavior was discovered in most common matter models encountered in general relativity, including pure gravity [2], null fluid [3] and, more generally, perfect fluid [4,5], as well as more exotic models. Despite the fact that the evolution equations are very complex and highly non-linear, the dynamics of the near-critical field evolution is relatively simple and, in some important aspects, universal. The critical solution, which depends on the matter model only, serves as an intermediate attractor in the phase space of solutions, and often has an additional symmetry called self-similarity. The mass of the black hole produced in supercritical evolution scales as a power law

$$M_{\mathrm{BH}}(p) \propto |p - p^{\times}|^{\beta}, \tag{1}$$

with parameter p describing initial data, and mass-scaling exponent β is dependent only on the matter model, but not on the initial data family. Universality of the near-critical behavior is explained by the fact that critical solutions generally have only one unstable perturbation mode [3–6]. As the near-critical field configuration evolves, all its perturbation modes decay, losing information about the initial data and bringing the solution closer to critical, except the one growing mode which will

---

[1] Email: andrei@phys.ualberta.ca

CP493, *General Relativity and Relativistic Astrophysics*, edited by C. P. Burgess and R. C. Myers
© 1999 American Institute of Physics 1-56396-905-X/99/$15.00

eventually drive the solution to black hole formation or dispersal, depending on its content in the initial data.

One of the few known closed form solutions related to critical phenomena is the Roberts solution, originally constructed as a counterexample to the cosmic censorship conjecture [7], and later rediscovered in the context of critical gravitational collapse [8,9]. It is a continuously self-similar solution of a spherically symmetric gravitational collapse of a minimally coupled massless scalar field. While not a proper attractor [10], it is still related to the numerical solution discovered by Choptuik [1], and can be used to illustrate dynamics of the critical collapse of the massless scalar field.

In this talk, we discuss the questions of how generic is the critical behavior with respect to initial data, and how discrete self-similarity arises in the critical collapse of a massless scalar field. To this end, we consider general perturbations of the Roberts solution in a gauge-invariant formalism. Due to the symmetries of the background, the linear perturbation equations decouple and the variables separate, so an exact analytical treatment is possible. We find that there are no growing non-spherical perturbation modes, so all the non-sphericity of the initial data decays in the collapse. Only the spherically symmetric growing modes will play a role in the critical behavior, and are responsible for the departure of the solution away from the Roberts one. For certain quite generic initial conditions, scaling invariance of the Roberts solution is broken by the growing perturbation, and discretely self-similar structure is formed dynamically at the late times. The results presented here are derived in more details in [11,12].

II THE ROBERTS SOLUTION

The starting point of our investigation is the Roberts solution, which will serve as a background for linear perturbation analysis. It is a solution describing gravitational collapse of a minimally coupled massless scalar field, described by the Einstein-scalar field equations

$$R_{\mu\nu} = 2\phi_{,\mu}\phi_{,\nu}, \quad \Box\phi = 0, \tag{2}$$

which is spherically symmetric and also continuously self-similar. The latter symmetry means that there exists a vector field ξ such that $\mathcal{L}_\xi g_{\mu\nu} = 2g_{\mu\nu}$ and $\mathcal{L}_\xi \phi = 0$, where $\mathcal{L}$ denotes Lie derivative. Under these assumptions the field equations can be solved analytically, which is most easily done in null coordinates [8,9,13]. Self-similar solutions form a one-parameter family, with the critical solution given by the metric

$$ds^2 = -2\,du\,dv + r^2\,d\Omega^2, \tag{3}$$

where

$$r = \sqrt{u^2 - uv}, \quad \phi = \frac{1}{2}\ln\left[1 - \frac{v}{u}\right]. \tag{4}$$

142

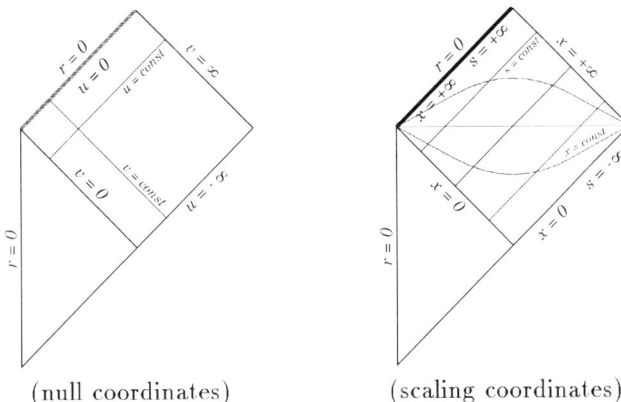

(null coordinates)   (scaling coordinates)

FIGURE 1. Global structure of the Roberts solution.

The global structure of the critical spacetime is shown in Fig. 1. The influx of the scalar field is turned on at the advanced time $v = 0$, so that the spacetime is Minkowskian to the past of this surface. The initial conditions for the field equations (2) are specified there by the continuity of the solution.

The evolution of perturbations of the Roberts solution is most easily followed in a coordinate system exploiting scale-invariance of the background, so that the self-similarity becomes apparent. Therefore, we introduce new coordinates, which we will call scaling coordinates, by $y = 1 - \frac{v}{u}$, $s = -\ln(-u)$. The scalar field ϕ does not depend on the scale variable s at all, and the metric coefficients depend on the scale only through the conformal factor e^{-2s}. The homothetic Killing vector ξ is simply $-\frac{\partial}{\partial s}$.

III   PERTURBATIVE ANALYSIS

Perturbation evolution is described by linearized Einstein-scalar field equations,

$$\delta R_{\mu\nu} = 4\phi_{(,\mu}\delta\phi_{,\nu)}, \quad \delta(\Box\phi) = 0 \tag{5}$$

with boundary conditions specified at $v = 0$ by continuity of matching with flat spacetime, and at the spatial infinity by well-behavedness of the perturbations there. Initial conditions are specified on the null surface $u = \text{const}$ by the initial shape of the wavepacket there.

To avoid complicated gauge issues of fully general perturbations, the above problem should be reformulated in gauge-invariant formalism, as only gauge-invariant quantities have inherent physical meaning. Once gauge-invariant quantities have been identified, one is free to convert between gauge-invariant perturbation amplitudes and their values in whatever gauge choice one desires. Without going

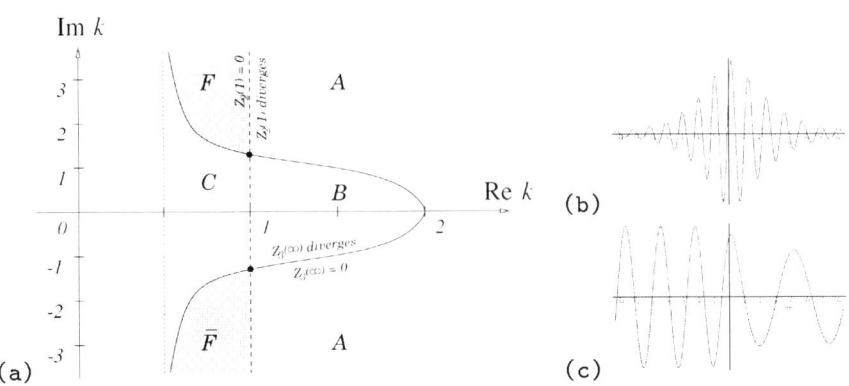

FIGURE 2. Perturbation spectrum of spherically-symmetric modes (a) and growing mode wave profiles (b) and (c). Shaded region F corresponds to allowed modes. Field profiles are plotted for $k = \frac{3}{4} + 3i$ (b) and $k = 1 + \sqrt{2}i$ (c).

into details of the calculation here, we will just say that it is possible to decouple perturbation equations, which results in master partial differential equation for gauge-invariant scalar field perturbation, and the rest of the gauge-invariant quantities can be derived from it.

The partial differential equations governing perturbation evolution can be further simplified by exploiting symmetries of the background to separate spatial and scale variables. Because the background spacetime is spherically symmetric, perturbations around it can be decomposed in spherical harmonics. Similarly, continuous self-similarity of the Roberts solution guarantees separation of the scale-dependence of the perturbation modes using Laplace transform. Thus the mode decomposition $\delta\phi = f(y)e^{ks}Y_{lm}(\theta, \varphi)$ reduces linear perturbation equations to the ordinary differential equations. One can show that the master equation for the scalar field reduces to hypergeometric equation

$$y(1-y)\ddot{f} + [c - (a+b+1)y]\dot{f} - abf = 0, \tag{6}$$

where parameters a, b, and c depend on the mode eigenvalues. As properties of the hypergeometric equation are extensively studied, it is possible to study linear perturbations of Roberts solution analytically.

An exact analysis of the perturbation eigenvalue problem [11] reveals that there are no growing non-spherical perturbation modes, in agreement with numerical results [14,15]. However, there are growing spherical perturbation modes. Their spectrum is continuous and occupies a big chunk of the complex plane [10], shown in Fig. 2. In view of these findings, the following picture of dynamics of scalar field evolution near self-similarity emerges: As we evolve generic initial data which is sufficiently close to the critical Roberts solution, non-spherical modes decay and the solution approaches the spherically symmetric one. Asymmetry of the initial

144

data does not play a role in the collapse. The growing spherical modes, on the other hand, drive the solution farther away from the continuously self-similar one. In this sense, the critical Roberts solution is an intermediate attractor for non-spherical initial data.

Additionally, using stationary phase approximation, one can show that the late-time evolution of the field for quite generic initial conditions is dominated by a single growing spherically symmetric mode $Z_2(y; k_0)e^{k_0 s}$, with $k_0 = 1 + \sqrt{2}i$ lying on the edge of the continuous spectrum [12]. The complex value of growth exponent gives rise to the most interesting physical effect: the perturbation developing on the scale-invariant background evolves to have a scale-dependent structure $e^s \cos(\mathrm{Im}\, k_0 s)$! The exponential growth of the amplitude of the perturbation will eventually be stopped by the non-linear effects, while the periodic dependence of the perturbation on scale will remain. The period of oscillation, obtained in the linear approximation, is

$$\Delta = \frac{2\pi}{\mathrm{Im}\, k_0} = 4.44. \tag{7}$$

This periodic dependence of the field perturbation on the scale is nothing else but the discrete self-similarity observed in numerical simulations. Thus, our simple analytical model of the critical collapse of the massless scalar field demonstrates how continuous self-similarity of the Roberts solution is dynamically broken to discrete self-similarity by the growing perturbations, reproducing the essential feature of numerical critical solutions. The value of echoing period above is within 25% of the numerical value $\Delta = 3.44$ measured by Choptuik [1].

REFERENCES

1. M. W. Choptuik, Phys. Rev. Lett. **70**, 9 (1993).
2. A. M. Abrahams and C. R. Evans, Phys. Rev. Lett. **70**, 2980 (1993).
3. C. R. Evans and J. S. Coleman, Phys. Rev. Lett. **72**, 1782 (1994).
4. T. Koike, T. Hara, and S. Adachi, Phys. Rev. Lett. **74**, 5170 (1995).
5. D. Maison, Phys. Lett. **B366**, 82 (1996).
6. T. Hara, T. Koike, and S. Adachi, gr-qc/9607010, 1996.
7. M. D. Roberts, Gen. Relativ. Gravit. **21**, 907 (1989).
8. P. R. Brady, Class. Quant. Grav. **11**, 1255 (1994).
9. Y. Oshiro, K. Nakamura, and A. Tomimatsu, Prog. Theor. Phys. **91**, 1265 (1994).
10. A. V. Frolov, Phys. Rev. D **56**, 6433 (1997).
11. A. V. Frolov, Phys. Rev. D **59**, 104011 (1999).
12. A. V. Frolov, to be published (1999).
13. A. V. Frolov, Class. Quant. Grav. **16**, 407 (1999).
14. C. Gundlach, Phys. Rev. D **57**, 7075 (1998).
15. J. M. Martín-García and C. Gundlach, Phys. Rev. D **59**, 064031 (1999).

Self-Similar Spherically Symmetric Perfect-Fluid Models

Martin Goliath

Stockholm University
Department of Physics
Box 6730, S-113 85 Stockholm
SWEDEN

Abstract. The purpose of this contribution is to present the different classes of solutions that exist for self-similar spherically symmetric spacetimes with a perfect fluid as matter source. In particular, there is a class of 'asymptotically Minkowski' solutions that was discovered only recently. This class of solutions appears to be of relevance in the study of critical phenomena in spherically symmetric perfect-fluid collapse.

The results presented here are part of an investigation [1,2], in which the work of Carr & Coley [3] has been combined with that of Goliath, Nilsson & Uggla [4,5].

SELF-SIMILARITY

Self-similarity plays an important role when trying to understand and model physical phenomena. It is a simplifying assumption that –by finding the appropriate variables– may enable, *e.g.*, the reduction of a system of partial differential equations to a system of ordinary differential equations. An excellent introduction to the subject is the book by Barenblatt [6].

The study of self-similarity in General Relativity was initiated by Cahill & Taub [7], who studied self-similar spherically symmetric perfect fluids. They assumed similarity under the coordinate transformation $(t, r) \to (at, ar)$, and that all physical quantities depend only on the *self-similar variable* $z = r/t$. This corresponds to a *self-similarity of the first kind*, see [6]. Furthermore they noted that this implies that the spacetime admits a homothetic vector field. The formalism for spatially self-similar spacetimes of this type was further investigated by Eardley [8].

Applications of self-similarity in General Relativity include: the evolution of primordial black holes [9–11]; primordial density perturbations [12]; and naked singularities in gravitational collapse [13,14]. Also, many spatially homogeneous cosmologies exhibit self-similar asymptotics, see, *e.g.*, [15]. More recently, the critical solution of spherically-symmetric collapse of a perfect fluid has been found to be self-similar [16]. For more examples of self-similarity in General Relativity, see, *e.g.*, [17].

CP493, *General Relativity and Relativistic Astrophysics*, edited by C. P. Burgess and R. C. Myers
© 1999 American Institute of Physics 1-56396-905-X/99/$15.00

SELF-SIMILAR SPHERICALLY SYMMETRIC MODELS

In what follows, self-similar spherically symmetric spacetimes will be considered. The matter content will be taken to be a perfect fluid with linear equation of state $p = (\gamma - 1)\mu$, where p is the pressure, μ is the energy density, and the equation-of-state parameter is in the range $1 < \gamma < 2$. The bifurcation values $\gamma = 1$ and $\gamma = 2$ have been excluded, as these models behave somewhat differently.

Different approaches

There are at least three natural structures to adapt the coordinates to:

The fluid. which results in the *comoving approach.* The surfaces of constant time coordinate t are then orthogonal to the fluid flow, and the spatial coordinates (r, θ, ϕ) are constant along each flow line. One of the advantages with this approach is that it is easy to extract the physical properties of the fluid.

The homothetic symmetry. which gives the *homothetic approach.* One of the coordinates then correspond to the homothetic vector ξ^a, and the line element can be written on the form [18] $ds^2(\xi, \eta) = e^{2\xi} d\tilde{s}^2(\eta)$, where η is the remaining coordinate dependence in $d\tilde{s}^2$. Whereas the original spacetime ds^2 is hypersurface self-similar, the unphysical spacetime $d\tilde{s}^2$ is hypersurface homogeneous [19,8]. This is useful, as there exists an extensive literature in which homogeneous models are considered. This makes it straight-forward to formulate the problem as a dynamical system.

The spherical symmetry. which leads to the *Schwarzschild approach.* The Schwarzschild radius R, which gives the area of the spherical symmetry surfaces according to $A = 4\pi R^2$, is then used as one of the coordinates. This approach is particularly useful when studying null geodesics [13].

These approaches are complementary, and in the work presented here, an investigation done in the comoving approach [3] is combined with the homothetic approach [4,5].

Physical quantities

The comoving approach guides us to look for physical quantities that are functions of the self-similar variable $z = r/t$. Consequently, we have considered the following four quantities:

The scale factor $S = R/r$, which indicates when a singularity forms ($S \to 0$ for finite z) and when the solution is dispersing ($S \to \infty$ for finite z).

The velocity function V, which is the speed of the fluid with respect to homothetic (constant z) symmetry surfaces. There are a number of significant values of V: $|V| = \sqrt{\gamma - 1}$ constitutes a *sonic surface* through which only a small subset of solutions may pass in a regular way. (Note that $\sqrt{\gamma - 1}$ is the speed of sound in the fluid.); $|V| = 1$ corresponds to an event horizon or cosmological particle horizon, and is important when studying naked singularities; $|V| \to \infty$ marks the formation of a naked singularity.

The density profile μt^2, which gives the matter distribution at a given co-moving time t.

The Mass function $2m/R$, which indicates the presence of an apparent horizon.

The state space

The homothetic approach results in a four-dimensional dynamical system, where the dependent variables are related by a constraint. The system is thus effectively three-dimensional. Furthermore, the state space can be made compact, which means that the system contains no 'infinities'. Consequently, we can draw three-dimensional pictures of the state space. Any solution corresponds to an orbit in the state space, each point on the orbit corresponding to a particular value of z. It turns out that all orbits are asymptotic to only a few *equilibrium points* of the system. This results in three main classes of solutions, each of which has a time-reversed counter-part:

Asymptotically Friedmann solutions. These are associated with a Friedmann singularity. There is a one-parameter set of such solutions, and they can be of two types: either they are *recollapsing*, and form a non-isotropic singularity (as opposed to the Friedmann singularity); else they are *ever-expanding* and have an infinitely dispersed final state. Solutions of the latter type have to pass a sonic surface, which results in bands of regular solutions. Also, the flat Friedmann solution itself belongs to this latter type. Consequently, this class of solutions can be viewed as inhomogeneous perturbations of the flat Friedmann model.

Asymptotically quasi-static solutions. There is a single orbit in the interior of the state space that corresponds to the static solution first discussed by Tolman [20]. Furthermore, there is a two-parameter set of solutions that mimic the static solution for part of their evolution. They are all associated with non-isotropic singularities, and can be of two types: either they recollapse to a singularity, or else they expand forever. The naked-singularity solutions studied by Ori & Piran [13] belong to the time-reverse of the latter type, *i.e.*, those solutions start out from an infinitely dispersed state and collapse.

'Asymptotically Minkowski' solutions. When $\gamma > 6/5$, there exist solutions that are 'asymptotically Minkowski' in the sense that the corresponding state-space orbits are asymptotic to a Minkowski equilibrium point. Upon approaching this point, the corresponding solutions become infinitely diluted. It should be emphasized that these solutions are not asymptotically flat in the usual sense, see further [1,2]. There are two subclasses with different asymptotic behavior, and each of these subclasses can be divided into two types: either they are *singular* and expand from a non-isotropic singularity; or else they are *regular* and are associated with a regular center.

148

THE CRITICAL SOLUTION

Critical phenomena in gravitational collapse were first studied by Choptuik [21], and their study remains an important field of research in General Relativity (for an overview see, *e.g.*, [22]). Evans & Coleman [16] studied the spherically symmetric collapse of a radiation fluid ($\gamma = 4/3$), and found that the critical solution has the following properties:

1) It is (continuously) self-similar.

2) It is everywhere regular. In particular it is regular at the center, and must pass any sonic surfaces in a regular way.

3) The solution consists of a collapsing region, surrounded by a dispersing exterior.

Subsequent studies [23,24] have used these criteria to investigate perfect-fluid collapse for general values of γ. One conclusion made was that the critical solution only existed when $\gamma \lesssim 1.89$. Recently, Neilsen & Choptuik [25] have demonstrated that the critical solution *does* exist all the way up to $\gamma = 2$. The work presented here supports their results. Furthermore, we have examined the critical solution in the whole range $1 < \gamma < 2$. The behavior in state space is as follows: starting from the regular center, the orbit corresponding to the critical solution pass a sonic surface in a regular way. The subsequent behavior depends on the equation of state: for $1 < \gamma \lesssim 1.28$, the critical orbit reaches a second sonic surface where the solution is irregular. This poses no problem, as the critical solution generally is matched to an asymptotically flat exterior already outside the first sonic surface [13,14]. For $1.28 \lesssim \gamma < 2$, the critical solution belongs to the 'asymptotically Minkowski' solutions, discussed above.

CONCLUSIONS

We have considered self-similar spherically symmetric models with a perfect-fluid matter source. By combining the homothetic and comoving approaches, we gain the advantages from both methods: the homothetic approach gives full control on what types of solutions there are, and the comoving approach provides the means to interpret them physically. Thus, combining them proves very useful.

Three main classes of solutions can be identified: the asymptotically Friedmann, asymptotically quasi-static and 'asymptotically Minkowski' solutions. The solutions of the latter class are physical only when $\gamma > 6/5$, and their existence was not appreciated until quite recently.

A particularly interesting example of a solution that can be investigated with the methods presented is the critical solution of spherically symmetric collapse of a perfect fluid.

149

REFERENCES

1. Carr B. J., Coley A. A., Goliath M., Nilsson U. S., and Uggla C., preprint gr-qc/9901031 (1999).
2. Carr B. J., Coley A. A., Goliath M., Nilsson U. S., and Uggla C., preprint gr-qc/9902070 (1999).
3. Carr B. J., and Coley A. A., to appear in *Phys. Rev. D*, preprint gr-qc/9901050 (1999).
4. Goliath M., Nilsson U. S., and Uggla C., *Class. Quantum Grav.* **15**, 167 (1998).
5. Goliath M., Nilsson U. S., and Uggla C., *Class. Quantum Grav.* **15**, 2841 (1998).
6. Barenblatt G. I., *Scaling, Self-Similarity, and Intermediate Asymptotics*, Cambridge: Cambridge University Press, 1996.
7. Cahill M. E., and Taub A. H., *Commun. Math. Phys.* **21**, 1 (1971).
8. Eardley D. M., *Commun. Math. Phys.* **37**, 287 (1974).
9. Carr B. J., and Hawking S. W., *Mon. Not. Roy. Astr. Soc.* **168**, 399 (1974).
10. Bicknell G. V., and Henriksen R. N., *Astrophys. J.* **219**, 1043 (1978).
11. Bicknell G. V., and Henriksen R. N., *Astrophys. J.* **225**, 237 (1978).
12. Carr B. J., and Yahil A., *Astrophys, J.* **360**, 330 (1990).
13. Ori A., and Piran T., *Phys. Rev. D* **42**, 1068 (1990).
14. Foglizzo T., and Henriksen R. N., *Phys. Rev. D* **48**, 4645 (1993).
15. Wainwright J., and Ellis G. F. R., *Dynamical Systems in Cosmology*, Cambridge: Cambridge University Press, 1997.
16. Evans C. R., and Coleman J. S., *Phys. Rev. Lett.* **72**, 1782 (1994).
17. Carr B. J., and Coley A. A., *Class. Quantum Grav.* **16**, R1 (1999).
18. Bogoyavlensky, O. I., *Methods in the Qualitative Theory of Dynamical Systems in Astrophysics and Gas Dynamics*, Berlin: Springer, 1985.
19. Defrise-Carter L., *Commun. Math. Phys.* **40**, 273 (1975).
20. Tolman R. C., *Proc. Natl. Acad. Sci. U. S.* **20**, 169 (1934).
21. Choptuik M. W., *Phys. Rev. Lett.* **70**, 9 (1993).
22. Gundlach C., *Phys. Rev. D* **55**, 695 (1997).
23. Maison D., *Phys. Lett. B* **366**, 82 (1996).
24. Koike T., Hara T., and Adachi S., *Phys. Rev. D* **59**, 104008 (1999).
25. Neilsen D. W., and Choptuik M. W., preprint gr-qc/9812053 (1998).

Radiative Falloff in Black-Hole Spacetimes — Part I

Eric Poisson

Department of Physics, University of Guelph, Guelph, Ontario, Canada N1G 2W1

Abstract. This is the first of a two-part contribution devoted to the evolution of a massless scalar field in various black-hole spacetimes; the second part follows this one and is authored by William G. Laarakkers. In this contribution we consider the evolution of a scalar field propagating in Schwarzschild-de Sitter spacetime. The field is non-minimally coupled to curvature through a coupling constant ξ. The spacetime possesses a cosmological horizon in addition to the usual event horizon. The presence of this new horizon affects the late-time evolution of the scalar field: The usual inverse power-law decay of pure Schwarzschild spacetime is replaced by a faster, exponential decay. The decay constant is proportional to the surface gravity of the cosmological horizon, and depends on the field's multipole order ℓ and the constant ξ.

INTRODUCTION

A theorem that establishes the uniqueness of the Schwarzschild black hole as the endpoint of gravitational collapse without rotation was proved by Werner Israel more than 30 years ago [1], and the mechanism by which the gravitational field eventually relaxes to the Schwarzschild form was elucidated by Richard Price more than 25 years ago [2]. Given the venerable age of this topic, it is surprising that more can be said about it today. Yet, many papers on radiative falloff have been written in the last few years [3–15]. Most of the new developments are concerned with rotating collapse, and how the gravitational field eventually relaxes to the Kerr form. The question we pursue in this two-part contribution is different. Focusing our attention on nonrotating black holes, we ask: How do the conditions far away from the black hole affect the relaxation process? In Part I we consider a black hole immersed in an inflationary universe. (This was first done by Brady *et al.* [14], and additional details can be found in Ref. [15].) In Part II, William G. Laarakkers will consider a black hole immersed in a spatially-flat, dust-filled universe.

CP493, *General Relativity and Relativistic Astrophysics,* edited by C. P. Burgess and R. C. Myers

SCHWARZSCHILD SPACETIME

Price's result [2] can be summarized as follows. As a nonspherical star undergoes gravitational collapse, the gravitational field becomes highly dynamical, and the escaping radiation interacts with the spacetime curvature surrounding the star. At late times, well after the initial burst of radiation was emitted, the gravitational field relaxes to a pure spherical state. If δg schematically represents the deviation of the metric from the Schwarzschild form, then $\delta g \sim t^{-(2\ell+2)}$, where ℓ is the multipole order of the perturbation; the dominant contribution to δg comes from the quadrupole ($\ell = 2$) mode.

The inverse power-law decay applies to many other situations involving radiation interacting with the curvature created by a massive object. The simplest model problem which exhibits this behaviour involves a massless scalar field in Schwarzschild spacetime. In this context, the background geometry is not affected by the field Φ, which satisfies the wave equation

$$(g^{\alpha\beta}\nabla_\alpha\nabla_\beta - \xi R)\Phi = 0, \tag{1}$$

where $g_{\alpha\beta}$ is the spacetime metric, R the Ricci scalar (which vanishes for Schwarzschild spacetime), and ξ a coupling constant. Because the spacetime is spherically symmetric, the field can be decomposed according to

$$\Phi = \sum_{\ell m} \frac{1}{r}\, \psi_\ell(t,r)\, Y_{\ell m}(\theta,\phi). \tag{2}$$

This leads to a decoupled equation for each wave function ψ_ℓ, and we can focus on a single mode at a time.

The problem is formulated as follows. A pulse of scalar radiation (described by ψ_ℓ) impinges on the black hole and interacts with the spacetime curvature, which creates a potential barrier fairly well localized near $r = 3M$. The wave pulse is partially reflected and transmitted, and at late times, a tail remains. At such times, the field falls off as $\psi_\ell \sim t^{-(2\ell+3)}$. This is Price's power-law decay, and this behaviour is displayed in Fig. 1.

A number of analytical and numerical studies of radiative dynamics [3–7] have revealed that the inverse power-law behaviour is not sensitive to the presence of an event horizon. In fact, power-law tails are a weak-curvature phenomenon, and it is the asymptotic structure of the spacetime at radii $r \gg 2M$ which dictates how the field behaves at times $t \gg 2M$. It is this observation that motivated our work: How is the field's evolution affected if the conditions at infinity are altered?

SCHWARZSCHILD-DE SITTER SPACETIME

To provide an answer to this question, we remove the black hole from its underlying flat spacetime and place it in de Sitter spacetime, which describes an

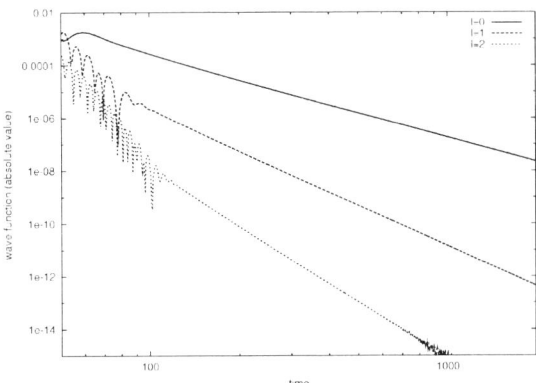

FIGURE 1. Absolute value of the wave function $\psi_\ell(t, r)$ as a function of time t, evaluated at $r = 10$ in Schwarzschild spacetime. We use units such that $2M = 1$. The cases $\ell = 0, 1, 2$ are considered, and the wave functions are plotted on a log-log scale. In such a plot, a straight line indicates power-law behaviour, and a change of sign in the wave function is represented by a deep trough. We see that the field's early behaviour is oscillatory, but that it eventually decays according to an inverse power law.

exponentially expanding universe. The Schwarzschild-de Sitter (SdS) spacetime has a metric given by

$$ds^2 = -f\, dt^2 + f^{-1}\, dr^2 + r^2\, d\Omega^2, \qquad f = 1 - 2M/r - r^2/a^2. \qquad (3)$$

Here, $a^2 = 3/\Lambda$, where Λ is the cosmological constant. (The SdS metric is a solution to the modified vacuum field equations, $G_{\alpha\beta} + \Lambda g_{\alpha\beta} = 0$, which imply $R = 4\Lambda = 12/a^2$.) The spacetime possesses an event horizon at $r = r_e \simeq 2M$ and a cosmological horizon at $r = r_c \simeq a$. We assume that $r_e \ll r_c$, so that the two length scales are cleanly separated.

We examine the time evolution of a scalar field in SdS spacetime; the field is still governed by Eq. (1), and it still admits the decomposition of Eq. (2). Figure 2 provides a comparison between the behaviour of ψ_ℓ in the two spacetimes (Schwarzschild and SdS). We see that at early times, the wave functions behave identically; the field has not yet become aware of the different conditions at $r \gg r_e$. At later times, however, deviations become apparent. For $\ell = 0$, the Schwarzschild behaviour $\psi_0 \sim t^{-3}$ is replaced by the wave function changing sign at $t \sim 260$, and settling down to a constant value at late times. For $\ell = 1$, the Schwarzschild behaviour $\psi_1 \sim t^{-5}$ is replaced by a faster decay which eventually becomes exponential.

The field's exponential decay is confirmed by monitoring its evolution up to times $t > r_c$. If $\xi = 0$, we find that $\psi_\ell \sim e^{-\ell\kappa_c t}$ at late times [14], where $\kappa_c \simeq 1/r_c$ is the surface gravity of the cosmological horizon.

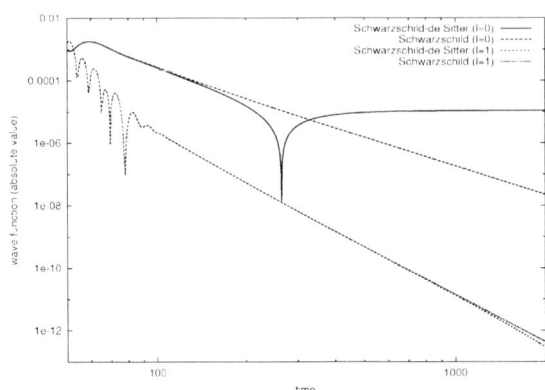

FIGURE 2. Absolute value of the wave function $\psi_\ell(t, r)$ as a function of time t, evaluated at $r = 10$ in Schwarzschild spacetime ($r_e = 1$) and SdS spacetime ($r_e = 1$ and $r_c = 2000$). The cases $\ell = 0, 1$ are considered, and the wave functions are plotted on a log-log scale. In both cases, $\xi = 0$.

A rich spectrum of late-time behaviours is revealed when ξ, the curvature-coupling constant, is allowed to be nonzero. Figure 3 shows the time-dependence of ψ_0 for several values of ξ. For ξ smaller than a critical value ξ_c, the field decays monotonically with a decay constant that increases with increasing ξ. When $\xi > \xi_c$, however, the wave function oscillates with a decaying amplitude. As ξ is increased away from the critical value ξ_c, the frequency of the oscillations increases, but the decay constant stays the same.

This qualitative change of behaviour as ξ goes through ξ_c is quite remarkable. It can be explained with a detailed analytical calculation that will not be presented here (see Ref. [15]). This calculation reveals that at late times, the field behaves as $\psi_\ell \sim e^{-p\kappa_c t}$, where

$$p = \ell + \frac{3}{2} - \frac{1}{2}\sqrt{9 - 16\xi} + O\left(\frac{r_e}{r_c}\right). \tag{4}$$

This relation implies that p becomes complex, and ψ_ℓ oscillatory, when $\xi > \xi_c \equiv 3/16$.

ACKNOWLEDGMENTS

The work presented here was carried out in collaboration with Patrick Brady, Chris Chambers, and Bill Laarakkers; additional details can be found in Ref. [15]. This work was supported by the Natural Sciences and Engineering Research Council.

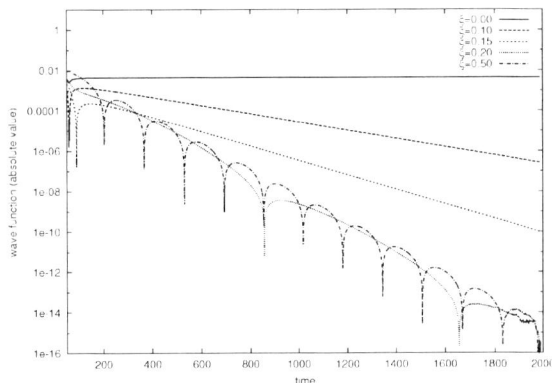

FIGURE 3. Absolute value of the wave function $\psi_0(t,r)$ as a function of time t, evaluated at $r = 10$ in SdS spacetime ($r_e = 1$ and $r_c = 100$). Several values of ξ are considered, in the interval between $\xi = 0$ and $\xi = \frac{1}{2}$. The wave functions are plotted on a semi-log scale, in which a straight line indicates exponential behaviour.

REFERENCES

1. W. Israel, Phys. Rev. **164**, 1776 (1967).
2. R.H. Price, Phys. Rev. D **5**, 2419 (1972); **5**, 2439 (1972).
3. C. Gundlach, R.H. Price, and J. Pullin, Phys. Rev. D **49**, 883 (1994); **49**, 890 (1994).
4. L.M. Burko and A. Ori, Phys. Rev. D **56**, 7820 (1997).
5. E.S.C. Ching, P.T. Leung, W.M. Suen, and K. Young, Phys. Rev. Lett. **74**, 2414 (1995); Phys. Rev. D **52**, 2118 (1995).
6. N. Andersson, Phys. Rev. D **55**, 468 (1997).
7. L. Barack, Phys. Rev. D **59**, 044016 (1999); 044017 (1999).
8. L. Barack and A. Ori, gr-qc/9902082; gr-qc/9907085.
9. S. Hod and T. Piran, Phys. Rev. D **58**, 024017 (1998); **58** 024018 (1998); **58** 024019 (1998); **58**, 044018 (1998).
10. S. Hod, Phys. Rev. D **58**, 104022 (1998); gr-qc/9902072; gr-qc/9902073; gr-qc/9907044; gr-qc/9907096.
11. W. Krivan, P. Laguna, and P. Papadopoulos, Phys. Rev. D **54**, 4728 (1996).
12. W. Krivan, P. Laguna, P. Papadopoulos, and N. Andersson, Phys. Rev. D **56**, 3395 (1997).
13. W. Krivan, gr-qc/9907038.
14. P.R. Brady, C.M. Chambers, W. Krivan, and P. Laguna, Phys. Rev. D **55**, 7538 (1997).
15. P.R. Brady, C.M. Chambers, W.G. Laarakkers, and E. Poisson, gr-qc/9902010.

Radiative Falloff in Black-Hole Spacetimes – Part II

Wm. G. Laarakkers

Department of Physics, University of Guelph, Guelph, Ont., N1G 3E1

Abstract. In part II of this talk the evolution of a massless scalar field in Schwarzschild-Einstein-de Sitter spacetime is studied numerically. The spacetime has two distinct regions: an inner black-hole region and an outer cosmological region. Early on in the evolution the field behaves as if it were in pure Schwarzschild spacetime, with each multipole of the field first exhibiting quasi-normal ringing followed by a power-law decay. However, later in the evolution the field learns of the existence of the cosmological region and changes behaviour. For the $l = 0$ mode, the field first changes sign due to the discontinuous negative potential at the boundary. The field then decays again with a power-law falloff, but with a slower decay rate than in the pure Schwarzschild case. For the $l > 0$ modes of the field the potential at the boundary is discontinuous and positive. The field therefore encounters a potential barrier at the boundary, and the part of the field that is reflected back from the barrier gives rise to an echo of its earlier quasi-normal oscillations.

THE SPACETIME

The background spacetime in which the scalar field's evolution is followed is the Schwarzschild-Einstein-de Sitter spacetime. Qualitatively, it can be described as follows. The idea is to start out with a spatially-flat, expanding, dust-filled universe. Then a ball of dust is "scooped" out, which leaves behind a spherical vacuum region. The dust that was removed is replaced by a Schwarzschild black hole, which is placed in the middle of the vacuous region. This produces a spacetime with two distinct regions. The inner (black hole) region is described by the Schwarzschild metric, and the outer (cosmological) region is described by the Friedman-Robertson-Walker (FRW) metric (see figure 1).

There are two important things to note about this spacetime. First, if the mass of the black hole is the same as the mass of the dust that was scooped out, the metric will be smooth across the boundary separating the two regions of the spacetime. Also, since the dust is pressureless it will not flow across the boundary, and the boundary itself will be co-moving with the universe.

Because the specific finite-difference equation used in the numerical work requires the use of null coordinates (see [1]), the metrics of the two regions must be put in double-null coordinate form. For the black hole region the metric is written as

$$ds^2 = -\left(1 - \frac{2M}{r}\right) du\, dv + r^2(d\theta^2 + \sin^2\theta\, d\phi^2). \tag{1}$$

CP493, *General Relativity and Relativistic Astrophysics*, edited by C. P. Burgess and R. C. Myers
© 1999 American Institute of Physics 1-56396-905-X/99/$15.00

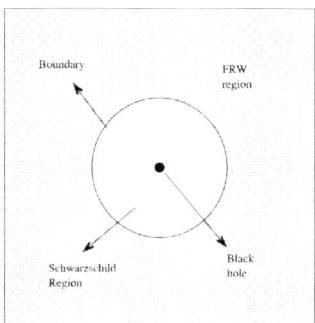

FIGURE 1. Schematic of the Schwarzschild-Einstein-de Sitter spacetime. Note that the boundary between the two regions is expanding outwards, co-moving with the universe.

Here, u and v are ingoing and outgoing null coordinates, and r is defined implicitly by $r + 2M \ln(r/2M - 1) = 1/2(v - u)$. In the cosmological region the metric takes the form:

$$\mathrm{d}s^2 = a^2(u^*, v^*) \left[-\mathrm{d}u^* \, \mathrm{d}v^* + \chi^2(\mathrm{d}\theta^2 + \sin^2\theta \, \mathrm{d}\phi^2) \right], \tag{2}$$

where u^* and v^* are ingoing and outgoing null coordinates of the FRW spacetime, different from u and v. The FRW radial coordinate is $\chi = \frac{1}{2}(v^* - u^*)$. The scale factor a is given by $a(u^*, v^*) = \frac{1}{16}C(u^* + v^*)^2$, where C is a constant that depends on the mass M of the black hole and the density of the dust.

The first task is to find one coordinate system that can describe both regions of the spacetime. This is required so that a single wave equation valid over the entire spacetime can be constructed. Since it is known that the metric is continuous across the boundary we can evaluate the metric induced on both sides of the boundary hypersurface, and set them equal. This construction allows us to find the ingoing Schwarzschild coordinate u as a function of the ingoing cosmological coordinate u^*, and the outgoing coordinate v as a function of v^*. Thus we now have a single coordinate system covering both regions of the spacetime.

THE WAVE EQUATION

The wave equation that governs the evolution of the scalar field is one without curvature coupling (equivalent to setting $\xi = 0$ (see Part I of talk and [2]). Thus the massless scalar field Φ obeys the equation

$$\Box \Phi = g^{\alpha\beta} \nabla_\alpha \nabla_\beta \Phi = 0. \tag{3}$$

The spherical symmetry of the problem allows us to decompose the field in terms of spherical harmonics, and then to evolve only the part of the field that depends on the null coordinates. Thus the field can be decomposed as

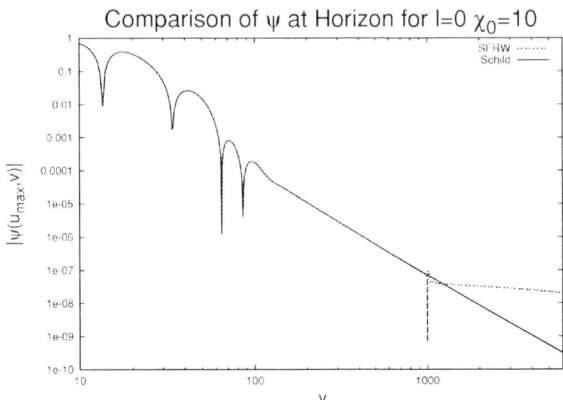

FIGURE 2. Absolute value of the field on the event horizon as a function of v, for $l = 0$. The solid line is the evolution of the the scalar wave in pure Schwarzschild spacetime. The dashed line is the evolution in the Schwarzschild-Einstein-de Sitter spacetime. The sharp dip near $v = 1000$ is where the field learns about the boundary Σ and changes sign. Before this point, the field decays as $\psi \sim v^{-3}$. After this point, the field decays as $\psi \sim v^{-1}$.

$$\Phi = \sum_{lm} \frac{1}{R} \psi_l Y_{lm}(\theta, \phi), \tag{4}$$

where $R = r$, $\psi_l = \psi_l(u, v)$ in the black hole region, and $R = a\chi$, $\psi_l = \psi_l(u^*, v^*)$ in the cosmological region. When all quantities are expressed in the starred coordinate system, each wavefunction ψ_l satisfies the equation

$$4 \frac{\partial^2 \psi}{\partial u^* \partial v^*} + V \psi = 0, \tag{5}$$

where the potential V takes a different form depending on which region of the spacetime the field lies:

$$V_{Schild} = \frac{du}{du^*} \frac{dv}{dv^*} f \left[\frac{l(l+1)}{r^2} + \frac{2M}{r} \right] \tag{6a}$$

$$V_{FRW} = \frac{4\, l(l+1)}{(v^* - u^*)^2} - \frac{8}{(v^* + u^*)^2}. \tag{6b}$$

RESULTS

The numerical code evaluates the field on the event horizon, on the boundary between the two regions of the spacetime, and at future null infinity. Our discussion here will be restricted to the value of the field on the event horizon for the $l = 0$ and $l = 1$ modes. The evolution was started at a "late" time, meaning that

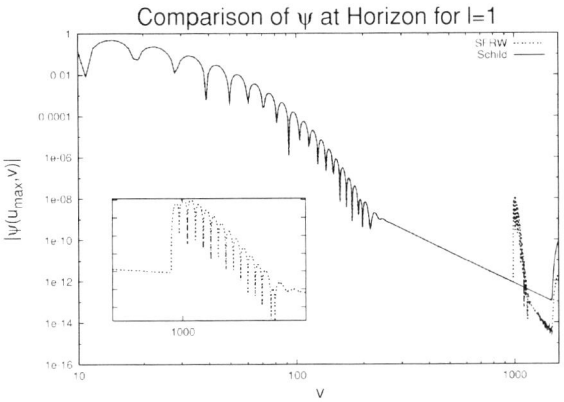

FIGURE 3. Absolute value of the field on the event horizon, as a function of v for $l = 1$. The solid line is the evolution in pure Schwarzschild spacetime. The dashed line is the evolution in the Schwarzschild-Einstein-de Sitter spacetime. The inset is a close-up of the region near $v = 1000$ where the field learns about the boundary Σ. Echoes of the quasi-normal oscillations in the field can be seen.

the boundary has expanded far enough that it can be clearly seen that the field initially behaves as it would in pure Schwarzschild spacetime. For both modes considered we see in figures 2 and 3 that the field first exhibits quasi-normal ringing followed by the well known power law decay (see, among others, [3]). However, at a certain time in the evolution, the field's behaviour deviates from the behaviour exhibited in the pure Schwarzschild case. The point at which the field changes behaviour corresponds to the time at which information about the existence of the cosmological region reaches the event horizon.

As the wave packet falls towards the event horizon (approximated by $u^* = u_{max}$, where u_{max} is the largest value of u^* in the numerical grid) it encounters the localized potential (dashed line - see figure 4). Part of the wave is transmitted through the barrier and reaches the event horizon, and part of the wave is back-scattered by the potential. The reflected wave heads out towards the cosmological region (to the right), where it encounters the boundary Σ. For the $l = 0$ mode the potential at the boundary is discontinuous and negative, and the field now changes sign. This sign change is the large dip in figure 2 (note that this is a log scale, so that as the field passes through $\psi = 0$ the logarithm goes to negative infinity). It is when this information reaches the event horizon that the evolution of the field deviates from its evolution in pure Schwarzschild spacetime. The field continues to decay with a power-law falloff, but the falloff is much slower than in the pure Schwarzschild case.

The discussion for the $l = 1$ case is similar, until the reflected wave reaches the boundary Σ. This is because the potential at the boundary is discontinuous and *positive* for $l > 0$ (see equation 6b). Therefore the field will be partially transmitted

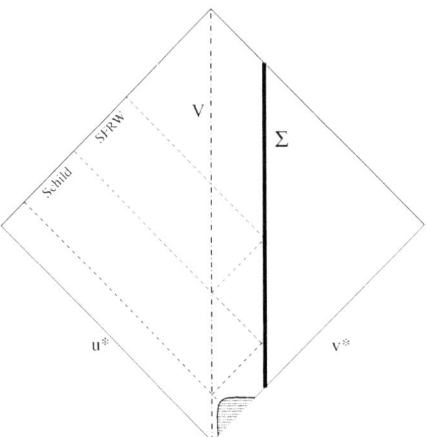

FIGURE 4. Evolution of the scalar field. Here the boundary between the two spacetimes is the line Σ, with the black hole region to the left and the cosmological region to the right. The line marked V represents the maximum of the potential. The dashed lines are the reflection and transmission of the wave pulse from the potential and the boundary. At the event horizon the field initially behaves as if it were in pure Schwarzschild spacetime (Schild), then evolves differently (SFRW).

through the barrier at the boundary and partially reflected off. The transmitted wave will make its way off to future null infinity. As for the part of the wave packet that has now been reflected twice, it will fall back towards the black hole where it once again encounters the localized potential. The part of the wave that manages to make it through the potential on its second encounter heads back towards the event horizon, carrying information about the existence of the boundary. This second encounter with the localized potential has the same effect on the packet as it did the first time—namely, the field again exhibits quasi-normal ringing (see inset of figure 3). This "echoing" phenomenon occurs only for the $l > 0$ modes of the field.

Acknowledgements: This work was carried out with Éric Poisson, and was supported by NSERC.

REFERENCES

1. C. Gundlach, R.H. Price, and J. Pullin, Phys. Rev. D **49**, 886 (1994).
2. P.R. Brady, C.M. Chambers, W.G. Laarakkers, and É. Poisson, *Radiative falloff in Schwarzschild-de Sitter spacetime*, gr-qc/9902010.
3. R.H. Price, Phys. Rev. D **5**, 2439 (1972).

Evolution of Evaporating Black Holes in a Higher Dimensional Inflationary Universe.

Manasse R. Mbonye

Physics Department. University of Michigan. Ann Arbor. Michigan 48109

Abstract. Spherically symmetric Black Holes of the Vaidya type are examined in an asymptotically de Sitter, higher dimensional spacetime. The various horizons are identified and located. The structure and dynamics of such horizons are studied.

INTRODUCTION

Several solutions to the Einstein equations of localized sources in higher dimensions have been obtained in the recent years. This includes the higher dimensional generalizations of the Schwarzschild and the Reisner-Nordstrom solutions [1], the Kerr solution [2] and the Vaidya solution [3]. Recently the metric of a radiating black hole in a de Sitter background, that is a generalization of the Mallett [4] metric, has been written down [5]. In the present work our aim is to demonstrate that the dynamics of a radiating black hole in a higher dimensional cosmological background can be sensibly discussed. First, we seek to identify and locate the various horizons. Then we study the structures and discuss the dynamics of such horizons. It is shown, at each stage, that all the results we obtain reduce to the well known Mallett [6] results as we go down to four dimensions.

In Section I we introduce the working metric. In Section II we derive equations for the horizons. We solve these equations and use the solutions to identify and locate the various horizons in the problem. In Section III we take up the issue of the structure of such horizons and study their dynamics. In Section IV we conclude the discussion.

CP493, *General Relativity and Relativistic Astrophysics,* edited by C. P. Burgess and R. C. Myers
© 1999 American Institute of Physics 1-56396-905-X/99/$15.00

I   THE METRIC

We wish to consider a radiating black hole introduced in an N dimensional de Sitter space-time. In advanced time, comoving, coordinates the line element is [5],

$$ds^2 = -\left[1 - \frac{2G_N m(v)}{nr^n} - \frac{2\Lambda}{(n+1)(n+2)}r^2\right]dv^2 + 2dvdr + d\Omega_{n+1}^2, \qquad (1.1)$$

where $n = N - 3$, $m(v)$, the mass, is a monotonically decreasing function of the advanced time coordinate v, G_N is the N-dimensional gravitational constant, Λ is the cosmological constant and $d\Omega_{n+1}^2 = d\theta_1^2 + \sin^2\theta_1 d\theta_2^2 + ... + \sin^2\theta_1 \sin^2\theta_2...\sin^2\theta_n d\theta_{n+1}^2$ is the line element on the $(n+1)$-sphere. The Luminosity $L_0 = -\frac{dm}{dv} < 1$, is measured in regions where $\frac{d}{dv}$ is time-like.

One can introduce a basis of N vectors at every point in this spacetime. Two such vectors β_a and l_a that span the radial-temporal subspace are $\beta_a = \delta_a^v$ and $l_a = -\frac{1}{2}\left[1 - \frac{2G_N m(v)}{nr^n} - \frac{2\Lambda}{(n+1)(n+2)}r^2\right]\delta_a^v + \delta_a^r$. The rest of the $(N-2)$ vectors are defined on the $(n+1)$-sphere and induce on the latter a tensor field γ_{ab} of the form

$$\gamma_{ab} = r^2\left(\delta_a^{\theta_1}\delta_b^{\theta_1} + \sin^2\theta_1\delta_a^{\theta_2}\delta_b^{\theta_2} + ... + \sin^2\theta_1\sin^2\theta_2...\sin^2\theta_n\delta_a^{\theta_{n+1}}\delta_b^{\theta_{n+1}}\right).$$

The vectors satisfy $\beta_a\beta^a = l_a l^a = 0$, $\gamma_{ab}k^b = \gamma_{ab}l^b = 0$, $\beta_a l^a = -1$.

II   LOCATION OF THE HORIZONS

The structure and dynamics of horizons of such non-static metrics can be approached from the non-perturbative description of deformation of relativistic membranes. The quantities that characterize how a variation in the symmetry of a membrane evolves are the expansion rate, θ, the shear rate, σ, and the vorticity (twist), ω. They obey the Rachaudhuri [7] equation,

$$\frac{d\theta}{dv} = \kappa\theta - (\gamma_c^c)^{-1}\theta^2 - \sigma_{ab}\sigma^{ab} + \omega_{ab}\omega^{ab} - R_{ab}l^a l^b \qquad (2.1)$$

where R_{ab} is the N-dimensional Ricci tensor, γ_c^c is the trace of the projection tensor for null geodesics and κ is to be identified as the surface gravity.

Spherically symmetric irrotational spacetimes, such as under consideration, are vorticity and the shear free. The structure and dynamics of the horizons are then only dependent on the expansion, θ. Following York [8] we note that to $O(L_0)$ the evolution of an apparent horizon (AH) is to satisfy the requirement that $\theta \simeq 0$, while that of an event horizons (EH) is to satisfy the requirement that $\frac{d\theta}{dv} \simeq 0$.

The Apparent Horizons
In our basis the expansion θ can be written as

$$\theta = \gamma^{ab}\nabla_a l_b \qquad (2.2)$$

From eq. (2.2) and the York condition $\theta \simeq 0$ we find that the (AHs) satisfy,

$$r^{n+2} - \frac{(n+1)(n+2)}{2\Lambda} r^n + \frac{(n+1)(n+2)}{n} \frac{G_N m(v)}{\Lambda} = 0. \tag{2.3}$$

While eq. (2.3) admits no simple solutions it is possible to cast it in a form for which approximate solutions are reasonable and justifiable. To this end we set $r = k(1 - \xi)$ where $k = \sqrt{\frac{(n+1)(n+2)}{2\Lambda}}$ to transform eq. (2.3) to the form

$$\xi (2 - \xi)(1 - \xi)^n - \beta(n) = 0., \tag{2.4}$$

where $\beta(n) = \frac{2G_N m(v)}{n} \left[\frac{2\Lambda}{(n+1)(n+2)} \right]^{\frac{n}{2}}$. In our model $\Lambda > 0$ and so $\beta(n) > 0$ so that for real and positive r values, $0 < \xi < 1$ and $0 < \beta < 1$. This justifies seeking lower order solutions by expanding the expression $(1 - \xi)^n$ in the first term of equation (2.4). Thus to $O(\xi^4)$ we find

$$\xi^4 + a\xi^3 + b\xi^2 + c\xi + d = 0, \tag{2.5}$$

where $a = -\frac{6n}{(n-1)(2n-1)}$, $b = \frac{6(2n+1)}{n(n-1)(2n-1)}$, $c = -\frac{12}{n(n-1)(2n-1)}$, and $d = \frac{6}{n(n-1)(2n-1)} \beta(n)$. We should mention that this expansion is strictly valid for $n > 2$, although the solutions have the right limits when $n = 1$. The solutions for the 5-dimensional ($n = 2$) case are exact. In general this approximation is good in the limit $\xi \to 0$.

Of the four solutions for eq. (2.5) the two physically interesting ones for our purposes are

$$\xi_\pm = -\frac{1}{4}a + \frac{1}{2}R(p,\varphi) \pm \frac{1}{2}D(p,\varphi), \tag{2.6}$$

where $R = \sqrt{\frac{a^2}{4} - b + \left(2\sqrt{\frac{-p}{3}}\right) \cos \frac{1}{3}\varphi}$, $D = \sqrt{\frac{3a^2}{4} - R^2(p,\varphi) - 2b + \frac{4ab - 8c - a^3}{4R(p,\varphi)}}$ and where $\left(\frac{\pi}{2} < \varphi < \pi\right) = \arccos\left(\frac{3q}{2\sqrt{\frac{-p^3}{3}}}\right)$, $p = \frac{1}{3}[(3ac - b^2) - 12d]$ and $q = \frac{1}{27}[9abc - 2b^3 - 27c^2 - 9(3a^2 + b)d]$.

Recalling that $r = k(1 - \xi)$, the solutions above give

$$r^\pm_{AH}(v) \simeq k \left\{ 1 \pm \frac{1}{4} [2D(p,\varphi) \mp 2R(p,\varphi) \pm a] \right\}. \tag{2.7}$$

In the limit $n \to 1$ one recovers the well known solutions [6]. Thus

$$\lim_{n=1} r^\pm_{AH}(v) = \pm \left(\frac{2}{\sqrt{\Lambda}}\right) \cos \frac{1}{3}\Psi_\pm, \tag{2.8}$$

where $\left(\frac{\pi}{2} < \Psi_-(v) < \pi\right) = \arccos\left[-3m(v)\sqrt{\Lambda}\right]$ and $\Psi_+ = (\Psi_- + \pi)$.

Consequently, we identify the loci of $r_{AH}^-(v)$ and $r_{AH}^+(v)$ in equations (2.7) as the black hole and de sitter apparent horizons (AH^-) and (AH^+), respectively, for a radiating black hole in an N dimensional background with a cosmological constant.

The Event Horizons

The event horizons are null surfaces. To $O(L_0)$ the evolution of these surfaces can be determined by applying to eq. (2.2) the second of the York conditions, $\frac{d\theta}{dv}|_{EH} \simeq 0$. One finds that the event horizons (EHs) in this problem satisfy

$$r^{n+2} - \frac{(n+1)(n+2)}{2\Lambda} r^n + \frac{(n+1)(n+2)}{n} \frac{G_N m^*(v)}{\Lambda} = 0. \tag{2.9}$$

where $m^*(v)$ is some effective mass given by $m^*(v) = m(v) - \frac{L_0}{\kappa}$

Eq. (2.9) is exactly of the same form as its counterpart equation (2.3) with the mass $m(v)$ replaced by the effective mass $m^*(v)$. Hence we can immediately write the solutions to (2.9) as

$$r_{EH}^\pm(v) \simeq k \left\{ 1 \pm \frac{1}{4} \left[2D^*(p,\varphi) \mp 2R^*(p,\varphi) \pm a^* \right] \right\}, \tag{2.10}$$

where $*$ means $m(v) \to m^*(v) = m(v) - \frac{L_0}{\kappa}$ and $\frac{1}{2}\pi < \varphi^* < \pi$.

In the limit $n \to 1$ one recovers the well known solutions [6]. Thus

$$\lim_{n=1} r_{EH}^\pm(v) = \pm \left(\frac{2}{\sqrt{\Lambda}} \right) \cos \frac{1}{3} \Psi_{(\pm)}^*, \tag{2.11}$$

where $\left(\frac{\pi}{2} < \Psi_{(-)}^*(v) < \pi \right) = \arccos \left[-3m^*(v)\sqrt{\Lambda} \right]$ and $\Psi_{(+)}^* = \left(\Psi_{(-)}^* + \pi \right)$

Consequently, we identify the loci of $r_{AH}^-(v)$ and $r_{AH}^+(v)$ in eq. (2.10) as the black hole and de sitter event horizons, i.e. (EH^-) and (EH^+) respectively, for a radiating black hole in an N-dimensional background with a cosmological constant.

The ordering of the horizons can now be made. One finds from our results that $EH^- < AH^- < AH^+ < EH^+$.

III STRUCTURE AND DYNAMICS OF THE HORIZONS

Structure of the Apparent Horizons

At the apparent horizons the expansion, θ, vanishes. One then finds that eq. (1.1) will, along with eqs. (2.7), induce on the surfaces r_{AH}^- and r_{AH}^+ metrics of the form

$$ds^2 \big|_{r=r_{AH}^\pm} = \mp \frac{k}{2} \alpha_\pm(p,\varphi) \left[\rho \frac{\sin \frac{1}{3}\varphi}{\sin \varphi} + \sigma \cos \frac{1}{3}\varphi \right] \frac{dm}{dv} dv^2 + d\Omega_{n+1}^2, \tag{3.1}$$

where $\alpha_\pm(p,\varphi) = (DR)^{-1}\left[2R + \frac{(4ac-8c-a^3)}{4R} \pm 1\right]$, $\rho = \frac{1}{3}\sqrt{\frac{-p}{3}}\frac{d}{dm}\left(\frac{3q}{2\sqrt{\frac{-p^3}{3}}}\right)$

and $\sigma = \frac{2d}{m(v)\sqrt{-3p}}$. Noting that α, ρ and $\alpha_\pm$ are all positive quantities then for $\frac{\pi}{2} < \varphi < \pi$, $\frac{dm}{dv} < 0$ contributes the only negative quantity in eqs. (3.1). We conclude, on this basis that the apparent horizon surface $r_{AH}^-(v)$ of an evaporating black hole in a higher dimensional de Sitter spacetime is timelike while the associated cosmological apparent horizon r_{AH}^+ is spacelike.

In the limit $n \to 1$, one finds that

$$ds^2\Big|_{r=r_{AH}^\pm} = \pm\frac{4\sin\left(\frac{1}{3}\Psi_\pm\right)}{\sin\Psi}\frac{dm}{dv}dv^2 + d\Omega_{n+1}^2. \tag{3.2}$$

Eqs. (3.2) are the known results [6] for the four dimensional case.

The dynamics of the Horizons

One can rewrite equations (3.1) in the form

$$ds^2\Big|_{r=r_{AH}^\pm} \simeq \pm 2L_0\Gamma_{(AH)n}^\pm dv^2 + d\Omega_{n+1}^2 \tag{3.3}$$

where $\Gamma_{(AH)n}^\pm = \frac{k}{2}\alpha_{(\pm)}(p,\varphi)\left[\rho\frac{\sin\frac{1}{3}\varphi}{\sin\varphi} + \sigma\cos\frac{1}{3}\varphi\right]$. We see then that to $O(L_0)$ the apparent horizons $r_{AH}^\pm$ move with velocities given by

$$\frac{dr_{AH}^\pm}{dv} \simeq \pm 2L_0\Gamma_{(AH)n}^\pm. \tag{3.4}$$

Similarly the motion of the event horizons $r_{EH}^\pm$ can be deduced from eqs. (2.9). One finds that the velocities of these surfaces are given by

$$\frac{dr_{EH}^\pm}{dv} \simeq \pm 2L_0\Gamma_{(EH)n}^\pm \tag{3.5}$$

where the $\Gamma_{(EH)n}^\pm$ are obtained by applying to $\Gamma_{(AH)n}^\pm$ the transformation $m(v) \to m^*(v) = m(v) - \frac{L_0}{\kappa}$ that turns AH quantities to EH quantities. For the range $\left(\frac{1}{2}\pi < \varphi, \varphi^* < \pi\right)$ considered $\Gamma_{(AH)n}^+$ and $\Gamma_{(EH)n}^+$ are positive so that for the observer at $r_{AH}^- < r < r_{AH}^+$ both EH^- and AH^- move with negative velocities $-2L_0\Gamma_{(EH)n}^-$ and $-2L_0\Gamma_{(AH)n}^-$, respectively. Such motion represents, in each case, a contraction of the respective black hole horizon. Conversely the cosmological horizons EH^+ and AH^+ are observed to expand at velocities $2L_0\Gamma_{(EH)n}^+$ and $2L_0\Gamma_{(AH)n}^+$, respectively.

As $(n \to 1)$ one recovers the known results [6] for the four dimensional case, and in this limit as $\Lambda \to 0$, we recover from the equations the standard relations $\lim_{\substack{n-1 \\ \Lambda-0}}\frac{dr_{AH}^-}{dv} = -2L_0$ and $\lim_{\substack{n-1 \\ \Lambda-0}}\frac{dr_{EH}^-}{dv} = -2L_0$.

IV CONCLUSION

In this discussion we have examined black holes of the Vaidya type in an spatially flat higher dimensional spacetime with a cosmological constant. We find four horizons identified as the event horizon, EH^- and the apparent horizon AH^- for the black hole and their cosmological counterparts, EH^+ and AH^+, respectively. We have, to good order of accuracy, located these horizons and deduced both their structure and dynamics. Our results reduce to the known ones, under various limits, including the four dimensional case [6]. It is seen then that the problem of the dynamics of a radiating blackhole in a higher dimensional cosmological background can be sensibly discussed.

An application of our results to the Hawking radiation problem will be the topic of a future discussion, elsewhere.

REFERENCES

1. A. Chados and S. Detweiler, Gen Relativ. Gravit. **14**, 879 (1982); G. W. Gibbons and D. L. Wiltshire, Ann. Phys. **167**, 201 (1986).
2. R. C. Myers and M. J. Perry, Ann. Phys. (NY) **172**, 304 (1986); P. O. Mazur, J. Math. Phys., **28**, 406 (1987).
3. B. R. Iyer and C. V. Vishveshwara, Pramana J. Phys., **32**, 749 (1989).
4. R. L. Mallett, Phys. Rev. D **31**, 416 (1985).
5. L. K. Patel and L S. Desai, Pramana J. Phys. **48**, 819 (1997); Zhen-Qiang Tan and You-Gen Shen, Il Nuovo Cimento **113 B**, 339 (1998).
6. R. L. Mallett, Phys. Rev D **33**, 2201 (1986).
7. A. Raychaudhuri, Phys. Rev. **98**, 1123 (1955); B. Carter, *General Relativity*, (edited by S. W. Hawking and I. Isreal) Cambridge Univ. Press, Cambrige, UK (1979)
8. J. W. York, Jr., in Quantum Theory of Gravity: Essays in Honor of the Sixtieth Birthday of Bryce S. DeWitt, edited by S. Christensen (Hilger, Bristol, 1984).

Metric Fluctuation Corrections to Black Hole Radiation

C. Barrabès[1] [1], V. Frolov[2] [2] and R. Parentani[3] [1]

[1] *Laboratoire de Mathématiques et Physique Théorique, CNRS UPRES A 6083, Université de Tours, 37200 Tours, France*
[2] *Theoretical Physics Institute, Department of Physics, University of Alberta, Edmonton, Canada T6G 2J1*

I   INTRODUCTION

In the original derivation of Hawking radiation [1] the spacetime metric of the collapsing body is treated classically and it is assumed to be insensitive to the emission of quanta of radiation. However the fluctuations of the black hole horizon geometry induced by quantized matter fields can modify the properties of Hawking radiation. To describe these fluctuations quantum mechanically and to determine their effects on Hawking radiation requires full quantum gravity. Nevertheless, it is not unreasonable to hope that the main modifications of Hawking radiation induced by metric fluctuations can be extracted from a much simpler framework in which the fluctuations of the metric are treated classically.

The model we shall use is inspired by that proposed by York [2]. In that model, the fluctuating geometry near the horizon of the black hole is represented by a Vaidya-type metric with a fluctuating mass. The spectrum of these fluctuations is characterized by the zero point fluctuations of quantum fields. We shall further simplify this model by considering only spherically symmetric fluctuations and by neglecting the scattering by the gravitational potential which occurs in the 4-dimensional Dalembertian. Then we determine how these fluctuations modify the energy flux and the asymptotic spectrum of s-waves.

II   MODEL

In the model proposed by York [2] the fluctuations of the black hole geometry are approximated by an incoming Vaidya metric with a fluctuating mass

[1] e-mail: barrabes@celfi.phys.univ-tours.fr
[2] e-mail: frolov@phys.ualberta.ca
[3] e-mail: parenta@celfi.phys.univ-tours.fr

CP493, *General Relativity and Relativistic Astrophysics*, edited by C. P. Burgess and R. C. Myers
© 1999 American Institute of Physics 1-56396-905-X/99/$15.00

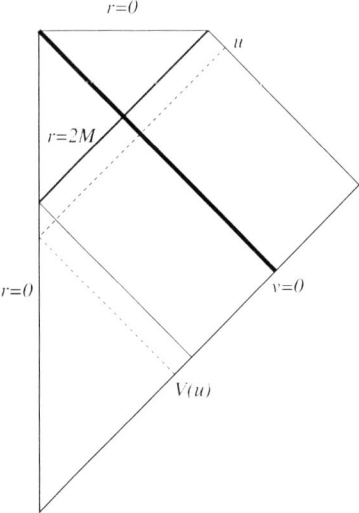

FIGURE 1. Conformal diagram for a black hole created by a collapse of a massive null shell. Solid dark line $v = 0$ represents the collapsing massive null shell.

$$ds^2 = -f(r, v)\,dv^2 + 2dv\,dr + r^2 dS_2^2\,, \tag{II.1}$$

where dS_2^2 is the metric of a unit 2-sphere and

$$f(r, v) = 1 - \frac{2m(v)}{r}\,, \tag{II.2}$$

$$m(v) = M[1 + \mu_0 \sin(\omega v)]\,\vartheta(v)\,. \tag{II.3}$$

This is the standard Vaidya metric in advanced time coordinates (v, r). The mass fluctuates with frequency ω and dimensionless amplitude μ_0 (for spherical modes one can take $\mu_0 = \alpha(m_{\text{Planck}}/M)$, where α is a pure number, and $\mu_0 \ll 1$ for black holes with mass $M \gg m_{\text{Planck}}$). The step function $\vartheta(v)$ in relation (II.3) indicates that the black hole results from the gravitational collapse of a massive (with mass M) null shell propagating along $v = 0$. Therefore inside the collapsing null shell the spacetime is flat.

The conformal diagram of the whole geometry (II.1) in the absence of fluctuations (that is for $\mu_0 = 0$) is schematically shown in Figure 1. The dashed line on this figure shows a radial null ray which reaches $\mathcal{J}^+$ at the moment of the retarded time u and which was sent from $\mathcal{J}^-$ at $v = V(u)$ of advanced time.

For the study of Hawking radiation we need to know the trajectory of radial null rays in the fluctuating black hole geometry (II.1). Out-going null rays obey the equation

$$f(r, v)\,dv = 2\,dr\,. \tag{II.4}$$

168

In order to solve this equation, we use a method of perturbations and write

$$r(v) = R(v) + \rho(v) + \sigma(v) + \ldots . \tag{II.5}$$

$R(v)$ is the solution of equation (II.4) in the absence of fluctuations, and $\rho(v)$ and $\sigma(v)$ are respectively the first and second order perturbation in μ_0. Higher order corrections are denoted by dots.

If u is the retarded time, the unperturbed trajectory $r = R(v; u)$ of a radial ray arriving at $\mathcal{J}^+$ at a given u is given by

$$u = v - 2R_* = \text{const} . \tag{II.6}$$

where

$$R_* = R - 2M + 2M \ln \frac{R - 2M}{2M} \tag{II.7}$$

is the usual tortoise radial coordinate. Notice that the retarded time u is a fixed parameter which specifies the unperturbed ray under consideration and $R = R(v; u)$.

Before describing an arbitrary null ray reaching $\mathcal{J}^+$ we discuss the particular solution corresponding to the event horizon in the fluctuating geometry. In the unperturbed geometry the event horizon corresponds to $R = 2M$. Starting with this solution we easily obtain the first and second order terms in (II.5) and the modified event horizon

$$r_{EH} \approx 2M \left[1 + \mu_0 \frac{\Omega \cos(\omega v) + \sin(\omega v)}{1 + \Omega^2} \right.$$
$$\left. + \mu_0^2 \frac{2\Omega^2(2 - \Omega^2)\cos(2\omega v) + \Omega(1 - 5\Omega^2)\sin(2\omega v)}{2(1 + \Omega^2)^2(1 + 4\Omega^2)} \right], \tag{II.8}$$

where $\Omega = \omega/\kappa$ and $\kappa = (4M)^{-1}$ is the unperturbed surface gravity of the black hole.

From this one can compute the modified value of the surface area $\mathcal{A}$ of the event horizon. When averaging over advanced time v, we find

$$\mathcal{A} \equiv 4\pi \overline{(r_{\text{hor}}^2(v))} \approx 16\pi M^2 \left[1 + \frac{\mu_0^2}{2(1 + \Omega^2)} \right] . \tag{II.9}$$

Similarly the average value of the surface gravity in the fluctuating geometry is

$$\bar{\kappa} \equiv \overline{\left(\frac{1}{2} \frac{\partial f}{\partial r} \right)}_{EH} \approx \kappa \left[1 + \frac{\mu_0^2}{2(1 + \Omega^2)} \right] . \tag{II.10}$$

We now consider the general case, that is we assume that $R \neq 2M$. To study quantum black hole radiation in the fluctuating geometry, we need to solve the wave equation in this geometry. As usual [1] we use the geometrical optics approximation.

Thus we only need to derive the relation, $v = V(u)$, between the advanced time v at which a radial null ray leaves $\mathcal{J}^-$ and the retarded time u at which it reaches $\mathcal{J}^+$ (see Fig. 1). To establish the relation $v = V(u)$ we first get the solution (II.5) for a ray propagating in the fluctuating geometry outside the collapsing massive null shell, see [3] for more details. Then we glue it to the solution inside the shell and introduce the reflection condition at $r = 0$. The result is the following

$$-\kappa V(u) \approx \kappa V_0 + e^{-\kappa u} \left[1 + A_1 \sin(\omega u + \phi_1) + A_2 \sin(2\omega u + \phi_2) + C\kappa u \right], \quad \text{(II.11)}$$

where ϕ_1 and ϕ_2 are functions of $\Omega = \omega/\kappa$ and

$$\kappa V_0 = 1 + \mu_0 \frac{\Omega}{1 + \Omega^2} + \mu_0^2 \frac{\Omega^2(2 - \Omega^2)}{(1 + \Omega^2)^2(1 + 4\Omega^2)}, \quad \text{(II.12)}$$

$$A_1 = \mu_0 \sqrt{\frac{2\pi}{\Omega(1 + \Omega^2)(e^{2\pi\Omega} - 1)}} \left[1 + 2\mu_0 \frac{\Omega^2 - 1}{\Omega(1 + \Omega^2)} \right], \quad \text{(II.13)}$$

$$C = -\frac{\mu_0^2}{2(1 + \Omega^2)}, \quad \text{(II.14)}$$

The coefficient A_2 in (II.11) also depends on Ω and is of second order in the amplitude μ_0 of the fluctuations, see [3].

III MODIFIED ENERGY FLUX AND ASYMPTOTIC SPECTRUM OF HAWKING RADIATION

Now we derive the s-mode contribution to Hawking radiation. In what follows, we shall neglect the scattering by the gravitational potential barrier which appears in the 4D Dalembertian. In other words, we use 2D approximation. To determine the importance of these interesting effects is complicated and goes beyond the scope of the present paper which is to describe the effects on Hawking radiation induced by the fluctuations of the geometry in the very close vicinity of the horizon. In this respect, we wish to emphasize that our classical metric has been choosen to mimic the near horizon quantum fluctuations and not the fluctuations of the height of the barrier around $r = 3M$. On physical grounds, in a self-consistent treatment, one might expect that the residual fluctuations around $3M$ be much smaller that the near horizon ones.

In the 2D simplified description, when the field is in its vacuum state before the formation of the black hole, the mean energy flux at $\mathcal{J}^+$ is

$$\frac{dE}{du} \equiv 4\pi r^2 \langle T_{uu} \rangle^{\text{ren}} = \frac{1}{12\pi} \left(\frac{dV}{du} \right)^{1/2} \frac{d^2}{du^2} \left[\left(\frac{dV}{du} \right)^{-1/2} \right]. \quad \text{(III.1)}$$

Here $V(u)$ is the function calculated in the previous section (II.11). Notice also, that u in this relation is the time which defines positive frequency at $\mathcal{J}^+$. It is convenient to split the expression of the energy flux into two parts

$$dE/du = (dE/du)^{\text{perm}} + (dE/du)^{\text{fluct}} . \tag{III.2}$$

where $(dE/du)^{\text{perm}}$ is the mean value of the flux

$$(dE/du)^{\text{perm}} \approx \frac{\kappa^2}{48\pi} \left[1 + \mu_0^2 \frac{\pi\Omega}{e^{2\pi\Omega} - 1} - 2C \right] . \tag{III.3}$$

and $(dE/du)^{\text{fluct}}$ is its fluctuating part.

$$(dE/du)^{\text{fluct}} \approx -\mu_0 \sqrt{\frac{\pi\Omega(1 + \Omega^2)}{2(e^{2\pi\Omega} - 1)}} \cos(\omega u + \varphi_1) + \mu_0^2 B_2 . \tag{III.1}$$

where B_2 is coefficient depending on Ω and oscillating at frequencies ω and 2ω, see [3]. The remarkable fact is that, to second order in μ_0, the correction term which is linear in u in (II.11) does not give any time-dependent contribution. It only gives an additional constant to $(dE/du)^{\text{perm}}$ in (III.3) which can be removed by making a "renormalization" of the surface gravity

$$\kappa \to \kappa_r = \kappa(1 - C) = \kappa \left[1 + \frac{\mu_0^2}{2(1 + \Omega^2)} \right] . \tag{III.5}$$

Hence the expression for $(dE/du)^{\text{perm}}$ can be identically rewritten as

$$(dE/du)^{\text{perm}} \approx \frac{\kappa_r^2}{48\pi} \left[1 + \mu_0^2 \frac{\pi\Omega}{\exp(2\pi\Omega) - 1} \right] . \tag{III.6}$$

It is interesting to note that the renormalized surface gravity κ_r which is introduced here coincides with the average value of the surface gravity $\bar{\kappa}$ defined by equation (II.10). As a consequence of this the Hawking temperature becomes

$$T = \frac{\kappa_r}{2\pi} . \tag{III.7}$$

The modifications in the black hole temperature and surface area in the presence of metric fluctuations raise the question about the modifications of black hole thermodynamics. If we identify the energy of the system E with the averaged mass of the black hole M. and the temperature of the black hole with (III.7), then the first law $dE = T\,dS$ and eqs.(II.9) leads to the averaged entropy

$$\bar{S} \approx \frac{\bar{A}}{4} \left(1 - \frac{\mu_0^2}{1 + \Omega^2} \right) . \tag{III.8}$$

Therefore, one looses the Bekenstein-Hawking relation $S = A/4$. If furthermore one takes for the amplitude of the fluctuations $\mu_0 = a m_{\text{Planck}}/M$ where a is dimensionless, the modification $s = \bar{S} - S$ of the entropy is

$$s = -\frac{2\pi a^2}{1 + \Omega^2} \qquad (\text{III.9})$$

and does not depend on the black hole mass.

In order to know how the asymptotic spectrum is modified by the fluctuating part of the metric we need to compute the Bogoliubov coefficients in the modified geometry. These coefficients are given by the overlap of the initial (infalling) modes which are specified on $\mathcal{J}^-$ and the final (outgoing) modes specified on $\mathcal{J}^+$. Both are solutions of the Dalembertian equation in the metric (II.1). For s-waves and under the neglection of the potential barrier, these modes satisfy the 2D equation $\partial_u \partial_v \phi = 0$. Thus the in-modes $\phi_\nu(v)$ and out-modes $\phi_\lambda(u)$ can be decomposed in terms of planes waves

$$\phi_\nu(v) = \frac{e^{-i\nu v}}{\sqrt{4\pi\nu}} \quad , \quad \phi_\lambda(u) = \frac{e^{-i\lambda u}}{\sqrt{4\pi\lambda}} \qquad (\text{III.10})$$

where ν is the energy measured on $\mathcal{J}^-$ and λ the energy measured on $\mathcal{J}^+$.

The scattering of in-modes in the time dependent geometry follows as usual from the 'reflection' condition on $r = 0$ wherein the Wronskian must vanish. This implies that the scattered in-modes are given by $\phi_\nu(V(u))$ where $V(u)$ in the perturbed geometry is given by (II.11). From the modified value of Bogoliubov coefficients [3].

From the modified value of Bogoliubov coefficients [3] one obtains that the mean (quantum averaged and time averaged) number of quanta which reach $\mathcal{J}^+$ per unit time is

$$\langle n_\lambda \rangle = \frac{1}{2\pi} \left\{ \frac{1}{e^{2\pi\lambda/\kappa_r} - 1} \left(1 - (\frac{A_1}{\kappa_r})^2 \frac{\lambda^2}{2} \right) \right.$$
$$\left. + (\frac{A_1}{2\kappa_r})^2 \lambda \left[\frac{\lambda - \omega}{e^{2\pi(\lambda-\omega)/\kappa_r} - 1} + \frac{\lambda + \omega}{e^{2\pi(\lambda+\omega)/\kappa_r} - 1} \right] \right\} \qquad (\text{III.11})$$

The spectrum of the energy flux is directly related to the spectrum of the particle number flux

$$\frac{dE}{du\, d\lambda} = \lambda \langle \bar{n}_\lambda \rangle \qquad (\text{III.12})$$

One sees that the frequency ω acts as a chemical potential, as in superradiance, as λ is replaced by $\lambda \pm \omega$.

IV CONCLUSION

Quantum fluctuations of the matter fields in the vicinity of a black hole horizon induce fluctuations of the spacetime metric which in turn should affect the spectrum of Hawking radiation. In our work, as a first attempt to characterize the modifications of Hawking radiation induced by metric fluctuations, we have used a model wherein these fluctuations are treated classically by an incoming Vaidya metric with a mass $m(v)$ oscillating around its mean value with amplitude μ_0 and frequency $\omega/2\pi$. For further simplicity we only considered spherical modes of the radiation field and we neglected the scattering by the gravitational potential barrier. Our results are the following:

-The expectation value of the outgoing flux of energy is no longer constant but fluctuates with frequencies given by harmonics of ω. Its time averaged is shifted by a quantity which is of second order in the amplitude μ_0 of the fluctuations and decreases for large ω.

-The spectrum of Hawking radiation is modified in essentially two ways. First the Hawking temperature is shifted by also a second order term in μ_0. Second the averaged flux contains additional terms with Bose thermal factors where the frequency of the metric fluctuations plays the role of a chemical potential and produces an effect similar to superradiance.

-The area of the event horizon and the surface gravity are renormalized. Furthermore the Bekenstein-Hawking relation between black hole entropy and area of the event horizon is modified.

Even though these results were obtained in an extremely simplified model in which the metric fluctuations were treated classically, we believe that they indicate what might be the impact of the quantum fluctuations of the near horizon geometry on black hole radiance.

Acknowledgements:This work was supported by NATO Grant CRG.972079. One of the authors (V.F.) is grateful to the Natural Sciences and Engineering Research Council of Canada and to the Killam Trust for their financial support.

REFERENCES

1. S.W. Hawking, Comm. Math. Phys. **43** (1975) 199; see also N.D. Birrel and P.C.W. Davies, *Quantum Fields in Curved Space*.Cambridge University Press, 1982, and V. Frolov and I. Novikov. *Black Hole Physics: Basic Concepts and New Developments*, Kluwer Academic Publ., 1998.
2. J. W. York, Jr., Phys. Rev. D **28** (1983) 2929.
3. C. Barrabès, V. Frolov and R. Parentani. Phys. Rev. D **59** (1999) 124010(1-14).

A Hamiltonian Approach to the Mass of Isolated Black Holes

Christopher Beetle and Stephen Fairhurst

Center for Gravitational Physics and Geometry
Department of Physics, The Pennsylvania State University
University Park, PA 16802

Abstract. Boundary conditions defining a *non-rotating isolated horizon* are given in Einstein–Maxwell theory. A spacetime representing a black hole which itself is in equilibrium but whose exterior contains radiation admits such a horizon. Inspired by Hamiltonian mechanics, a (quasi-)local definition of isolated horizon mass is formulated. Although its definition does not refer to infinity, this mass takes the standard value in a Reissner–Nordström solution. Furthermore, under certain technical assumptions, the mass of an isolated horizon is shown to equal the future limit of the Bondi energy.

I INTRODUCTION

In their standard form, the zeroth and first laws of black hole mechanics apply only to the relatively small class of stationary black hole solutions. These solutions seem too limited to describe physically realistic situations since the requirement of stationarity precludes the presence of radiation even far from the horizon. As an example, consider the gravitational collapse depicted in figure 1. Although the horizon will presumably reach equilibrium at late times, there will also be gravitational and other radiation near null infinity. One would hope the familiar laws of black hole mechanics continue to apply in such situations. Recently, using the framework of *isolated horizons* [1,2], this expectation has been shown to be correct: The zeroth and first laws extend to a broad class of spacetimes containing both radiation and black holes which are, however, isolated from that radiation. This contribution focuses on the definition of black hole mass in this expanded, non-stationary setting and collects the results presented in both of our talks at the Conference.

The key idea of the isolated horizon framework is to replace the global construct of an event horizon with a set of boundary conditions applied locally at the surface of the black hole. These boundary conditions model a portion of the horizon which is isolated, i.e., across which there is no flux of gravitational radiation or matter.

CP493, *General Relativity and Relativistic Astrophysics*, edited by C. P. Burgess and R. C. Myers

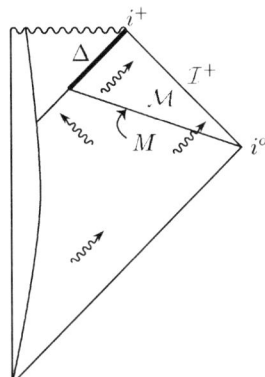

FIGURE 1. A typical gravitational collapse. The portion Δ of the horizon at late times is isolated. The spacetime $\mathcal{M}$ of interest is the triangular region bounded by Δ, $\mathcal{I}^+$ and a partial Cauchy slice M.

However, since the boundary conditions are applied *only* at the horizon, the exterior of an isolated horizon will generically contain both. For technical simplicity, we confine our attention here to non-rotating isolated horizons in Einstein–Maxwell theory.

To formulate the laws of black hole mechanics for isolated horizons, one needs definitions of the "extrinsic parameters" of a black hole — particularly its mass M and surface gravity κ — which depend *only* on the structure available at the horizon. In the static context, M is taken to be the ADM mass and κ to be the acceleration at the event horizon of that static Killing field which is unit at infinity. Although these parameters are associated with the black hole, they cannot be constructed solely from the near-horizon geometry; they are genuinely global concepts. In the non-static context, the ADM energy *cannot* be identified with the mass of the black hole since it will also include contributions from the energy contained in radiative fields far from the horizon. Similarly, without a static Killing field, the usual prescription for normalizing the null generator of the horizon is no longer applicable and the usual definition of κ therefore fails. Drawing motivation from the Hamiltonian formulation of an isolated horizon, this paper shows how the problem of defining its mass can be overcome.

The details of our construction, along with a more complete set of references, may be found in [2].

II   ISOLATED HORIZONS

An isolated horizon is defined by a set of boundary conditions which capture the essential local structure of a stationary event horizon. The physical situation we wish to model is illustrated by the example of figure 1. The late stages of the collapse pictured here should describe a non-dynamical, isolated black hole. However, a realistic collapse will generate gravitational radiation which must either be scattered back into the black hole or radiated to infinity. Physically, one expects most of the back-scattered radiation will be absorbed rather quickly and, in the absence of outside perturbations, the black hole will "settle down" to a steady-state configuration. Although the presence of radiation elsewhere in spacetime implies $\mathcal{M}$ cannot be stationary, the portion Δ of the event horizon at late times will describe an isolated black hole. The boundary conditions are designed to represent precisely such situations. They are also satisfied at the event horizon of a static, Reissner–Nordström black hole.

We are now in a position to state our definition. For clarity, we shall give a fairly complete statement of the boundary conditions, even though some of the details will not be needed here. A *non-rotating, isolated horizon* is a three-dimensional hypersurface Δ in spacetime which satisfies the following four conditions:

(a) Δ is null, topologically $S^2 \times \mathrm{R}$, and equipped with a preferred foliation by two-spheres S_Δ transverse to its null normal ℓ^a. We will denote the second null normal to the foliation by n^a and partially fix the normalizations by setting $\ell^a n_a = -1$ and requiring the pull-back to Δ of n_a to be curl-free.

(b) Δ is a non-rotating, non-expanding, future boundary of spacetime. Specifically, ℓ^a is twist-, shear- and expansion-free and n^a is twist- and shear-free with strictly negative expansion $\theta_{(n)}$ which is constant on each spherical leaf of the preferred foliation of Δ.[1]

(c) All equations of motion are satisfied at Δ.

(d) The flux densities of the electric and magnetic fields are both constant on each two-sphere S_Δ.

Let us summarize the consequences of the boundary conditions which are directly relevant here.

First, the notion of an isolated horizon is intrinsically local since it is defined through a set of boundary conditions. This makes the definition well-suited to situations such as that of figure 1. Although the boundary conditions imply the surface Δ is isolated from infalling matter and radiation, the geometry of the exterior region $\mathcal{M}$ remains largely unconstrained. Even *at* Δ, the *outgoing* radiative modes (i.e., those flowing along Δ) are undetermined.

Second, a non-rotating isolated horizon in Einstein–Maxwell theory is characterized by three parameters; its radius r_Δ, electric charge Q_Δ and magnetic monopole moment P_Δ. Much of the intrinsic structure of Δ depends only on this remarkably small set of parameters. In particular, the Newman–Penrose components Ψ_2

---

[1] The spherical symmetry of $\theta_{(n)}$ guarantees the uniqueness of the preferred foliation.

and ϕ_1 [3], describing respectively the "Coulombic parts" of the gravitational and electromagnetic fields at Δ, are functions of these variables alone.

Third, the boundary conditions guarantee all three parameters are constant in time. The radius is time-independent due to the vanishing expansion of Δ and the imposition of the Maxwell equations at Δ implies the same for the electromagnetic charges. It is important to note that this time-independence is a result of *boundary conditions* and not of dynamics. Consequently, this result holds in *any* history compatible with the boundary conditions and not just "on-shell."

Before moving on to our discussion of mass, let us briefly consider the surface gravity. As discussed in the Introduction, the standard formulation of black hole mechanics defines surface gravity as the acceleration of the appropriately normalized null generator of the horizon. This normalization is possible since the horizon-generating Killing field can be fixed to an unit time translation at infinity. However, a generic isolated horizon spacetime possesses no such Killing field. To overcome this problem, we must provide a means of fixing the normalization of ℓ^a on *any* isolated horizon; then κ may again be defined as the acceleration of the properly normalized ℓ^a. Recall that ℓ^a itself is null and free of shear, expansion and twist, whence we cannot normalize it through its intrinsic properties. However, since $\theta_{(n)}$ is strictly negative by condition (b), we *can* normalize n^a by setting its expansion to some given value. This in turn will fix the normalization of ℓ^a since $\ell^a n_a = -1$.

Fortunately, a preferred value of $\theta_{(n)}$ exists. In a Reissner–Nordström solution, with ℓ^a taken to be the restriction to Δ of the properly normalized static Killing field, one finds $\theta_{(n)} = -2/r_\Delta$. Since we want to include this family of solutions in our analysis, let us require n^a (and hence ℓ^a) *always* be normalized such that $\theta_{(n)} = -2/r_\Delta$. This convention, which is *not* a consequence of the boundary conditions, provides a definition of surface gravity for a generic isolated horizon: The value of κ is simply the acceleration of the properly normalized ℓ^a.

III   THE HAMILTONIAN AND BLACK HOLE MASS

A key question faced by any set of boundary conditions is whether they allow the formulation of a well-defined action principle. For boundary conditions enforcing asymptotic flatness in spacetimes with no interior boundaries, it is well known that an action principle can be found by adding a boundary term at infinity to the standard bulk action. The situation is similar when one allows an isolated horizon at an interior boundary. A well-defined action principle can then be obtained by also including a boundary term at Δ [4]. However, as we will see now, there is some subtlety in using this action principle to formulate a Hamiltonian description of an isolated horizon system.

In passing to the Hamiltonian framework, one performs a Legendre transform of the action. To do so, let us foliate the spacetime $\mathcal{M}$ with partial Cauchy surfaces M which extend to spatial infinity and whose inner boundaries are the preferred two spheres S_Δ. Next, fix a smooth vector field t^a, transverse to the leaves M, which tends to a ℓ^a at the horizon[2] and to an unit time translation orthogonal to M near infinity. Then, after the Legendre transform, the Hamiltonian corresponding to evolution along t^a can be written in the form

$$H_t = \int_M (\text{constraints}) + H_\infty - H_\Delta. \tag{1}$$

The quantities H_∞ and H_Δ represent the surface terms in the Hamiltonian at infinity and Δ respectively. The term at infinity is the usual one arising from the asymptotically flat boundary conditions imposed there. It can be expressed in terms of components of the Weyl curvature as

$$H_\infty = \lim_{R \to \infty} \oint_{S_R} \left[-\frac{R}{4\pi G} \Psi_2 \right]\, {}^2\epsilon. \tag{2}$$

On any solution to the field equations, this surface term will equal the ADM energy of spacetime.

On the other hand, due to an ambiguity in the action caused by the variational principle, the horizon boundary term H_Δ in the Hamiltonian is initially an *arbitrary* function of the parameters r_Δ, Q_Δ and P_Δ. Since every allowable history must satisfy our boundary conditions, the variation of these parameters must be time-independent. However, the action principle allows only variations which vanish at the initial and final time slices. It follows that δr_Δ, δQ_Δ and δP_Δ vanish everywhere in the Lagrangian formalism. Therefore, one may add *any* function of these parameters to the action and still have a well-posed variational principle. As a result of this freedom, H_Δ will not be fixed by the Lagrangian.

The ambiguity in H_Δ is resolved within the Hamiltonian formalism. The reason for this lies in subtle differences between the Lagrangian and Hamiltonian variations. In the Hamiltonian framework, the phase space consists of fields on a fixed spacelike three-manifold M which describe horizons of arbitrary radius and charges. Consequently, there *are* tangent vectors δ to phase space which change the values of r_Δ, Q_Δ and P_Δ. Requiring the consistency of Hamilton's equations under such variations in phase space determines the Hamiltonian *uniquely*. The resulting horizon surface term in the Hamiltonian is

$$H_\Delta = \oint_{S_\Delta} \left[-\frac{r_\Delta}{4\pi G} \Psi_2 + \frac{Q_\Delta - i P_\Delta}{2\pi r_\Delta} \phi_1 \right]\, {}^2\epsilon, \tag{3}$$

where the Newman–Penrose components Ψ_2 and ϕ_1 are determined in terms of r_Δ, Q_Δ and P_Δ by the boundary conditions.

Remarkably, in a static Reissner–Nordström solution, one finds the horizon surface term (3) is exactly equal to that at infinity (2). Consequently, the full Hamiltonian vanishes in a static solution. This feature is not accidental: There is a

---

[2] Here, ℓ^a denotes the null generator of Δ normalized according to the prescription given at the end of the previous section. Physically, this vector field defines the "rest frame" of the black hole.

general argument from symplectic geometry which requires the vanishing of H_t on a stationary solution.

Now let us turn to the definition of black hole mass. In physical theories, energy is defined as the on-shell value of the Hamiltonian generating an appropriate time translation. In general relativity, the bulk term in the Hamiltonian consists solely of constraints and therefore vanishes on-shell. Thus, the energy of a system is given by the surface terms in its Hamiltonian. For example, in asymptotically flat spacetime, the Hamiltonian surface term at infinity is precisely equal to the ADM energy when the constraints are satisfied. In analogy, we *define* the mass of an isolated horizon to be the horizon surface term in the Hamiltonian:

$$M_\Delta := H_\Delta = \oint_{S_\Delta} \left[-\frac{r_\Delta}{4\pi G} \Psi_2 + \frac{Q_\Delta - iP_\Delta}{2\pi r_\Delta} \phi_1 \right] {}^2\epsilon. \qquad (4)$$

This formula for black hole mass resembles the expression (2) for ADM mass, but includes an additional contribution from the electromagnetic field. Without this contribution, M_Δ would *not* give the correct mass for a Reissner–Nordström black hole. For a generic isolated horizon, the expression (4) for M_Δ can be regarded as "the mass of the black hole together with its Coulombic hair." That is, it includes the energy associated with static fields emanating from Δ, but *not* contributions due to radiative excitations outside Δ.

IV RADIATIVE ENERGY AND BLACK HOLE MASS

As discussed in the Introduction, the ADM energy cannot be used to measure the mass of a non-stationary black hole since it also includes contributions from the radiative modes of the fields. This raises the question of whether there exists a precise relation among the ADM energy, the black hole mass and the radiative energy. One would expect such a relation to be simply

$$(\text{ADM energy}) = (\text{black hole mass}) + (\text{radiative energy}). \qquad (5)$$

We will now show that, under certain technical assumptions, this relation holds within the framework of isolated horizons. This result is on a somewhat different footing from the rest of the calculations presented here due to the additional assumptions needed to complete the proof. Nevertheless, from a physical point of view, the relation (5) serves to strengthen our intuition regarding isolated horizon mass and provides further justification for our definition.

Consider an isolated horizon, such as Δ in figure 1, which extends to future time-like infinity i^+. Since no radiation can cross Δ, all the radiation in $\mathcal{M}$ must "register on" future null infinity $\mathcal{I}^+$. It is therefore not surprising that the radiative data can be encoded in fields *on* $\mathcal{I}^+$ [5]. Furthermore, these fields on $\mathcal{I}^+$ admit a phase space structure [6]. That is, there exists a symplectic structure $\Omega_{\mathcal{I}}$ on the space of data at null infinity and a Hamiltonian $H_{\mathcal{I}}$ which generates time evolution in that space. Moreover, on any solution to the field equations, the value of $H_{\mathcal{I}}$ is precisely equal to the total radiative energy. This fact is the key to the proof of (5) for isolated horizon systems.

We now have two phase spaces; the isolated horizon phase space (Ω, H_t) discussed in the previous section and the asymptotic phase space $(\Omega_{\mathcal{I}}, H_{\mathcal{I}})$ introduced here. Both of these describe the radiative modes of the fields in $\mathcal{M}$, and hence it is reasonable to expect they should also be "equivalent" in some appropriate sense. This expectation turns out to be correct. There is a natural map from the isolated horizon phase space to the phase space at null infinity such that $\Omega \mapsto \Omega_{\mathcal{I}}$. Using this fact, we can apply Hamilton's equations in each phase space to find

$$\delta H_t = \Omega(\delta, X_H) = \Omega_{\mathcal{I}}(\delta, X_H) = \delta H_{\mathcal{I}}. \tag{6}$$

It follows that the isolated horizon Hamiltonian and its counterpart at null infinity differ at most by a constant. In the previous section, however, we showed the isolated horizon Hamiltonian H_t vanishes on a Reissner–Nordström solution. Furthermore, these static solutions contain no radiation, whence the Hamiltonian at $\mathcal{I}^+$ must also vanish. Therefore, the "constant of integration" for (6) is zero and the two Hamiltonians are *equal* on any solution to the equations of motion. Since H_t equals the difference between the ADM energy and the black hole mass by (1) and $H_{\mathcal{I}}$ equals the radiative energy in spacetime, the relation (5) follows.

Therefore, *the mass of an isolated horizon is equal to the ADM energy of spacetime minus the energy contained in radiation outside the horizon*. It is well known that the difference of the ADM energy and the flux of energy through $\mathcal{I}^+$ is the future limit of the Bondi energy. Hence, isolated horizon mass is the future limit of the Bondi energy. Equivalently, M_Δ can be thought of as the mass remaining in spacetime after all radiation has escaped to infinity.

V SUMMARY

We have presented the definition of a non-rotating isolated horizon. Within the surrounding framework, we have shown how the definitions of mass (and surface gravity) can be extended to a non-static context. Although it has not been discussed here, these definitions enable one to formulate and prove the zeroth and first laws of black hole mechanics for generic isolated horizons [2].

The mass M_Δ is given by the horizon surface term in the *unique* consistent Hamiltonian for an isolated horizon system. It is manifestly (quasi-)local to the horizon and correctly reproduces the mass of Reissner–Nordström black holes. Furthermore, given certain technical assumptions, one can show the ADM energy of a spacetime containing a single isolated horizon is precisely the sum of M_Δ and the total energy contained in the radiative modes of the fields. In this case, M_Δ is equal to the future limit of the Bondi energy.

Acknowledgements   We would like to express our gratitude to Abhay Ashtekar for his collaboration in the work reported here. The authors were supported in part

180

by NSF grants PHY95-14240 and INT97-22514 and by the Eberly research funds of the Pennsylvania State University.

REFERENCES

1. A. Ashtekar, A. Corichi and K. Krasnov. Isolated horizons: the classical phase space. *Adv. Theor. Math. Phys.* in press; gr-qc/9905089.
2. A. Ashtekar, C. Beetle and S. Fairhurst. Isolated horizons: a generalization of black hole mechanics. *Class. Quantum Grav.* **16** (1999) L1-L7.
   A. Ashtekar, C. Beetle and S. Fairhurst. Mechanics of isolated horizons. Penn State preprint; gr-qc/9907068.
3. R. Penrose and W. Rindler. *Spinors and Spacetime*, Volume 1. Cambridge University Press, Cambrindge, 1984.
4. A. Ashtekar, J. Baez, A. Corichi and K. Krasnov. Quantum geometry and black hole entropy. *Phys. Rev. Lett.* **80** (1998) 904-907.
5. A. Ashtekar. Radiative degrees of freedom of the gravitational field in exact general relativity. *J. Math. Phys.* **22** (1981) 2885-2895.
6. A. Ashtekar and M. Streubel. Symplectic geometry of radiative modes and conserved quantities at null infinity. *Proc. R. Soc. Lond.* A **376** (1981) 585-607.

Quasilocal Energy
and Naked Black Holes

Ivan Booth and Robert Mann

Department of Physics,University of Waterloo, Waterloo, Ontario, N2L 3G1

Abstract.
    We extend the Brown and York notion of quasilocal energy to include coupled elec-
tromagnetic and dilaton fields and also allow for spatial boundaries that are not orthogonal to the foliation of the spacetime. We investigate how the quasilocal quantities
measured by sets of observers transform with respect to boosts. As a natural application of this work we investigate the naked black holes of Horowitz and Ross calculating
the quasilocal energies measured by static versus infalling observers.

I INTRODUCTION

The definition of energy in general relativity continues to be an area of active
research. It is widely accepted that while one cannot localize gravitational energy
and therefore define a gravitational stress energy tensor one can define a notion
of the total energy in a spacetime (for example the ADM or Bondi masses). In
between these two extremes one can define energy quasilocally – that is define the
amount of energy contained in a finite volume of spacetime.

One popular definition of quasilocal energy (QLE) was proposed by Brown and
York in 1993 [1]. As we shall see in the next section their approach derives a
notion of quasilocal energy by Hamiltonian methods from the standard Einstein-
Hilbert action. Since its proposal this quasilocal energy has found application
in gravitational thermodynamics and the study of the production of pairs of black
holes. It has been shown to reduce to the ADM and Bondi masses in the appropriate
limits as well as a Newtonian notion of gravitational energy for certain specific
examples. References for these may be found in [2].

In this paper, we extend this notion of quasilocal energy to include coupled electromagnetic and dilaton fields and allow for general motions of the set of observers.
We see that the QLE transforms in a Lorentzian way under boosts of the observers.
In the last section we find a natural application for this work in the naked black
holes first studied by Horowitz and Ross [3]. Such black holes have small curvature invariants yet observers falling into them experience massive tidal forces as
they approach the event horizon. We calculate the quasilocal energies measured by

CP493, *General Relativity and Relativistic Astrophysics*, edited by C. P. Burgess and R. C. Myers
© 1999 American Institute of Physics 1-56396-905-X/99/$15.00

these observers and see that they distinguish between naked and standard "clothed" black holes.

II   QUASILOCAL ENERGY

In classical mechanics the action I of a point particle is the time integral of its kinetic energy minus its potential energy. To wit,

$$I = \int dt (p\dot{q} - H)$$

where p is the particle momentum, q is its position, and H is its Hamiltonian/potential energy. Taking the first variation of this action we obtain the equations of motion of the particle.

Now the action for gravity is well known (the standard Einstein-Hilbert action) so it is natural to extend and reverse this procedure to define a Hamiltonian for gravity. Following the procedure of Brown and York [1] which was generalized to allow for moving observers and coupled dilaton and matter fields in [4,2] the quasilocal energy contained by a spacelike two-surface of observers Ω who are being evolved in spacetime by a timelike vector field T^α is

$$E = \int_\Omega d^2 x \sqrt{\sigma}\,(\varepsilon + \varepsilon^m - \underline{\varepsilon}).$$

Defining u^α as the (unit-normalized) component of T^α perpendicular to Ω and n^α as the outward pointing spacelike unit-normal vector perpendicular to both Ω and T^α, the *geometrical* and *matter* quasilocal energy densities are respectively

$$\varepsilon = -\frac{1}{16\pi}\sigma^{\alpha\beta}\mathcal{L}_n\sigma_{\alpha\beta} \quad \text{and} \quad \varepsilon^m = \frac{1}{4\pi}(n_\alpha B^\alpha)(u^\beta A^\star_\beta).$$

$\mathcal{L}_n$ is the Lie derivative in the direction n^α and so ε measures how much the area of Ω would change if the surface was Lie-dragged along that vector field. $B^\alpha = -e^{2a\phi}\star F_{\alpha\beta}u^\beta$ is the magnetic field seen by the observers defining Ω. $\star F_{\alpha\beta}$ is the dual of the regular electromagnetic field tensor, $A^\star_\alpha$ is a vector potential generating that dual field, ϕ is the dilaton field, and a is the coupling constant between the dilaton and EM fields. Note that we are working with the dual here because we ultimately wish to study magnetic black holes. Conversely if we wished to study electric black holes we would use $F_{\alpha\beta}$. This is explained in detail in [2].

For asymptotically flat spacetimes the reference term $\underline{\varepsilon}$ is defined so that E will vanish for the observers undergoing arbitrary motion in flat space. The simplest way to do this is to embed Ω in Minkowski space and then define a vector field $\underline{T}^\alpha$ over Ω so that $\underline{T}^\alpha\underline{T}_\alpha = T^\alpha T_\alpha$, $\mathcal{L}_{\underline{T}}\sigma_{\alpha\beta} = \mathcal{L}_T\sigma_{\alpha\beta}$, and $\underline{T}^\alpha\sigma^\beta_\alpha = T^\alpha\sigma^\beta_\alpha$. Then the reference quasilocal energy density is $\underline{\varepsilon} = -\frac{1}{16\pi}\sigma^{\alpha\beta}\mathcal{L}_n\sigma_{\alpha\beta}$.

Note that the matter term ε^m is manifestly gauge dependent. The gauge choice is essentially a choice of where to set the zero of this gauge energy. In the following we will make use of two different gauge choices. First, setting the (magnetic) Coulomb potential $-u^\alpha A_\alpha^\star = 0$ over Ω and therefore $\varepsilon^m = 0$ we define the *geometrical* energy E_{Geo}. Second choosing $-u^\alpha A_\alpha^\star = 0$ at the black hole horizon we define the *total* energy E_{tot}.

Simple transformation laws are defined for the QLE's with respect to motion of the observers. Consider two sets of observers who instantaneously coincide on a surface Ω. Here we take them as a "static" set being evolved by the timelike unit vector $T^\alpha = u^\alpha$ with normal vector n^α to Ω and a "moving" set being evolved by $T^{\star\alpha} = u^{\star\alpha}$. Then the moving set are seen to have velocity $v = -\frac{T^{\star\alpha} n_\alpha}{T^{\star\alpha} u_\alpha}$ in the direction n^α by the static set. Defining $\gamma = (1 - v^2)^{-1/2}$, it is not hard to show that

$$\varepsilon^\star = \gamma(\varepsilon + v j_\vdash) \quad \text{and} \quad \varepsilon^{m\star} = \gamma(\varepsilon + v j_\vdash^m).$$

$j_\vdash = -\frac{1}{16\pi}\sigma^{\alpha\beta}\mathcal{L}_u\sigma_{\alpha\beta}$ and so represents the (local) rate of change of the area of Ω as measured by the observers with respect to proper time. It can be thought of as a momentum flow through the surface. $j_\vdash^m$ is a matter term that can be set to zero by an appropriate gauge choice (which we'll make here for simplicity).

Things are slightly complicated because the reference terms transform with respect to a different velocity. Looking back at the defining conditions for the reference term we see that by construction $\underline{j}_\vdash = j_\vdash$. Then, it is not surprising that the surface of observers would have to travel at a different speed in the reference spacetime than they do in the original one to keep this rate of change the same. Thus,

$$\underline{\varepsilon}^\star = \underline{\gamma}(\underline{\varepsilon} + \underline{v} j_\vdash)$$

where $\underline{v}$ is defined in an analogous way to v.

With these transformation laws it is also easy to see that $\varepsilon^{\star 2} - j_\vdash^{\star 2}$ is a constant, independent of the boost. This is analogous to the special relativity relation $E^2 - p^2 c^2 = m^2$ (which is a constant) and will be of use in the later calculations.

III   NAKED BLACK HOLES

Naked black holes are a subclass of the low-energy-limit string theory solutions with metric

$$ds^2 = -F(r)dt^2 + \frac{dr^2}{F(r)} + R(r)^2(d\theta^2 + \sin^2\theta d\varphi^2),$$

where $F(r) = \frac{(r-r_+)(r-r_-)}{R^2}$ and $R(r) = r\left(1 - \frac{r_-}{r}\right)^{a^2/(1+a^2)}$. r_+ is the location of the black hole horizon and (for the coupling constant $a \neq 0$) r_- gives the position of the singularity behind the horizon. There are also dilaton and Maxwell fields defined by $\phi = -\frac{a}{1+a^2}\ln\left(1 - r_-/r\right)$ and $F = G_0 \sin\theta d\theta \wedge d\varphi$. If $a = 0$ this solution reduces to a magnetically charged Reissner-Nordström (RN) black hole. For the purposes of this short paper we shall also assume that if $a \neq 0$ then $a \approx 1$. Very small values of a cause complications; we deal with these elsewhere [2]. The (ADM) mass and magnetic charge of these solutions are given by $M = \frac{r_+}{2} + \frac{1-a^2}{1+a^2}\frac{r_-}{2}$ and $G_0 = \left(\frac{r_+ r_-}{1+a^2}\right)^{1/2}$ respectively.

If $R_+ = R(r_+) \gg 1$ (that is R_+ is much larger than the Planck length) then these black holes have a very large surface area and a correspondingly large mass (in the appropriate Planck units). All of the curvature invariants are small outside of the horizon, and static observers measure very small curvature components. For example consider a static spherical set of observers surrounding a black hole. Then $u^\alpha = F^{-1/2}\partial_t$ and $n^\alpha = F^{1/2}\partial_r$. Contracting u^α and any unit tangent vector $X^\alpha \subset T\Omega$ with the Riemann tensor, $\mathcal{R}_{\alpha\beta\gamma\delta}u^\alpha X^\beta u^\gamma X^\delta \propto R_+^{-2} \ll 1$ near to the horizon.

If, however, these holes are also extremely close to being extreme with $\delta \equiv (1 - r_-/r_+)^{\frac{1}{1+a^2}} \ll a/R_+$, then observers who are falling into these holes tell a very different story about the curvature components. Consider a spherical set of observers who started out with velocity zero at some very large r and then fell towards the black hole along a radial geodesic. Then $u^{\star\alpha} = \gamma(u^\alpha + vn^\alpha)$ where $v = -(1 - F)^{1/2}$ is the radial velocity of the infalling observers as seen by the static ones, and $\gamma = (1 - v^2)^{-1/2}$ is the standard Lorentz factor from special relativity. Again taking $X^\alpha \subset T\Omega$, $\mathcal{R}_{\alpha\beta\gamma\delta}u^{\star\alpha}X^\beta u^{\star\gamma}X^\delta \propto a^2(R_+\delta)^{-2} \gg 1$. That is they see extremely large, Planck scale curvature components. The resulting huge geodesic deviation laterally crushes them. Horowitz and Ross [3] dubbed this subclass of Maxwell-dilaton holes *naked* because Planck scale curvature components could be seen outside of their horizons.

A   QLE of naked black holes

We use our methods to calculate the static and boosted quasilocal energies of these black holes. Static observers measure

$$E_{Geo} = R - \sqrt{(r - r_+)(r - r_-)}\dot{R} \quad \text{and} \quad E_{Tot} = R\left(1 - \sqrt{\frac{r - r_+}{r - r_-}}\right),$$

where $\dot{R} = \frac{dR}{dr}$. Note that both of these go to $R_+ \gg 1$ at the horizon. This is not surprising since both measure the quasilocal energy and a large R_+ corresponds to a large mass black hole.

Next consider the infalling measurements. The radial velocity of the infalling observers in the original spacetime is $v = -(1 - F)^{1/2}$ with respect to the static observers. By contrast, the shell of observers have to travel at $\underline{v} = -\dot{R}(1 - F)^{1/2}/(1 + \dot{R}^2(1 - F))^{1/2}$ in the reference spacetime if they want their surface area to change at the same rate. Therefore at the horizon r_+,

$$E_{Geo}^{\star} = \left\{ \begin{array}{ll} C_1 R_+ \delta \ll 1 & \text{for naked black holes} \\ C_2 R_+ \gg 1 & \text{for "clothed" holes} \end{array} \right\} \quad \text{and} \quad E_{Tot}^{\star} = \frac{R_+}{\delta} \gg \gg 1,$$

where C_1 and C_2 are constants that are on the order of unity. Thus we see that our extension of the QLE formalism detects the difference between regular "clothed" and naked black holes. Note however that while static/infalling observers see small/large curvatures they measure large/small geometric QLE's.

B   Why do naked black holes behave this way?

These small/large measurements can be understood physically in the following manner. As was noted by Horowitz and Ross for a naked black hole $R_+\delta$ is more-or-less the time left to an infalling observer before she reaches the singularity at r_-. This time is very small and so the singularity is "just behind" the outer horizon. At $r = r_-$ the surface area of an $r =$ constant shell of observers goes to zero. This means that naked holes will have a very large $j_\vdash$ at r_+ since the area is very large but will soon be zero. By contrast for an RN hole, the area would only go to zero at $r = 0$ which is not so "close" to the horizon. Thus $j_\vdash$ is not so large.

Thus, we can see why observers falling into a naked black hole experience the huge lateral crushing forces. As a shell of them travelling on geodesics cross the horizon the surface area of that shell is rapidly decreasing and so the crushing lateral forces are to be expected. By contrast for "clothed" black holes the rate of change of the area is much smaller and so are the corresponding lateral forces.

The relative sizes of $j_\vdash$ in the two cases also explain the geometrical QLE observations. As we saw earlier, $\varepsilon^2 - j_\vdash^2$ is a constant independent of boosts. Therefore if ε and $\underline{\varepsilon}$ are the geometric and reference QLE densities for the static observers, ε^* and $\underline{\varepsilon}^*$ are their boosted counterparts, and recalling that $\underline{j}_\vdash = j_\vdash$,

$$\varepsilon^* - \underline{\varepsilon}^* = \sqrt{j_\vdash^2 + \varepsilon^2} - \sqrt{j_\vdash^2 + \underline{\varepsilon}^2} \approx \frac{\varepsilon^2 - \underline{\varepsilon}^2}{2j_\vdash}$$

for $j_\vdash$ much larger than ε and ε^*. Thus as $j_\vdash$ becomes larger and larger, $\varepsilon^* - \underline{\varepsilon}^*$ becomes smaller and smaller. Physically, though ε^* and $\underline{\varepsilon}^*$ are boosted to be very large, at the same time the difference between them becomes smaller and smaller. The relativistic effect of the motion overpowers that of the gravity.

By contrast E_{Tot}^* includes matter terms. These terms are also boosted to be very large but there is no term in the reference spacetime to counterbalance them as there was for the geometric terms. Therefore in E_{Tot}^* the matter terms dominate over the geometrical ones and so the total infalling quasilocal energy is large.

ACKNOWLEDGEMENTS

This work was supported by the Natural Sciences and Engineering Council of Canada.

REFERENCES

1. J.D. Brown and J.W. York, *Phys. Rev. D* **47**, 1407–1419 (1993).
2. I.S. Booth and R.B. Mann, gr-qc/9907072.
3. G.T. Horowitz and S.F. Ross, *Phys. Rev. D* **56**, 2180–2187 (1997); *Phys. Rev. D.* **57**, 1098–1107 (1998).
4. I.S. Booth and R.B. Mann, *Phys. Rev. D* 064010, (1999).

III. COSMOLOGY

Preheating and Supergravity

Lev Kofman

Canadian Institute for Theoretical Astrophysics
University of Toronto, M5S 3H8 Canada

Abstract. In this talk recent developments of the theory of preheating after inflation are briefly reviewed. In inflationary cosmology, the particles constituting the Universe are created after inflation due to their interaction with moving inflaton field(s) in the process of reheating. In inflationary models motivated by supergravity, both bosons and fermions are created. In the bosonic sector, the leading channel of particle production is the non-perturbative regime of parametric resonance dominated by those bosons which are created exponentially fast with the largest characteristic exponent. In the fermionic sector, the leading channel corresponds to the regime of parametric excitation of fermions, which respects Pauli blocking but differs significantly from the perturbative expectation. In supergravity we also have to consider production of gravitinos and moduli fields, which are cosmologically dangerous relics. We discuss the derivation of the gravitino equations in curved space-time with moving background scalars. We describe recent results on the production of gravitinos from preheating, which may put strong constraints on the inflationary models.

Preheating after Inflation

According to the inflationary scenario, the Universe initially expands quasi-exponentially in a vacuum-like state without entropy or particles. At the stage of inflation, all energy is contained in a classical slowly moving inflaton field ϕ. The fundamental Lagrangian $\mathcal{L}(\phi, \chi, \psi, A_i, h_{ik}, ...)$ contains the inflaton part with the potential $V(\phi)$ and other fields which give subdominant contributions to gravity. The Friedmann equation for the scale factor $a(t)$ and the Klein-Gordon equation for $\phi(t)$ determine the evolution of the background fields. In most of the models, soon after the end of inflation, an almost homogeneous inflaton field $\phi(t)$ coherently oscillates with a very large amplitude of the order of the Planck mass around the minimum of its potential. This scalar field can be considered as a coherent superposition of inflatons with zero momenta. The amplitude of oscillations gradually decreases not only because of the expansion of the universe, but also because energy is transferred to particles created by the oscillating field. At this stage we shall recall the rest of the fundamental Lagrangian which includes all the fields interact-

CP493, *General Relativity and Relativistic Astrophysics,* edited by C. P. Burgess and R. C. Myers
© 1999 American Institute of Physics 1-56396-905-X/99/$15.00

ing with inflaton. These interactions lead to the creation of many ultra-relativistic particles from the inflaton. Gradually, the inflaton field decays and transfers all of its energy non-adiabatically to the created particles. In this scenario all the matter constituting the universe is created from this process of reheating. If the creation of particles is sufficiently slow, the particles would simultaneously interact with each other and come to a state of thermal equilibrium at the reheating temperature T_r. This gradual reheating can be treated with the perturbative theory of particle creation and thermalization. However, typically particle production from coherently oscillating inflatons occurs not in the perturbative regime but in the non-perturbative regime of parametric excitation. Indeed, let us consider a simple toy model of chaotic inflation with the quadratic potential $V(\phi) = \frac{1}{2}m_\phi\phi^2$ and $\mathcal{L}_{int} = -\frac{1}{2}g^2\phi^2\chi^2$ describing the interaction between the inflatons and other massless Bose particles χ. The quantum scalar field $\hat{\chi}$ in a flat FRW background has the eigenfunctions $\chi_k(t)\,e^{-i\mathbf{k}\mathbf{x}}$ with comoving momentum $\mathbf{k}$. The temporal part of the eigenfunction obeys the equation

$$\ddot{\chi}_k + 3\frac{\dot{a}}{a}\dot{\chi}_k + \left(\frac{\mathbf{k}^2}{a^2} - \xi R + g^2\phi^2\right)\chi_k = 0 \tag{1}$$

with vacuum-like initial conditions: $\chi_k \simeq \frac{e^{-ikt}}{\sqrt{2k}}$ in the far past. The coupling to the curvature ξR will not be important in the presence of the interaction (but would lead to gravitational preheating in the absence of the interaction). In this model, the inflaton field $\phi(t)$ coherently oscillates as $\phi(t) \approx \Phi(t)\sin(m_\phi t)$, with the amplitude $\Phi(t) = \frac{M_p}{\sqrt{3\pi}} \cdot \frac{1}{m_\phi t}$ decreasing as the universe expands. The smallness of g^2 alone does not necessarily lead to the perturbative excitation of χ_k modes. To check whether the interaction term $g^2\phi^2$ in eq. (1) is perturbative or not, it is convenient to use a new time variable $z = mt$ and the essential dimensionless coupling parameter $q = \frac{g^2\Phi^2}{m^2}$. Scalar metric fluctuations in this model are compatible with cosmology if the inflaton mass is $m \simeq 10^{-6}M_p$; therefore, it is expected that $q \simeq 10^{10}g^2 \gg 1$ for not negligibly small g^2. In fact, a consistent setting for the problem of χ-particle creation from the ϕ-inflaton requires $q \gg 1$ even without additional assumptions about g^2 [2]. Indeed, it is known that if we have two scalars ϕ and χ, then the latest stage of inflation will be driven by the lightest scalar. The square of the effective mass of the χ-field includes a term $g^2\phi^2$. Inflation is driven by the ϕ-field if its square mass m^2 is smaller than $g^2\phi^2$. This leads to the condition $q \gg 1$, i.e. to the creation of χ-particles in the resonance regime.

Supergravity and the Early Universe

To make the next step beyond toy models of particles interactions with the inflaton, we have to choose the fundamental Lagrangian $\mathcal{L}(\phi, \chi, \psi, A_i, h_{ik}, ...)$. We may expect that the low-energy physics of the early universe will be described by the general four-dimensional $N = 1$ supergravity-Yang-Mills-matter theory [1]. A rather lengthy $N = 1$ phenomenological supergravity Lagrangian [1] begins with the terms

$$
\begin{aligned}
e^{-1}\mathcal{L} = {}& -\frac{1}{2}M_P^2 R - \hat{\partial}_\mu \Phi^i \hat{\partial}^\mu \Phi_i + e^K \left(D^i W D_i W - 3\frac{WW^*}{M_P^2} \right) \\
& - \bar{\chi}_j \, \mathcal{D}\chi^i - \bar{\chi}^i \, \mathcal{D}\chi_j - \left(e^{K/2} D^i D^j W \bar{\chi}_i \chi_j + h.c. \right) \\
& - \frac{1}{2}\bar{\psi}_\mu R^\mu + \left(\frac{1}{2}e^{K/2}W \bar{\psi}_{\mu R}\gamma^{\mu\nu}\psi_{\nu R} + \bar{\psi}_{\mu L}\,\hat{\partial}\Phi^i\gamma^\mu\chi_i + \bar{\psi}_R \chi_i e^{K/2}D^i W + h.c. \right) \\
& + \ldots
\end{aligned}
\tag{2}
$$

A particular choice of the form of the Lagrangian is motivated and notations are given in [3]. In Eq. (2) we choose the minimal Kähler potential $K = \frac{\Phi_i \Phi^i}{M_P^2}$, where Φ^i is the complex conjugate of Φ_i. The last term of the 1st line is the scalar potential $V(\Phi_i)$. The equations of motion based on the first line should describe inflation, which is a challenging problem by itself. For simplicity we will take a prototype model of the superpotential $W = \sqrt{\lambda}\Phi^3/3$, which for $|\Phi| \ll M_P$ leads to the effective potential $\lambda\Phi^4$. We will illustrate some effects of preheating in supergravity with this model.

Preheating of Bosons

Let us briefly recall the basics of the bosonic preheating. Consider the creation of χ-particles due to the $g^2\chi^2\phi^2$ interaction. To get a feeling for the way in which χ-particles are created from inflaton oscillations, we have to understand the character of solutions of eq. (1) for the mode functions $\chi_k(t)$. For preheating of bosons we need the first line of the Lagrangian (2). Consider the model with the inflaton potential $V(\phi) = \frac{1}{4}\lambda\phi^4$ in an expanding universe. The problem of particle creation in this theory can be reduced to a similar problem in Minkowski space-time. This can be realized with the conformal transformation of the scalar field $\phi \to \varphi = a\phi$ and with the conformal time variable $\tau = \sqrt{\lambda}\tilde{\varphi}\int \frac{dt}{a(t)}$. Therefore, $\frac{1}{4}\lambda\phi^4$ theory is sometimes dubbed the conformal theory. In conformal variables, the Klein-Gordon equation for $\varphi(\tau)$ is reduced to an equation in flat space-time. Its solution is $\varphi(\tau) \approx \tilde{\varphi}\, f(\tau)$, where the amplitude of the oscillations $\tilde{\varphi}$ is constant until the backreaction of created particles is taken into account. The time-dependence of the oscillations in this theory is not sinusoidal, but given by an elliptic function $f(\tau) = cn\left(\tau, \frac{1}{\sqrt{2}}\right)$. Eq. (1) for quantum fluctuations χ_k can be simplified in this theory. Using a conformal transformation of the mode function $X_k(t) = a(t)\chi_k(t)$ in Eq. (1) we obtain

$$
X_k'' + \left(\kappa^2 + qf^2(\tau) \right)X_k = 0 \ ,
\tag{3}
$$

---

[1] When I show a viewgraph with the full supergravity Lagrangian at the cosmology conferences, somehow people start to laugh hysterically.

where $\kappa^2 = \frac{k^2}{\lambda \bar\varphi^2}$, and $q = \frac{g^2}{\lambda}$. The equation for fluctuations does not depend on the expansion of the universe and is completely reduced to the similar problem in Minkowski space-time. This is a special feature of the conformal theory $\frac{1}{4}\lambda\phi^4 + \frac{1}{2}g^2\phi^2\chi^2$. The mode equation (3) belongs to the class of Lamé equations. The combination of parameters $q = g^2/\lambda$ ultimately defines the structure of the parametric resonance in this theory. This means that the condition of a broad parametric resonance does not require a large initial amplitude of the inflaton field, as for the quadratic potential. The strength of the resonance depends non-monotonically on the value of the q parameter. The stability/instability chart of the Lamé equation (3) in the variables $\left(\kappa^2, \frac{g^2}{\lambda}\right)$ was constructed in [6]. To see how the general theory works, let us consider parametric resonance of inflaton fluctuations fluctuations ϕ_k due to the self-interaction of inflaton field in $\lambda\phi^4$ theory. Using conformal transformation of the mode functions $\varphi_k(\tau) = a(t)\phi_k(t)$, the equation for φ_k can be reduced to the general equation (3) with the particular value $q = 3$. The equation for fluctuations X_k when $q = 3$ can be solved analytically [6]. The resonance in the "inflaton" direction ϕ is weak; the maximal value of the characteristic exponent of the fluctuations $\phi_k \propto e^{\mu_1 \tau}$ is $\mu \approx 0.036$. Let us however consider a supersymmetric version of the conformal theory. In the first line of the Lagrangian (2) we will put the superpotential $W = \sqrt{\lambda}\Phi^3/3$, which at $\phi \ll M_P$ gives us a scalar potential with the conformal properties $V(\Phi) = \frac{1}{4}(\Phi^2)^2$. In supersymmetric theories, all scalar are complex. The scalar field Φ will have two components $(\phi, \bar\phi)$. Let us assume that inflaton direction corresponds to the real component ϕ, $\text{Re } \Phi = \frac{\phi}{\sqrt{2}}$. and initially $\bar\phi = 0$, $\text{Im } \Phi = 0$. The equation for the mode function of fluctuations in the direction $\bar\phi$ can be obtained with the conformal transformation $\bar\varphi_k(\tau) = a(t)\bar\phi_k(t)$ and can be reduced reduces to the general equation (3) with the parameter $q = 1$. Again, in this case the problem can be solved analytically [6]. The factor $q = 1$ instead of 3 in eq. (3) makes a big difference, which manifests the subtlety of the parametric resonance. The resonance in the direction $\bar\phi$ is much stronger and broader than the resonance in the inflaton direction, $\bar\phi_k \propto e^{\bar\mu \tau}$, with $\bar\mu \approx 0.147$. Thus, the supergravity generalization of bosonic preheating in the conformal theory makes a big difference in the bosonic preheating due to the self-interaction. Note that the character and strength of the parametric resonance for the general equation (3) depends on the shape of the effective inflaton potential $V(\phi)$. Moreover, the investigation of the resonance in an expanding universe typically cannot be reduced to the study of the regular stability/instability chart. If theory is not conformal, (say due to the mass term $m^2\phi^2$ in $V(\phi)$) the parameter q in an expanding universe is time-dependent. For the broad resonance case $q \gg 1$ this parameter can jump over a number of instability bands within a single oscillation of the inflaton field, and the concept of stability/instability bands is inapplicable here. Parametric resonance in this case is a stochastic process [5].

Parametric Excitations of Fermions

A simple model for the inflaton's interaction with Fermi particles χ is a Yukawa term $h\bar{\chi}\phi\chi$. For instance, consider second line of eq. (2) and ignore mixing between χ and ψ_μ (which corresponds to the rigid SUSY limit). For our toy model with $W = \frac{1}{3}\sqrt{\lambda}\Phi^3$ and $\phi \ll M_P$, the "mass" term of the chiral fermion χ is equal to $\sqrt{2\lambda}\phi$, which corresponds to the inflaton-fermion interaction $\sqrt{2\lambda}\bar{\chi}\phi\chi$. For fermions, the Pauli exclusion principle prohibits the occupation number from exceeding 1. For this reason, it has been silently assumed that fermions are created in the three-legs perturbative process $\phi \to \bar{\chi}\chi$ where individual inflatons decay independently into pairs of ψ-particles. Let us, however, consider the Dirac equation for a massless quantum Fermi field $\chi(t, \vec{x})$:

$$[\gamma^\mu \nabla_\mu + h\phi(t)]\chi = 0 , \tag{4}$$

where ∇_μ is the derivative with the spin connection. We are using the representation of gamma matrices where $\gamma_0 = diag(i, i, -i, -i)$. Here, similar to the bosonic case, the inflatons producing fermions also act not as individual particles but as a coherently oscillating field $\phi(t)$. Let us consider more general model $\frac{1}{4}\lambda\phi^4 + h\bar{\psi}\phi\psi$. This is a conformal theory in the sense that the problem of fermion production by the inflaton ϕ in an expanding universe can be reduced to equations in Minkowski space-time. Indeed, let us perform a conformal transformation of the involved fields, $\varphi \equiv a\phi$ and $\Psi \equiv a^{3/2}\chi$, and use a conformal time variable, τ as in the previous section. The equation for the eigenfunctions of the quantum fluctuations in this theory can be reduced to a second-order equation for an auxillary field $X(\tau, \vec{x})$, so that $\Psi = [\gamma^\mu \nabla_\mu + h\varphi]X$. The eigenmodes of the auxillary field have the form $X_k(\tau)e^{+i\mathbf{k}\cdot\mathbf{x}}R_r$, where the R_r are eigenvectors of the Dirac matrix γ^0 with eigenvalue $+1$. The temporal part of the eigenmode obeys an oscillator-like equation with a complex frequency which depends periodically on time

$$X_k'' + \left(\kappa^2 + qf^2 - i\sqrt{q}f'\right)X_k = 0 . \tag{5}$$

The dimensional comoving momentum k enters the equation in the combination $\frac{k^2}{\lambda\bar{\varphi}^2} \equiv \kappa^2$; therefore, the natural units of momentum are $\sqrt{\lambda}\bar{\varphi}$. The background oscillations enter in the form $f(\tau)$ of the previous section, having unit amplitude. The imaginary part of the frequency in Eq. (5) guarantees the Pauli blocking for the occupation number n_k. The results for n_k can be formulated as follows [4]. Even though the Yukawa interaction contains a small factor h, one cannot use the perturbation expansion in h. This is because the frequency of the background field oscillations is proportional to another small parameter $\sqrt{\lambda}$. The combination of the coupling parameters $\frac{h^2}{\lambda} \equiv q$ ultimately determines the strength of the effect. The growth of fermionic modes occurs in the non-perturbative regime of parametric excitation. The modes get fully excited with occupation numbers $n_k \simeq 1$ within tens of oscillations of the field ϕ, and the width of the parametric excitation of fermions in momentum space is about $q^{1/4}\sqrt{\lambda}\phi_0$. For instance, in the case of $h = \sqrt{2\lambda}$, $q = 2$, the modes will be excited in about ten oscillations, and the width will be about $\sqrt{\lambda}\phi_0$.

Equations for the Gravitino

The rest of my talk is based on the recent project [3][2]. Let us now consider the third line in the supergravity Lagrangian (2), which describes the gravitino field ψ_μ. In a general background metric and in the presence of complex scalar fields with non-vanishing VEV's, the equation for the gravitino has on the left hand side the kinetic part $R^\mu \equiv \epsilon^{\mu\nu\rho\sigma}\gamma_5\gamma_\nu\mathcal{D}_\rho\psi_\sigma$, and a rather lengthy right hand side. We will use the long derivative $\mathcal{D}_\mu$ with the spin connection and Christoffel symbols, for which $\mathcal{D}_\mu\gamma_\nu = 0$. Apart of varying gravitino mass m, the right hand side contains a chiral connection and various mixing terms like those in the 3rd line of (2). For a self-consistent setting of the problem, the gravitino equation should be supplemented by the equations for the fields mixing with gravitino, χ_i from (2), as well as by the equations determining the gravitational background and the evolution of the scalar fields. Let us make some simplifications. We consider the supergravity multiplet and a single chiral multiplet containing a complex scalar field Φ with a single chiral fermion χ. This is a simple non-trivial extension which allows us to study the gravitino with a non-trivial FRW cosmological metric supported by the scalar field. A nice feature of this model is that the chiral fermion χ can be gauged to zero so that the mixing between ψ_μ and χ in (2) is absent. We also can choose the non-vanishing VEV of the scalar field (inflaton) in the real direction, as in the previous sections. First we will derive the equation for a spin $3/2$ field in a curved background metric with non-vanishing VEVs for the scalar fields.

1 Spin 3/2 Field Equations in External gravitational and Scalar Fields

From (2) we can obtain the equation for the gravitino field (in case of single chiral multiplet χ and vanishing Im Φ) $R^\mu = m(\tau)\gamma^{\mu\nu}\psi_\nu$, where $\gamma^{\mu\nu} = \frac{1}{2}\left(\gamma^\mu\gamma^\nu - \gamma^\nu\gamma^\mu\right)$ and gravitino mass $m = m(\phi(\tau))$ is given by $m = e^{K/2}\frac{W}{M_P^2}$. This equation can be transformed into the form

$$\mathcal{D}\!\!\!\!/\,\psi_\mu + m\psi_\mu = \left(\mathcal{D}_\mu - \frac{m}{2}\gamma_\mu\right)\gamma^\nu\psi_\nu. \tag{6}$$

The gravitino equation (6) is a curved spacetime generalization of the familiar gravitino equation $(\partial\!\!\!/ + m_0)\psi_\mu = 0$ in a flat metric, where m_0 is a constant gravitino mass. The generalization of the first constraint equation $\partial^\mu\psi_\mu = 0$ can be obtained from the equality $\gamma_\mu R^\mu = 2\gamma^{\mu\nu}\mathcal{D}_\mu\psi_\nu$ and reads

---

[2] All credit for the correct results below go to [3]. All incorrect deviations is my fault.

$$\mathcal{D}^{\mu}\psi_{\mu} - \mathcal{D}\gamma^{\mu}\psi_{\mu} + \frac{3}{2}m\gamma^{\mu}\psi_{\mu} = 0 . \tag{7}$$

The generalization of the second constraint equation $\gamma^{\mu}\psi_{\mu} = 0$ can be obtained from the equality $\mathcal{D}_{\mu}R^{\mu} = -\frac{1}{2}G_{\nu\rho}\gamma^{\nu}\psi^{\rho}$ (dropping the torsion term) and is

$$\frac{3}{2}m^2\gamma^{\mu}\psi_{\mu} + (\gamma^{\mu}\partial_{\mu}m)\gamma^{\nu}\psi_{\nu} - (\partial_{\mu}m)\psi^{\mu} = -\frac{1}{2}G_{\mu\nu}\gamma^{\mu}\psi^{\nu} . \tag{8}$$

In our case $m = m(\tau)$ and from (8) one can find an algebraic relation between $\gamma^0\psi_0$ and $\gamma^i\psi_i$:

$$\gamma^0\psi_0 = \hat{A}\gamma^i\psi_i . \tag{9}$$

Here $\hat{A}$ is a matrix which will play a crucial role in our description of the interaction of the gravitino with the varying background fields. If ρ and p are the background energy-density and pressure, we have $G_0^0 = M_P^{-2}\rho$, $G_k^i = -M_P^{-2}p\,\delta_k^i$, and one can represent the matrix $\hat{A}$ as follows:

$$\hat{A} = \frac{p - 3m^2 M_P^2}{\rho + 3m^2 M_P^2} + \gamma_0\frac{2m'a^{-1}M_P^2}{\rho + 3m^2 M_P^2} = A_1 + \gamma_0 A_2 . \tag{10}$$

We shall solve the equation (6) using the constraint equations in the form (8) and (9). We use a plane-wave ansatz $\psi_{\mu} \sim e^{i\mathbf{k}\cdot\mathbf{x}}$ for the space-dependent part. Then ψ_i can be decomposed into its transverse part ψ_i^T, and to the longitudinal part ψ_i^L which is defined by the trace $\gamma^i\psi_i$. Two degrees of freedom of ψ_{μ} are associated with the transverse part ψ_i^T, which correspond to helicity $\pm 3/2$ and two degrees of freedom are associated with $\gamma^i\psi_i$ (or ψ_0) which correspond to helicity $\pm 1/2$. For the helicity $\pm 3/2$ states we have to derive the equation for ψ_i^T. We apply decomposition $\psi = \psi_i^T + \psi_i^L$ to the master equation (6) for $\mu = i$ and obtain

$$\left(\gamma^{\mu}\partial_{\mu} + \frac{a'}{2a}\gamma^0 + ma\right)\psi_i^T = 0 . \tag{11}$$

The transformation $\psi_i^T = a^{-1/2}\Psi_i^T$ reduces the equation for the transverse part to the free Dirac equation with a time-varying mass term ma, c.f. eq. (4). In the previous section we explained how to treat this type of equation. The essential part of Ψ_i^T is given by the time-dependent part of the eigenmode of the transversal component $X_T(\eta)$, which obeys second-order equation (c.f. (5)):

$$X_T'' + \left(k^2 + (ma)^2 - i(ma)'\right)X_T = 0 . \tag{12}$$

The corresponding equation for gravitino with helicity $1/2$ is more complicated. We have to find $k^i\psi_i$ and $\gamma^i\psi_i$. The equation for the components $k^i\psi_i$ can be obtained from the constraint equation (7). The equation for $\gamma^j\psi_j$ can be derived from (6). Using ψ_0 from (9), we get an equation for $\gamma^i\psi_i$

$$\left(\partial_\eta + \hat{B} - i\mathbf{k} \cdot \gamma\gamma_0\hat{A}\right)\gamma^i\psi_i = 0\,, \tag{13}$$

where $\hat{B} = -\frac{3a'}{2a}\hat{A} - \frac{ma}{2}\gamma_0(1 + 3\hat{A})$. The time-dependent factor of the spinor $\gamma^i\psi_i$, which we denote as $f_k(\tau)$, obeys a second-order differential equation. By the substitution $f_k(\tau) = E(\eta)X_L(\eta)$, with $E = (-A^*)^{1/2}\exp\left(-\int^\eta d\eta\, ReB\right)$, the equation for the function $f_k(\tau)$ is reduced to the final oscillator-like equation for the time-dependent mode function $X_L(\eta)$:

$$X_L'' + \left(|A|^2k^2 + \Omega_L^2 - i\Omega_L'\right)X_L = 0\,. \tag{14}$$

Here $a^{-1}\Omega_L = \frac{i}{2}\partial_\tau \ln A^* + \frac{3a'}{2a}A_2 + \frac{1}{2}ma(-1 + 3A_1)$. For an arbitrary background FRW metric $a(\tau)$ and background scalar field $\phi(\tau)$, equation (14) may lead to ill-defined physics. For instance, if $|A| > 1$, it describes noncasual propagation of spin 3/2 particles. In the context of preheating, the background fields $\phi(\tau)$ and $a(\tau)$ are oscillating. Naively, one could expect that $|A|$ is oscillating too. In this case one would reach a pathological conclusion that the strongest parametric excitation of spin 3/2 eigenmodes will be with the highest momenta k. Noncausality and other defects of spin 3/2 fields interacting with an external electromagnetic field are well known [7]. It is interesting to compare the equations for a spin 3/2 field in an external electromagnetic field and in external gravitational/scalar fields. Then we will show how the apparent defects of (14) are resolved in supergravity.

2   Spin 3/2 Field Equations in an External Electromagnetic Field

The equations for a charged spin 3/2 field interacting with an external electromagnetic field A_μ with field strength tensor $F_{\mu\nu}$ in flat space-time were derived in [7]. After simple manipulations, they can be re-written in the form of the equation of motion

$$\hat{\not{D}}\psi_\mu + m_0\psi_\mu = \left(\hat{D}_\mu - \frac{m_0}{2}\gamma_\mu\right)\gamma^\nu\psi_\nu\,, \tag{15}$$

and two constraint equations

$$\hat{D}^\mu\psi_\mu - \hat{\not{D}}\gamma^\mu\psi_\mu + \frac{3}{2}m_0\gamma^\mu\psi_\mu = 0\,, \tag{16}$$

$$\frac{3}{2}m_0^2\gamma^\mu\psi_\mu = -\frac{1}{2}\tilde{F}^{\mu\nu}\gamma_\mu\psi_\nu\,, \tag{17}$$

where $\tilde{F}^{\mu\nu} = e\gamma^5\epsilon^{\mu\nu\rho\sigma}F_{\rho\sigma}$. Here $\hat{D}_\mu \equiv \partial_\mu - ieA_\mu$. The system of equations (15), (16), (17) for flat case with EM background is similar to the equations (6), (7), (8) for cosmological problem up to definition of derivatives $\hat{D}_\mu$ instead of $\mathcal{D}_\mu$, a constant mass m_0 instead of the altering mass $m(\tau)$, and $\tilde{F}^{\mu\nu}$ instead of $G^{\mu\nu}$. It is interesting that a spin 3/2 field interacting with external fields can be described in the unified way. In [7] it was shown that for some configurations of the electromagnetic field, the propagation of spin 3/2 field violates causality.

3 Self-consistent Gravitino Problem in Supergravity

Contrary to the inconsistent setting of the problem of a spin $3/2$ field in an arbitrary EM or gravitational field, the spin $3/2$ gravitino field in supergravity should be consistent. In this section we will show that the gravitino equations in an expanding universe with moving scalars are consistent. We will use a model with a single chiral multiplet. Let us concentrate on the matrix $\hat{A}$ given by (10). In the models where the energy-momentum tensor is determined by the energy of a classical scalar field and Φ depends only on time we have $\rho = |\dot{\Phi}|^2 + V$, $p = |\dot{\Phi}|^2 - V$. The scalar potential is $V(\Phi) = e^K |\mathcal{D}W|^2 - 3m^2 M_P^2$. Also, we have $m' = ae^{K/2}\mathcal{D}W\dot{\Phi}/M_P^2$. Therefore, the matrix $\hat{A}$ can be rewritten in terms of $\dot{\Phi}$ and $e^{K/2}\mathcal{D}W$ only

$$\hat{A} = \frac{|\dot{\Phi}|^2 - \left|e^{K/2}\mathcal{D}W\right|^2}{|\dot{\Phi}|^2 + |e^{K/2}\mathcal{D}W|^2} + \gamma_0 \frac{2\dot{\Phi}e^{K/2}\mathcal{D}W}{|\dot{\Phi}|^2 + |e^{K/2}\mathcal{D}W|^2} . \tag{18}$$

From this form of $\hat{A}$ it follows that

$$|A|^2 \equiv A_1^2 + A_2^2 = 1 \tag{19}$$

for an arbitrary superpotential W. Thus A can be represented as $A = -\exp\left(2i \int_{-\infty}^{t} dt\, \mu(\eta)\right)$. Using the Einstein equations, one obtains $\mu = \mathcal{D}\mathcal{D}W + \Delta$, where the correction $\Delta = \mathcal{O}(M_P^{-1})$ is given in [3]. The expression for μ becomes much simpler and its interpretation is more transparent if the amplitude of oscillations of the field Φ is much smaller than M_P. In the limit $\Phi/M_P \to 0$ one has $\mu = \partial_\Phi \partial_\Phi W$. This coincides with the mass of both fields of the chiral multiplet (the scalar field and spin $1/2$ fermion) in rigid supersymmetry. When supersymmetry is spontaneously broken, the chiral fermion, goldstino, is 'eaten' by the gravitino which becomes massive and acquires helicity $\pm 1/2$ states in addition to the helicity $\pm 3/2$ states of the massless gravitino. With this form of $\hat{A}$ the gravitino equation (14) becomes

$$X_L'' + \left(k^2 + \Omega_L^2 - i\Omega_L'\right) X_L = 0 . \tag{20}$$

with $a^{-1}\Omega_L = \mu - \frac{3}{2}H \sin 2\int \mu dt - \frac{1}{2}m\left(1 + 3\cos 2 \int \mu dt\right)$. Equation (20) is consistent and describes the creation of gravitinos from preheating. The solution of the consistency problem in the equation (20) is that the symmetries imprinted to the Supergravity provide $|A| = 1$. But it does not mean that A is a constant. The matrix $\hat{A}$ does not become constant even in the limit $M_P \to \infty$. The phase of $\hat{A}$ rotates when the background scalar field oscillates. The amplitude and sign of A change two times within each oscillation. Consequently, the relation between $\gamma^0\psi_0$ and $\gamma^i\psi_i$ also oscillates during the field oscillations. This means that the gravitino with helicity $1/2$ (which is related to ψ_0) remains coupled to the changing background even in the limit $M_P \to \infty$. In a sense, the gravitino with helicity $1/2$

remembers its goldstino nature. This is the main reason why gravitino production in this background in general is not suppressed by the gravitational coupling. The main dynamical quantity which is responsible for the gravitino production in this scenario will not be the small changing gravitino mass $m(t)$, but the mass of the chiral multiplet μ, which is much larger than m.

Problem of Gravitino Over-Production from Preheating

As an example, consider the model with the superpotential $W = \sqrt{\lambda}\Phi^3/3$. The parameter μ for this model is given by $\mu = \sqrt{2\lambda}\phi$. It rapidly changes in the interval between 0 and $\sqrt{2\lambda}\phi_0$. Initially it is of the same order as H and m, but then H and m rapidly decrease as compared to μ, and therefore the oscillations of μ remain the main source of the gravitino production. In this case the production of gravitinos with helicity $1/2$ is much more efficient than that of helicity $3/2$. The theory of production of gravitinos with helicity $1/2$ in this model is similar to the theory of production of spin $1/2$ fermions with mass $\sqrt{2\lambda}\phi$ by the coherently oscillating scalar field in the theory $\lambda\phi^4/4$, which we considered above. Growth of helicity $1/2$ gravitino modes (20) occurs in the non-perturbative regime of parametric excitation. The modes get fully excited with occupation numbers $n_k \simeq 1$ within about ten oscillations of the field ϕ, and the width of the parametric excitation of fermions in momentum space is about $\sqrt{\lambda}\phi_0$. This result violates the cosmological constraints on the abundance of gravitinos with mass $\sim 10^2$ GeV by 4 orders of magnitude [3]. The most dangerous gravitino over-production (by 14 orders of magnitude) occurs in the class of inflationary models where $V(\phi)$ does not have a minimum and where preheating is gravitational (NO models) [2]. Thus the investigation of the non-thermal gravitino production in the early universe may serve as a useful tool helping us to discriminate among various versions of the cosmological theory.

REFERENCES

1. E. Cremmer, S. Ferrara, L. Girardello and A. Van Proeyen, *Nucl. Phys.* **B212** (1983).
2. G. Felder, L.A. Kofman, and A.D. Linde, hep-ph/9903350 and *Phys. Rev.* **D 59**, 123523 (1999), hep-ph/9812289.
3. R. Kallosh, L. Kofman, A.D. Linde, and A. Van Proeyen, hep-th/9907124
4. P. Greene and L. Kofman, *Phys. Lett.* **448**, 6 (1999).
5. L. Kofman, A. Linde and A. Starobinsky, *Phys. Rev. Lett.* **73**, 3195 (1994), hep-th/9405187; L. Kofman, A. Linde and A. Starobinsky, *Phys. Rev.* **D56**, 3258 (1997), hep-ph/9704452;
6. P. B. Greene, L. Kofman, A.D. Linde, and A.A. Starobinsky, *Phys. Rev.* D **56**, 6175-6192 (1997), hep-ph/9705347.
7. G. Velo and D. Zwanziger, *Phys. Rev.* **186**, 1337 (1969)

Scale Invariance and Cosmology

E. I. Guendelman

Physics Department, Ben Gurion University, Beer Sheva, Israel

Abstract. The possibility of mass in the context of scale-invariant, generally covariant theories, is discussed. Scale invariance is considered in the context of a gravitational theory where the action, in the first order formalism, is of the form $S = \int L_1 \Phi d^4 x + \int L_2 \sqrt{-g} d^4 x$ where Φ is a density built out of degrees of freedom independent of the metric. For global scale invariance, a "dilaton" ϕ has to be introduced, with non-trivial potentials $V(\phi) = f_1 e^{\alpha\phi}$ in L_1 and $U(\phi) = f_2 e^{2\alpha\phi}$ in L_2. This leads to non-trivial mass generation and a potential for ϕ which is interesting for new inflation. Scale invariant mass terms for fermions lead to a possible explanation of the present day accelerated universe and of cosmic coincidences. Although the scale symmetry is spontaneously broken there is no Goldstone boson. This surprising effect is due to the fact that in spite of the fact that there is a locally conserved current, no globally conserved dilatation charge exists due to the singular infrared behavior of the spatial components of such a current.

The concept of scale invariance appears as an attractive possibility for a fundamental symmetry of nature. In its most naive realizations, such a symmetry is not a viable symmetry, however, since nature seems to have chosen some typical scales. Even if we decide to spontaneously break this symmetry, this also leads to unsatisfactory consequences, since we end up with unobseved Goldstone bosons.

Here we will find that scale invariance can nevertheless be incorporated into realistic, generally covariant field theories and spontaneously broken in a way that Goldstone bosons do not appear. However, scale invariance has to be discussed in a more general framework than that of standard generally relativistic theories, where we must allow in the action, in addition to the ordinary measure of integration $\sqrt{-g} d^4 x$, another one, $\Phi d^4 x$, where Φ is a density built out of degrees of freedom independent of the metric.

For example, given 4-scalars φ_a (a = 1,2,3,4), one can construct the density

$$\Phi = \varepsilon^{\mu\nu\alpha\beta} \varepsilon_{abcd} \partial_\mu \varphi_a \partial_\nu \varphi_b \partial_\alpha \varphi_c \partial_\beta \varphi_d \tag{1}$$

One can allow both geometrical objects to enter the theory and consider[1]

$$S = \int L_1 \Phi d^4 x + \int L_2 \sqrt{-g} d^4 x \tag{2}$$

CP493, *General Relativity and Relativistic Astrophysics*, edited by C. P. Burgess and R. C. Myers

Here L_1 and L_2 are φ_a independent. There is a good reason not to consider mixing of Φ and $\sqrt{-g}$, like for example using $\frac{\Phi^2}{\sqrt{-g}}$. This is because (2) is invariant (up to the integral of a total divergence) under the infinite dimensional symmetry $\varphi_a \to \varphi_a + f_a(L_1)$ where $f_a(L_1)$ is an arbitrary function of L_1 if L_1 and L_2 are φ_a independent. Such symmetry (up to the integral of a total divergence) is absent if mixed terms are present.

We will study now the dynamics of a scalar field ϕ interacting with gravity as given by the action (2) with[2,3]

$$L_1 = \frac{-1}{\kappa} R(\Gamma, g) + \frac{1}{2} g^{\mu\nu} \partial_\mu \phi \partial_\nu \phi - V(\phi), L_2 = U(\phi) \tag{3}$$

$$R(\Gamma, g) = g^{\mu\nu} R_{\mu\nu}(\Gamma), R_{\mu\nu}(\Gamma) = R^\lambda_{\mu\nu\lambda}, R^\lambda_{\mu\nu\sigma}(\Gamma) = \Gamma^\lambda_{\mu\nu,\sigma} - \Gamma^\lambda_{\mu\sigma,\nu} + \Gamma^\lambda_{\alpha\sigma}\Gamma^\alpha_{\mu\nu} - \Gamma^\lambda_{\alpha\nu}\Gamma^\alpha_{\mu\sigma}. \tag{4}$$

In the variational principle $\Gamma^\lambda_{\mu\nu}, g_{\mu\nu}$, the measure fields scalars φ_a and the scalar field ϕ are all to be treated as independent variables.

If we perform the global scale transformation ($\theta = $ constant)

$$g_{\mu\nu} \to e^\theta g_{\mu\nu} \tag{5}$$

then (2), with the definitions (3), (4), is invariant provided $V(\phi)$ and $U(\phi)$ are of the form

$$V(\phi) = f_1 e^{\alpha\phi}, U(\phi) = f_2 e^{2\alpha\phi} \tag{6}$$

and φ_a is transformed according to $\varphi_a \to \lambda_a \varphi_a$ (no sum on a) which means $\Phi \to \left(\prod_a \lambda_a \right) \Phi \equiv \lambda \Phi$ such that $\lambda = e^\theta$ and $\phi \to \phi - \frac{\theta}{\alpha}$. In this case we call the scalar field ϕ needed to implement scale invariance "dilaton".

Let us consider the equations which are obtained from the variation of the φ_a fields. We obtain then $A^\mu_a \partial_\mu L_1 = 0$ where $A^\mu_a = \varepsilon^{\mu\nu\alpha\beta} \varepsilon_{abcd} \partial_\nu \varphi_b \partial_\alpha \varphi_c \partial_\beta \varphi_d$. Since $\det (A^\mu_a) = \frac{4^{-4}}{4!} \Phi^3 \neq 0$ if $\Phi \neq 0$. Therefore if $\Phi \neq 0$ we obtain that $\partial_\mu L_1 = 0$, or that $L_1 = M$, where M is constant. This constant M appears in a self-consistency condition of the equations of motion that allows us to solve for $\chi \equiv \frac{\Phi}{\sqrt{-g}}$

$$\chi = \frac{2U(\phi)}{M + V(\phi)}. \tag{7}$$

To get the physical content of the theory, it is convenient to go to the Einstein conformal frame where

$$\overline{g}_{\mu\nu} = \chi g_{\mu\nu} \tag{8}$$

and χ given by (7). In terms of $\overline{g}_{\mu\nu}$ the non Riemannian contribution (defined as $\Sigma^\lambda_{\mu\nu} = \Gamma^\lambda_{\mu\nu} - \{^\lambda_{\mu\nu}\}$ where $\{^\lambda_{\mu\nu}\}$ is the Christoffel symbol), disappears from the equations, which can be written then in the Einstein form ($R_{\mu\nu}(\bar{g}_{\alpha\beta})$ = usual Ricci tensor)

$$R_{\mu\nu}(\bar{g}_{\alpha\beta}) - \frac{1}{2}\bar{g}_{\mu\nu}R(\bar{g}_{\alpha\beta}) = \frac{\kappa}{2}T_{\mu\nu}^{eff}(\phi) \tag{9}$$

where

$$T_{\mu\nu}^{eff}(\phi) = \phi_{,\mu}\phi_{,\nu} - \frac{1}{2}\bar{g}_{\mu\nu}\phi_{,\alpha}\phi_{,\beta}\bar{g}^{\alpha\beta} + \bar{g}_{\mu\nu}V_{eff}(\phi), V_{eff}(\phi) = \frac{1}{4U(\phi)}(V+M)^2. \tag{10}$$

If $V(\phi) = f_1 e^{\alpha\phi}$ and $U(\phi) = f_2 e^{2\alpha\phi}$ as required by scale invariance, we obtain from (10)

$$V_{eff} = \frac{1}{4f_2}(f_1 + Me^{-\alpha\phi})^2 \tag{11}$$

Since we can always perform the transformation $\phi \to -\phi$ we can choose by convention $\alpha > O$. We then see that as $\phi \to \infty, V_{eff} \to \frac{f_1^2}{4f_2}$ = const. providing an infinite flat region. Also a minimum is achieved at zero cosmological constant for the case $\frac{f_1}{M} < O$ at the point $\phi_{min} = \frac{-1}{\alpha}ln \mid \frac{f_1}{M} \mid$. Finally, the second derivative of the potential V_{eff} at the minimum is $V''_{eff} = \frac{\alpha^2}{2f_2} \mid f_1 \mid^2 > O$ if $f_2 > O$,

There are many interesting issues that one can raise here. The first one is of course the fact that a realistic scalar field potential, with massive excitations when considering the true vacuum state, is achieved in a way consistent with the idea of scale invariance.

The second point to be raised is that since there is an infinite region of flat potential for $\phi \to \infty$, we expect a slow rolling new inflationary[4] scenario to be viable, provided the universe is started at a sufficiently large value of the scalar field ϕ.

Furthermore, one can consider this model as suitable for the present day universe rather than for the early universe, after we suitably reinterpret the meaning of the scalar field ϕ. This can provide a long lived almost constant vacuum energy for a long period of time, which can be small if $f_1^2/4f_2$ is small. Such small energy density will eventually disappear when the universe achieves its true vacuum state.

Notice that a small value of $\frac{f_1^2}{f_2}$ can be achieved if we let $f_2 >> f_1$. In this case $\frac{f_1^2}{f_2} << f_1$, i.e. a very small scale for the energy density of the universe is obtained by the existence of a very high scale (that of f_2) the same way as a small fermion mass is obtained in the see-saw mechanism[5] from the existence also of a large mass scale. In what follows, we will take $f_2 >> f_1$.

So far we have studied a theory which contains the metric tensor $g_{\mu\nu}$, the measure fields φ_a (a=1,2,3,4) and the "dilaton" ϕ, which makes global scale invariance possible in a non-trivial way. All of the above fields have some kind of geometrical significance, but if we are to describe the real world, the list of fields and/or particles has to be enlarged.

Taking, for example, the case of a fermion ψ, where the kinetic term of the fermion is chosen to be part of L_1

$$S_{fk} = \int L_{fk} \Phi d^4 x \tag{12}$$

$$L_{fk} = \frac{i}{2} \overline{\psi} [\gamma^a V_a^\mu (\overrightarrow{\partial}_\mu + \frac{1}{2} \omega_\mu^{cd} \sigma_{cd}) - (\overleftarrow{\partial}_\mu + \frac{1}{2} \omega_\mu^{cd} \sigma_{cd}) \gamma^a V_a^\mu] \psi \tag{13}$$

there V_a^μ is the vierbein, $\sigma_{cd} = \frac{1}{2}[\gamma_c, \gamma_d]$, the spin connection ω_μ^{cd} is determined by variation with respect to ω_μ^{cd} and, for self-consistency, the curvature scalar is taken to be (if we want to deal with ω_μ^{ab} instead of $\Gamma_{\mu\nu}^\lambda$ everywhere)

$$R = V^{a\mu} V^{b\nu} R_{\mu\nu ab}(\omega), R_{\mu\nu ab}(\omega) = \partial_\mu \omega_{\nu ab} - \partial_\nu \omega_{\mu ab} + (\omega_{\mu a}^c \omega_{\nu cb} - \omega_{\nu a}^c \omega_{\mu cb}). \tag{14}$$

Global scale invariance is obtained provided ψ also transforms, as in $\psi \to \lambda^{-\frac{1}{4}} \psi$. Mass term consistent with scale invariance exist,

$$S_{fm} = m_1 \int \overline{\psi} \psi e^{\alpha\phi/2} \Phi d^4 x + m_2 \int \overline{\psi} \psi e^{3\alpha\phi/2} \sqrt{-g} d^4 x. \tag{15}$$

If we consider the situation where $m_1 e^{\alpha\phi/2} \overline{\psi}\psi$ or $m_2 e^{3\alpha\phi/2} \overline{\psi}\psi$ are much bigger than $V(\phi) + M$, i.e. a high density approximation, we obtain that instead of (7) that the consistency condition is[3] $(3m_2 e^{3\alpha\phi/2} + m_1 e^{\alpha\phi/2} \chi) \overline{\psi}\psi = 0$, which means $\chi = -\frac{3m_2}{m_1} e^{\alpha\phi}$. Using this in (15), we obtain, after going to the conformal Einstein frame, which involves, also a transformation of the fermion fields, necessary so as to achieve Einstein-Cartan form for both the gravitational and fermion equations. These transformations are, $\overline{g}_{\mu\nu} = \chi g_{\mu\nu}$ (or $\overline{V}_\mu^a = \chi^{\frac{1}{2}} V_\mu^a$) and $\psi' = \chi^{-\frac{1}{4}} \psi$ and they lead to a mass term,

$$S_{fm} = -2m_2 (\frac{|m_1|}{3|m_2|})^{3/2} \int \sqrt{-\overline{g}} \overline{\psi}' \psi' d^4 x \tag{16}$$

The ϕ dependence of the mass term has disappeared, i.e. masses are constants.

There is one situation where the low density of matter can also give results which are similar to those obtained in the high density approximation, in that the coupling of the ϕ field disappears and that the mass term becomes of a conventional form in the Einstein conformal frame.

This is the case, when we study the theory for the limit $\phi \to \infty$. Then $U(\phi) \to \infty$ and $V(\phi) \to \infty$. In this case, taking $m_1 e^{\alpha\phi/2} \overline{\psi}\psi$ and $m_2 e^{3\alpha\phi/2} \overline{\psi}\psi$ much smaller than $V(\phi)$ or $U(\phi)$ respectively, therefore one can see that (7) is a good approximation and since also M can be ignored in the self consistency condition (7) in this limit, we get then, $\chi = \frac{2f_2}{f_1} e^{\alpha\phi}$. If this is inserted in (15), we get $S_{fm} = m \int \sqrt{-\overline{g}} \overline{\psi}' \psi' d^4 x$, where

$$m = m_1 \left(\frac{f_1}{2f_2} \right)^{\frac{1}{2}} + m_2 \left(\frac{f_1}{2f_2} \right)^{\frac{3}{2}} \qquad (17)$$

Comparing (16) and (17) and taking m_1 and m_2 of the same order of magnitude, we see that the mass of the Dirac particle is much smaller in the region $\phi \to \infty$, for which (17) is valid, than it is in the region of high density of the Dirac particle relative to $V(\phi) + M$, as displayed in eq. (16), if the "see-saw" assumption $\frac{f_1}{f_2} << 1$ is made.

Therefore if space is populated by these diluted Dirac particles of this type, the mass of these particles will grow substantially if we go to the true vacuum state valid in the absence of matter, i.e. $V + M = 0$, as dictated by V_{eff} given by eq. (11).

The presence of matter pushes therefore the minimum of energy to a state where $V + M > 0$. The real vacuum in the presence of matter should not be located in the region $\phi \to \infty$, which minimizes the matter energy, but maximizes the potential energy V_{eff} and not at $V + M = 0$, which minimizes V_{eff}, and where particle masses are big, but somewhere in a balanced intermediate stage. Clearly how much above $V + M = 0$ such true vacuum is located must be correlated to how much particle density is there in the Universe. A non zero vacuum energy, which must be of the same order of the particle energy density, has to appear and this could explain the "accelerated universe" that appears to be implied by the most recent observations, together with the "cosmic coincidence", that requires the vacuum energy be of the same order of magnitude to the matter energy[6].

It is worthwhile to point out that in the models with scale invariance dicussed here there is no Goldstone boson, when we look at the excitations arround the true vacuum with zero cosmological constant. The basic reason that Goldstone's theorem does not apply is that although there is a global symmetry, which leads, according to Noether's theorem to a locally conserved current, the spatial components of such current have an infrared singular behavior, leading to flux leaking through infinity and to a non conservation of the would be dilaton charge[7].

Let us see that this is indeed the case and for this purpose, let us ignore the fermions. Since there is a symmetry according to Noether's theorem, there is a conserved current given by (since the variation of the lagrangian density vanishes under the scale symmetry),

$$j^{\mu} = \frac{\partial L}{\partial(\partial_{\mu}\varphi_a)} \delta\varphi_a + \frac{\partial L}{\partial(\partial_{\mu}\phi)} \delta\phi \qquad (18)$$

since in the first order formalism $\frac{\partial L}{\partial(\partial_{\mu}g_{\alpha\beta})} = 0$ and $\delta\Gamma^{\lambda}_{\mu\nu} = 0$ under the scale symmetry defined before.

Let us now consider what we should take for $\delta\varphi_a$. As part of the dilatation symmetry, we have that $\varphi_a \to \lambda_a\varphi_a$ (no sum on a) and since $\left(\prod_a \lambda_a \right) \equiv \lambda = e^{\theta}$, we have, taking a transformation infinitesimally close to the identity, i.e. $\lambda_a = 1 + \epsilon_a$, with $\epsilon_a << 1$ and all ϵ_a equal, so that $\epsilon_a = \theta/4$ and since also $\delta\phi = -\frac{\theta}{\alpha}$, that the conserved dilatation current is,

$$j_\theta^\mu = -\frac{\theta}{\alpha}\Phi\partial^\mu\phi + \theta\varepsilon^{\mu\nu\alpha\beta}\varepsilon_{abcd}\varphi_a\partial_\nu\varphi_b\partial_\alpha\varphi_c\partial_\beta\varphi_d L_1 \equiv \theta j_D^\mu \tag{19}$$

To see the basic reasons why the dilatation current has an infrared singular behavior, let us consider the spatial behavior of the φ_a fields for the case of a simple spatially flat Robertson-Walker solution of the form

$$ds^2 = -dt^2 + R^2(t)(dx^2 + dy^2 + dz^2), \phi = \phi(t) \tag{20}$$

We see also from the constraint (7) that $\chi = \chi(t)$. Then, since $\chi = \chi(t) = \frac{\Phi}{R^3(t)}$, we get that,

$$\Phi = R^3(t)\chi(t) = \varepsilon^{\mu\nu\alpha\beta}\varepsilon_{abcd}\partial_\mu\varphi_a\partial_\nu\varphi_b\partial_\alpha\varphi_c\partial_\beta\varphi_d \tag{21}$$

This can be solved by taking

$$\varphi_1 = x, \varphi_2 = y, \varphi_3 = z, \varphi_4 = -\frac{1}{4!}\int \chi(t')R^3(t')dt' \tag{22}$$

For this case, with a time dependent scalar field $\phi(t)$ and with φ_a given above, the spatial components of the current j_D^μ, diverge linearly as $x^i \to \infty$ ($x^1 = x, x^2 = y, x^3 = z$). In fact $j_D^i \to Mx^i\chi(t)R^3(t)$ as $x^i \to \infty$

Such current does indeed give flux at infinity. The current grows linearly with distance, so that the total flux is proportional to the volume enclosed and obviously the total dilatation charge is not conserved here.

In the context of theories with additional measure Φ, there are other instances where Goldstone's theorem can fail. For example, take the model, without scale invariance where, in the context of the framework of equations (2), (3) and (4), we take $U(\phi) = \Lambda = $ constant and $V(\phi) = J\phi$. The model has a symmetry up to the integral of a total divergence, $\phi \to \phi + c, c = $ constant. In this case, since $V_{eff} = \frac{1}{4\Lambda}(J\phi + M)^2$, we see again that no Goldstone boson is present in the particle spectrum. Working out the conserved current associated with this symmetry, we see that it is $j_{shift}^\mu = \Phi\partial^\mu\phi + J\varepsilon^{\mu\nu\alpha\beta}\varepsilon_{abcd}\varphi_a\partial_\nu\varphi_b\partial_\alpha\varphi_c\partial_\beta\varphi_d$ which has a singular infrared behavior, exactly for the same reasons the dilatation current has (i.e. because of the singular behavior of the φ_a fields at spatial infinity).

Notice that the potential $V_{eff} = \frac{1}{4\Lambda}(J\phi + M)^2$ is contained in the class of potentials being discussed here, i.e. $V_{eff} = \frac{1}{4f_2}(f_1 + Me^{-\alpha\phi})^2$, for the limit $\alpha \to 0, \alpha M \to$ constant, $\alpha^2 M, \alpha^3 M, ... \to 0$, so that $V_{eff} = \frac{1}{4f_2}(f_1 + M - M\alpha\phi)^2$, so that if $f_1 + M$ is kept fixed in this limit, we obtain a purely quadratic potential. The flat region has, in this limit been pushed out and has gone away.

I would like to thank J. Bekenstein, A. Davidson, A.Guth, A. Kaganovich, M.Paranjape and P.Mannheim for conversations on the subjects discussed here.

REFERENCES

1. For a review (on other issues, not including the scale invariant models discussed here) and further references to published papers (in Phys. Rev. D., Mod. Phys. Lett.A, etc.) see, E.I.Guendelman and A.B.Kaganovich, gr-qc/9905029, to appear in Phys. Rev. D.

2. E.I. Guendelman, Mod. Phys. Lett. A14 (1999) 1043.

3. E.I. Guendelman, gr-qc/9901067.

4. A.D. Linde, Phys. Lett, 108B, (1982) 389; A. Albrecht and P.J. Steinhardt, Phys. Rev. Lett, 48, (1982) 1220.

5. For a review see B.Kayser, with F. Gibrat-Debu and F. Perrier, The physics of massive neutrinos, World Scientific, Singapore (1989).

6. For a review see articles by C.J.Hogan, R.P.Kirshner and N.B.Suntzeff in Sci. Amer. 280 (1999), 28 and by L.Krauss, Sci. Amer. 280, (1999) 34.

7. E.I.Guendelman, gr-qc/9906025.

The Bright Side of Dark Matter[1]

Ariel Edery[2]

Department of Physics, McGill University

3600 University Street

Montreal, PQ, Canada, H3A 2T8

Abstract. We show that it is not possible in the absence of dark matter to construct a four-dimensional metric that explains galactic observations. In particular, by working with an effective potential it is shown that a metric which is constructed to fit flat rotation curves in spiral galaxies leads to the wrong sign for the bending of light i.e. repulsion instead of attraction. Hence, without dark matter the motion of particles on galactic scales cannot be explained in terms of geodesic motion on a four-dimensional metric. This reveals a new bright side to dark matter: it is indispensable if we wish to retain the cherished equivalence principle.

It has been known for over sixty years, since the work of Zwicky [1] in the early 1930s, that there are significant discrepancies between the luminous and dynamical mass in large astronomical systems such as galaxies and clusters of galaxies. The luminous mass in galaxies is considerably less than the dynamical mass inferred from applying Newtonian gravity to the motion of gas and stars orbiting the galaxies. In clusters of galaxies the luminous mass is again considerably less than the dynamical mass inferred from applying the virial theorem to the motion of the galaxies (keeping in mind that correctly applying the virial theorem depends on whether the clusters are in equilibrium). Gravitational lensing also reveals a mass discrepancy: the dynamical mass inferred from applying General Relativity to the bending of light in clusters is considerably greater than the luminous mass. To date, the most detailed evidence for the mass discrepancy in galaxies is the extended rotation curves in spiral galaxies determined from the observed 21 cm line emission of neutral hydrogen; outside the optical disc the rotation curves remain flat instead of falling off in a Keplerian fashion. There seems to be only two possible explanations for the mass discrepancy in astronomical systems. Either there is large amounts of non-luminous matter (or dark matter) that clusters on galactic scales or the Newtonian inverse-square law used to infer the dynamical mass breaks down on galactic scales (which implies General Relativity breaks down on those scales too). At the present time the dark matter paradigm is by far the most favored option.

---

[1] received an "honorable mention" in the 1999 Gravity Research Foundation Essay Competition.

[2] E-mail: edery@hep.physics.mcgill.ca

CP493, *General Relativity and Relativistic Astrophysics*, edited by C. P. Burgess and R. C. Myers

© 1999 American Institute of Physics 1-56396-905-X/99/$15.00

Nonetheless, the direct experimental detection of either baryonic or nonbaryonic dark matter in the amounts inferred from observations is presently lacking. The nature of the dark matter is still unknown. A few authors have therefore considered alternative gravity as a possible solution to the mass discrepancy; the best known of these is Milgrom's Modified Newtonian Dynamics (MOND) [2]. In MOND, Newtonian dynamics are modified at low accelerations typical of orbits on galactic scales. It has been reasonably successful at fitting galactic rotation curves using only one extra parameter. The main drawback with MOND is that it is a non-relativistic theory and as such cannot make any predictions on cosmology, the deflection of light, etc. Attempts at constructing a relativistic theory based on MOND have not been successful [3]. Other proposed alternatives include Bekenstein and Sanders' scalar-tensor theory [4]. However, it was found that the scalar field contributed a negative deflection of light which reduced the overall deflection instead of increasing it. To date, no satisfactory alternative to General Relativity has been found that can explain the galactic observations (in our context "galactic observations" means the galactic rotation curves and the observed bending of light). After over sixty years the problem of the mass discrepancy in astronomical systems is still with us and stands as one of the great unsolved problems in astrophysics.

In this letter we shed light on the problem by showing that there is a serious price to pay for choosing alternative gravity over dark matter: the equivalence principle must be violated. We arrive at this conclusion by first addressing the following question: without introducing dark matter can one construct a metric which is a modification of the Schwarzschild metric and explain galactic phenomenology? We show that such a metric does not exist. By analyzing the motion of particles with an effective potential it is shown that a metric constructed to explain flat rotation curves in spiral galaxies leads to light repulsion! Hence, without dark matter galactic observations cannot be explained in terms of geodesic motion on a four-dimensional metric. Simply put, dropping dark matter for alternative gravity implies doing away with the equivalence principle.

We now prove our conjecture that without dark matter a metric which explains flat rotation curves leads to the wrong sign for the bending of light. We assume that the total mass M of a spiral galaxy is luminous and is found inside the optical disc (we neglect the mass of the gaseous component which usually makes a negligible contribution compared to the total luminous mass). Therefore, outside the optical disc the mass within a sphere of radius r is constant and equal to M. Under the usual inverse r potential associated with the Schwarzschild metric the velocity of the gas orbiting the outskirts of the galaxy would not remain constant but decrease as a function of r. Therefore, we expect on galactic scales a potential that actually increases with distance in order to explain the flat rotation curves.

Consider a static spherically symmetric metric $ds^2 = B(r)dt^2 - A(r)dr^2 - r^2d\Omega^2$ with $B(r) \equiv 1+2\phi(r)$ (we will specify the function $A(r)$ later). The velocity v of gas in circular orbits around a galaxy is independent of the function $A(r)$ and is given by $v^2 = r\,\phi'(r)$ [5]. Since the velocity of the gas is constant (flat) on the outskirts of spiral galaxies, the function $\phi(r)$ on galactic scales is given by $C\,ln(r/b)$ where

C and b are positive parameters adjusted for each galaxy. A logarithmic function on galactic scales is therefore best suited to explain the flat rotation curves in the absence of dark matter. The dimensionless constant C is of order v^2/c^2 and since spiral galaxies have flat rotation velocities v in the range 60 to 300 km/s [7], the value of C roughly ranges from 10^{-8} to 10^{-6}. Actually the constant C is constrained by the Tully-Fisher relation to be proportional to the square root of the mass M of the galaxy [6] and this has been shown to fit rotation curves quite well [7]).

Of course, we need to recover the results of the Schwarzschild metric on solar system scales (such as the deflection of light from the sun, the precession of the perihelia of Mercury, etc.). We therefore let $B(r) = 1 - 2GM/r + 2C\ln(r/b)$ (the constant C is small enough so that the logarithmic term is negligible on solar system scales and similarly the GM term is negligible when r is large i.e. on galactic scales). We also require to a very high degree the same relation between $A(r)$ and $B(r)$ as in the Schwarzschild metric i.e. $A(r) = B^{-1}(r)$. Deviations from this relation spoil the classical solar system tests. The important point is that a metric which reduces to the Schwarzschild metric when r is small but which does not have the relation $A(r) = B^{-1}(r)$ cannot in general reproduce the solar system results. To see this consider a metric with $B(r) = 1 - 2GM/r + 2f(r)$ and $A(r) = 1/(1 - 2GM/r + g(r))$ where $f(r)$ and $g(r)$ are functions that are small compared to GM/r on solar system scales. Moreover, to reproduce galactic rotation curves the function $f(r)$ will be an increasing function of r and will approach infinity asymptotically (this is certainly the case for the logarithmic function we obtained but we wish here to be more general and not specify $f(r)$ exactly). The deflection of light with these two functions can be calculated using the standard integral-angle formula [5] and yields:

$$\Delta\varphi = 2\int_0^{\pi/2} \sqrt{A(r)\,B(r)}$$
$$\left(1 + \frac{GM}{r_0}\frac{(1 - \sin^3\theta)}{\cos^2\theta} - f(r_0) + (f(r) - f(r_0))\tan^2\theta\right)d\theta - \pi \qquad (1)$$

where r_0 is the point of closest approach, $r = r_0/\sin\theta$ and terms inside the brackets must be much smaller than the constant one to obtain any realistic deflection (of order arc seconds). The integration from 0 to $\pi/2$ represents light moving from r_0 to infinity. Note that the GM term and the $f(r)$ terms are separated inside the brackets i.e. there are no cross terms. If the product $A(r)\,B(r)$ is equal to one then the straight line angle π is cancelled out exactly and one obtains the deflection $4GM/r_0$ as in the Schwarzschild case plus a modification proportional to $f(r_0)$ (which by definition plays a significant role only when r_0 is on galactic scales and hence does not spoil the "Schwarzschild" result). However, if $A(r)\,B(r) \neq 1$ this changes the situation in a significant way. First, if $g(r)$ does not approach infinity asymptotically the deflection is infinite. Secondly, even if $g(r)$ approaches infinity asymptotically and the deflection is finite it will not reproduce the solar system results if $A(r)\,B(r)$ is not equal to one. The straight-line angle π will in general no longer cancel out as before and a constant deflection term will appear (which has not been observed on solar scales). Moreover, the result $4GM/r_0$ will no longer be reproduceable and cross terms involving combinations of GM/r_0, $f(r_0)$ and $g(r_0)$ will now appear. Since the solar deflection result $4GM/r_0$ has been confirmed to an accuracy of 1% using radio-interferometric methods [8] this leaves very little room for any deviation from a $A(r) = B^{-1}(r)$ relation.

We now show that the deflection of light on galactic scales is negative. One way to proceed is to calculate the integral for the deflection angle and show that the extra term is negative (this has already been done for the case where $\phi(r)$ is a linear function of r [9,10]). However, such a calculation is not necessary nor illuminating. It proves much more instructive to work with an effective potential. As in the analysis of the Schwarzschild metric one can write the geodesic equations of motion as "Newtonian" equations of motion with an effective potential $V(r)$ [9,11]. The derivative of the potential can be interpreted as the "radial force" acting on the particles and therefore its sign reveals whether a particle is attracted or repelled by the source. Armed with a potential we will have no need to calculate the full deflection angle and this will enable us to carry out a clear and general analysis. The effective potential for a metric with $B(r) = 1 + 2\phi(r)$ and $A(r) = B^{-1}(r)$ is given by [9]

$$V(r) = \phi(r) \left(\frac{J^2}{r^2} + E \right) \tag{2}$$

where J and E are constants of the motion. The above potential is a velocity-dependent potential. For particles moving non-relativistically (NR for short), such as gas in circular orbits around a spiral galaxy, one has the condition $E \approx 1$ and $J^2/r^2 << 1$ yielding the potential $V_{NR}(r) \approx \phi(r)$. For light, E is zero and the potential is $V_{light}(r) = \phi(r)(J^2/r^2)$. We now substitute the logarithmic function $\phi(r) = C \ln(r/b)$ into these two potentials. For the case of light, we can relate the constant b to the point of closest approach r_0 by using the fact that the derivative of $V_{light}(r)$ is a maximum at r_0. This leads to $b = r_0 e^{-5/6}$. The derivative of $V_{light}(r)$ can now be readily calculated and is found to be negative corresponding to repulsion. Note that $V_{light}(r)$ decreases with r. In contrast, the NR potential, which is equal to the logarithmic function, increases with r and its derivative is positive corresponding to attraction. We therefore see that the logarithmic function required to explain flat rotation curves in spiral galaxies leads to light repulsion. It is worthwhile to note that the deflection of light is a well defined finite quantity for the given logarithmic function $\phi(r)$ precisely because the potential for light decreases with r. If the potential for light had been an increasing function of r the deflection of light would yield a nonsensical divergent result i.e. scattering states would simply not exist. Hence, we see immediately that for functions $\phi(r)$ that increase faster than r^2 one cannot even talk of a deflection. Note that for a function $\phi(r) = -GM/r$ the derivative of both potentials is positive as required on solar system scales.

211

It is important to note that a negative deflection is not particular to a logarithmic function and this is where using an effective potential enables us to make a more general and powerful conclusion. We see immediately that the derivative of $V_{light}(r)$ is negative for any function $\phi(r)$ that increases slower than r^2 (which is well within the requirement for fitting galactic rotation curves). Hence, if we had allowed deviations from a strict logarithmic function in the fitting of galactic rotation curves the result would not change: the deflection of light would still be negative. One thing is clear: an increasing function $\phi(r)$ is required to explain the flat rotation curves and this inevitably leads to light repulsion. There is no way around this general and striking result. We have shown that, in the absence of dark matter, it is not possible to construct a four-dimensional metric that explains galactic phenomenology. Therefore, any attempt to explain the mass discrepancy in astronomical systems using alternative gravity instead of dark matter comes at the price of having to abandon the equivalence principle. Dark matter can now be seen in a new light: it is indispensable if we wish to hold on to this pillar of gravitational theory.

ACKNOWLEDGEMENTS

I wish to thank Robert Myers for his useful comments and McGill University for their financial support.

REFERENCES

1. F. Zwicky, Helv. Phys. Acta, **6**, 110 (1933).
2. M. Milgrom, Astrophys. J. **270**, 365,371,384 (1983).
3. J. D. Bekenstein, Phys. Lett. B **202**, 497 (1988).
4. J. D. Bekenstein and R. H. Sanders, Astrophys. J. **429**, 480 (1994).
5. S. Weinberg, *Gravitation and Cosmology* (Wiley, New York, 1972).
6. J. D. Bekenstein and M. Milgrom, Astrophys. J. **286**, 7 (1984).
7. R.H. Sanders, Astrophys. J. **473**, 117 (1996).
8. E. B. Fomalont and R. A. Sramek, Phys. Rev. Lett. **36**, 1475 (1976).
9. A. Edery and M. B. Paranjape, Phys. Rev. D **58** 024011(1998).
10. M. A. Walker, Astrophys. J. **430**, 463 (1994).
11. R. Wald, *General Relativity* (University of Chicago Press, Chicago, 1984).

Radial Variations of G on the Galactic Scale

C. Gauthier

Université de Moncton, Moncton, N.-B., E1A 3E9 Canada

Abstract. We consider the idea of replacing the gravitational constant G by a function depending on the surrounding medium and whose variation would be noticeable on the galactic scale. Such a function in newtonian dynamics allows one to explain the flat region of the rotation velocity curves of many galaxies, without the dark matter hypothesis. This study is done within a unified field theory of the Kaluza-Klein type where we give an interpretation of gravitation which includes general relativity.

L'HYPERESPACE-TEMPS

Admettons que les propriétés de notre univers soient déterminées par les caractéristiques géométriques d'un hyperespace-temps de dimension $4+d$, que l'on note V^{4+d}. La structure géométrique de V^{4+d} est supposée de la forme $M^4 \times (S^d(K)/\Gamma)$, où M^4 représente l'espace-temps. $S^d(K)$ est la sphère de dimension impaire d et de faible courbure K, alors que Γ est l'un de ses groupes discrets et cycliques d'invariance ayant un ordre très élevé. La variété V^{4+d} devient ainsi un fibré de groupe Γ et de base M^4. La multiconnexité de l'espace supérieur $S^d(K)/\Gamma$ fait qu'il existe une jauge discontinue sur M^4, que l'on nomme jauge supérieure [1].

La rigidité géométrique de $S^d(K)$ s'applique à chacun des polytopes de la d-sphère multiconnexe. Cependant, cette multiconnexité rend possible le passage direct entre deux polytopes non contigus au-dessus d'une trajectoire dans M^4. Cette propriété fait que V^{4+d} a une flexibilité φ qui s'harmonise à sa structure géométrique. Nous pouvons ainsi admettre que $\varphi = |\varphi| \exp(i\theta)$ fait montre d'une densité énergétique déterminée par un potentiel du type:

$$W(\varphi, R) := (2 + \cos N\theta)(\varphi^4 - 2B^2 R^{-2}\varphi^2), \qquad (1)$$

où l'angle $\theta \in [0, 2\pi[$ est mesuré par rapport à une valeur initiale donnée, R désigne la courbure scalaire de V^{4+d} et B est une constante correspondant à sa flexibilité minimale. Si $R \neq 0$, ce potentiel a N minimums absolus en:

$$\varphi_k := BR^{-1} \exp(2k\pi i/N), \quad k = 0, 1, \ldots, N-1.$$

CP493, *General Relativity and Relativistic Astrophysics*, edited by C. P. Burgess and R. C. Myers
© 1999 American Institute of Physics 1-56396-905-X/99/$15.00

LES PARTICULES FONDAMENTALES

Le degré de liberté relatif à la localisation de l'angle $\theta = 0$ en (1) entraîne l'existence d'une jauge locale et continue dans M^4 dont le groupe d'invariance est $U(1)$: nous l'appelons jauge du zéro. La variation angulaire entre l'angle zéro dans le potentiel W associé à la structure de base autocohérente d'un observateur et celui d'une structure de base qui n'a pas été observée détermine une onde scalaire appelée onde toriidale. Ces ondes ont des propriétés similaires à celles des ondes de de Broglie. Leur vitesse de propagation dans M^4 égale celle de la lumière et chaque onde a une fréquence qui est réglée par la vitesse de giration de la structure de base qui lui est associée, par rapport à la structure de base de l'observateur. Quand une onde toriidale participe à une interaction, la brisure spontanée des symétries des jauges du zéro et supérieure la fait se manifester comme un point dans M^4 et avec une énergie-impulsion déterminée. Nous appelons toriion ce qui se propage dans M^4 sous la forme d'une onde toriidale et s'y manifeste sous la forme d'un point.

LA MASSE

La quantité de mouvement associée aux ondes toriidales est à la base de notre interprétation de la gravitation. Pour l'expliquer, considérons un corps que nous définissons comme une agrégation de toriions dans une région limitée de M^4. Nous supposons que les ondes toriidales remplissent M^4 isotropiquement et que les corps leur sont presque perméables. Lors d'une interaction entre une onde toriidale et un corps, ce dernier subit l'impulsion fournie par l'onde. Ceci nous amène à identifier la masse inertielle d'un corps à la section efficace totale des toriions qui le composent vis-à-vis des ondes toriidales.

Dans tout référentiel inertiel en mouvement vers un corps, on observe la même section efficace pour ce corps, car toute section transversale au mouvement n'est pas affectée par ce mouvement. De plus, ce corps aura toujours la même vitesse par rapport aux ondes toriidales, car celles-ci se propagent à la vitesse de la lumière. Cette section efficace est donc un invariant relativiste pour les ondes venant par l'avant ou par l'arrière, et ces ondes sont les seules que nous considérons.

La section efficace représentant la masse d'un corps peut être vue comme une mesure de la capacité qu'a ce corps à rendre effective l'énergie associée aux ondes toriidales qui rencontrent ce corps. La section efficace d'un corps vis-à-vis des ondes toriidales devient alors une sorte de mesure de l'énergie que peut produire ce corps. Même si elles n'utilisent pas les mêmes unités, la masse et la section efficace d'un corps vis-à-vis des ondes toriidales mesurent donc la même propriété de ce corps.

Appliquant aux corps le principe de conservation de l'énergie-impulsion, on a qu'au moment où une onde toriidale est absorbée par un corps, celui-ci émet une onde toriidale capable de produire la même énergie-impulsion que l'onde absorbée. Il y a donc un équilibre entre le nombre d'ondes absorbées et le nombre d'ondes émises par chaque corps.

LA GRAVITATION

Soient deux corps dans une région minkowskienne de M^4. Chacun d'eux joue pour l'autre le rôle d'un écran à une partie des ondes toriidales formant le milieu environnant. Les deux corps reçoivent ainsi plus d'ondes venant de l'extérieur que de la région située entre eux. Plus les corps seront près l'un de l'autre et plus ce déséquilibre sera grand. Il en résulte une poussée tendant à rapprocher les corps l'un de l'autre, qui se traduit par une force d'attraction entre eux. On peut montrer que cette force est décrite par la loi newtonienne de l'attraction universelle si:

$$G = \frac{4h}{\pi c} \int_0^\infty \nu n(\nu) d\nu \,, \tag{2}$$

où h est la constante de Planck, c est la vitesse de lumière et $n(\nu)$ est le flux des ondes toriidales de fréquence ν dans une direction spatiale orientée de M^4 [2].

Cette correspondance entre la loi de Newton et la force engendrée par l'interaction d'ondes toriidales avec les corps s'étend de façon à retrouver le formalisme de la relativité générale. En effet, lorsqu'un corps se déplace dans M^4, l'onde toriidale de chacun des toriions qui le composent demeure associée au minimum radial de W. Le fait que la courbure scalaire de l'espace supérieur soit constante entraîne que les mouvements de ce corps s'effectuent selon une trajectoire qui est dans un M^4 dont la courbure scalaire correspond en chaque point de M^4 au minimum radial de W. Ceci justifie la pertinence du principe variationnel de Hilbert et, par là, les équations d'Einstein pour M^4.

L'interaction des ondes toriidales avec les corps permet enfin d'expliquer l'inertie. La manifestation de tout toriion sous la forme d'un point dans M^4 fait effectivement en sorte que l'impulsion associée à cette manifestation se produit toujours dans le direction de M^4 déterminée par la tangente à sa trajectoire, selon les minimuns radiaux de W, au point d'interaction. Ce phénomène coïncide avec l'inertie.

LA FONCTION $G(r)$

L'équilibre entre le nombre des ondes toriidales qu'absorbe un corps et le nombre des ondes que celui-ci émet dans le milieu environnant s'applique à tout corps ou ensemble de corps, comme les galaxies ou les amas et superamas galactiques. L'échange continuel d'ondes toriidales entre les corps à l'intérieur d'un tel ensemble fait que le milieu y a une densité d'ondes toriidales qui est supérieure à celle du milieu loin de tout ensemble de corps. Le flux local des ondes toriidales sera sensiblement différent de sa valeur loin de tout ensemble de corps seulement à l'intérieur et près d'un ensemble de corps ayant une masse totale suffisamment grande pour y modifier de façon appréciable la densité des ondes toriidales du milieu. Nous admettrons que cette masse minimale est de l'ordre de celles des galaxies.

L'expression (2) montre que la valeur de G dépend du flux local des ondes toriidales. La force d'attraction causée par les ondes toriidales sera donc plus forte à l'intérieur qu'à l'extérieur des ensembles de corps. Il en va de même pour l'inertie. Cependant, l'inertie liée au mouvement orbital autour d'un tel ensemble sera constante à l'extérieur de l'ensemble, car le flux des ondes toriidales arrivant tangentiellement à l'orbite n'est pas affecté par la présence de l'ensemble.

À l'extérieur d'un ensemble de corps E, le flux local des ondes toriidales varie avec la distance r au barycentre de E. Soit $G_E(r)$ la fonction décrivant les valeurs de la 'constante' gravitationnelle correspondant au flux toriidal associé à E. Si G_∞ désigne la valeur de $G_E(r)$ très loin de tout ensemble de corps, alors $G_E(r) < G_\infty$ pour les valeurs de r se rapportant aux points près de E, à cause des ondes toriidales émises par E.

Si E a la forme d'un disque mince D_a de rayon a, on obtient que

$$G_{D_a}(r) = \int_0^\infty \frac{2hm_{D_a}(\nu, r)}{\pi^2 c\left(1 - \frac{r}{\sqrt{a^2+r^2}}\right)} \left\{ \nu \frac{\exp\left[\frac{h\nu r}{c}\left(1 - \frac{r}{\sqrt{a^2+r^2}}\right)\right] - \frac{r}{\sqrt{a^2+r^2}}}{\exp\left[\frac{h\nu r}{c}\left(1 - \frac{r}{\sqrt{a^2+r^2}}\right)\right] - 1} - \frac{c}{hr} \right\} d\nu \,,$$

où $m_{D_a}(\nu, r)$ est le flux des ondes toriidales de fréquence ν passant par un élément de volume situé à une distance r, mesurée à partir du centre du disque, et selon une direction passant par D_a [2]. Si E a la forme d'une sphère S_b de rayon b, on a:

$$G_{S_b}(r) = \frac{2}{\pi b} \int_0^\infty \int_{r-b}^{r+b} m_{D_{\sqrt{b^2-(r-x)^2}}}(\nu, |x|) F_{b,\nu}(x) dx d\nu \,.$$

où

$$F_{b,\nu}(x) := \begin{cases} -\dfrac{h\nu}{c} \dfrac{\exp\left[\frac{xh\nu}{c}\left(1 + \frac{x}{\sqrt{b^2-r^2+2rx}}\right)\right] + \frac{x}{\sqrt{b^2-r^2+2rx}}}{\exp\left[xp(\nu)\left(1 + \frac{x}{\sqrt{b^2-r^2+2rx}}\right)\right] - 1} + \dfrac{1}{x} \,, & \text{si} \quad x < 0 \\[3em] \dfrac{h\nu}{c} \dfrac{\exp\left[\frac{xh\nu}{c}\left(1 - \frac{x}{\sqrt{b^2-r^2+2rx}}\right)\right] - \frac{x}{\sqrt{b^2-r^2+2rx}}}{\exp\left[\frac{xh\nu}{c}\left(1 - \frac{x}{\sqrt{b^2-r^2+2rx}}\right)\right] - 1} - \dfrac{1}{x} \,, & \text{si} \quad x \geq 0 \,. \end{cases}$$

En modélisant une galaxie lenticulaire par une sphère de rayon $a = 1$ centrée au centre d'un disque de rayon $b = 10$, on trouve:

$$G_{S_1 D_{10}}(r) = \frac{1}{2}[G_{S_1}(r) + G_{D_{10}}(r)] \,.$$

LE PROBLÈME DE LA MASSE MANQUANTE

Selon la mécanique newtonienne, la masse $M(r)$ de la partie d'une galaxie située à l'intérieur d'un rayon r est donnée par

$$M(r) = r[v(r)]^2/G \,, \tag{3}$$

où $v(r)$ est la vitesse tangentielle d'un corps sur une orbite circulaire de rayon r autour du centre de la galaxie. L'équation (3) montre que $v(r)$ devrait être

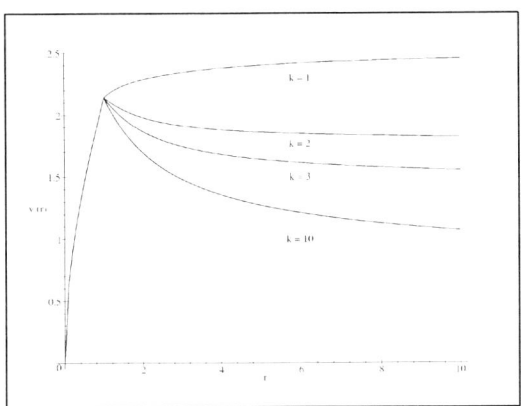

FIGURE 1. La vitesse de rotation $v(r)$ pour $S_1 D_{10}$, en unités arbitraires, en termes de la distance r, où l'unité est le rayon de S_1.

proportionnelle à $r^{-\frac{1}{2}}$. Cependant, des observations montrent que plusieurs galaxies ont des vitesses de rotation qui demeurent constantes au-delà d'un certain rayon.

Si l'on remplace G par $G_{S_1 D_{10}}(r)$ dans (3), il s'ensuit que

$$
v(r) = \begin{cases}
\left[\frac{4\pi G_{S_1 D_{10}}(r)}{r} \int_0^r \rho^2 \delta_{S_1}(\rho)d\rho \right]^{\frac{1}{2}}, & \text{si} \quad 0 \leq r \leq 1 \\[2ex]
\left[\frac{2\pi G_{S_1 D_{10}}(r)}{r} \left(2 \int_0^1 \rho^2 \delta_{S_1}(\rho)d\rho + \int_1^r \rho \delta_{D_{10}}(\rho)d\rho \right) \right]^{\frac{1}{2}}, & \text{si} \quad 1 < r \leq 10.
\end{cases}
$$

où $\delta_{S_1}(r)$ et $\delta_{D_{10}}(r)$ sont les densités respectives de S_1 et D_{10}.

Les courbes de la figure 1 donnent la fonction $v(r)$ avec $\delta_{S_1}(r) = 1/r$ et $\delta_{D_{10}}(r) = 1/kr$ pour $k = 1, 2, 3, 10$. On peut noter que ces courbes représentent des vitesses de rotation relativement constantes au-delà d'un certain rayon.

CONCLUSION

L'invariance de G par rapport au temps n'a été vérifiée que pour des objets en orbite circulaire autour du centre d'une galaxie [3]. Ces observations n'interdisent donc pas une possible variation de la valeur de G par rapport à la distance au centre d'une galaxie.

Malgré la simplicité de notre modèle, les résultats précédents montrent que la substitution de la constante G par une fonction $G(r)$ permet d'expliquer le plat dans les courbes des vitesses de rotation de plusieurs galaxies. À noter que le remplacement de G par $G(r)$ pour les grands ensembles de corps fait que ceux-ci ne peuvent avoir de satellite se déplaçant sur une orbite de forte excentricité.

REFERENCES

1. Gauthier C., *Nuovo Cimento* **110A**, 1997, 149.
2. Gauthier C., Inertia and gravitation from extra space multiconnectivity: a new perspective on the missing mass problem, *Gravitation and Cosmology*, à paraître.
3. Thorsett S.E., *Phys. Rev. Lett.* **77**, 1996, 1432.

A Critical Examination of Sciama's Heavy Neutrino Hypothesis

S. S. Seahra, J. M. Overduin
W. W. Duley and P. S. Wesson

Department of Physics, University of Waterloo,
Waterloo, ON, Canada, N2L 3G1

Abstract. Following recent results from the Super-Kamiokande experiment, we re-examine the decaying neutrino hypothesis of Sciama, including for the first time the effects of absorption by intergalactic dust. We find that the theory can (likely) be ruled out on observational grounds.

I   INTRODUCTION

Cosmic ray showers observed at the Super-Kamiokande observatory [1] have lent strong new support to the hypothesis of nonzero neutrino masses. This is of great interest to propenents of Sciama's theory, where heavy neutrinos (of mass ~ 30 eV) decay and contribute to the extragalactic background light (EBL) at ultraviolet wavelengths [2]. In a previous study (Overduin & Wesson [3]; hereafter "OW") we showed that the strength of this neutrino decay signal was at or near upper limits on EBL intensity. But now there are tighter new *observational limits* on the intensity of the EBL, based on a re-analysis of data from the Voyager spacecraft [4]. Taken together, we believe that these developments warrant a new look at the question of decaying neutrinos and the EBL. As a new feature to our model, we consider for the first time the effects of intergalactic dust on the neutrino decay signal.

II   CONTRIBUTION OF NEUTRINO DECAY TO THE EBL

The spectral intensity of the neutrino decay signal at observed wavelength λ_0 (including the effects of extinction) is given by eqs. (12) and (17) of OW:

$$I_\lambda(\lambda_0) = I_0 \int_0^{z_{\max}} \exp\left[-\frac{1}{2}\left(\frac{\lambda_0/(1+z) - \lambda_\gamma}{\sigma}\right)^2 - \tau(\lambda_0, z)\right](1+z)^{-9/2}dz, \quad (1)$$

CP493, *General Relativity and Relativistic Astrophysics*, edited by C. P. Burgess and R. C. Myers

where z refers to redshift and we have assumed an Einstein-de Sitter cosmology, as befits the fact that neutrinos in Sciama's theory provide enough dark matter to close the Universe.

The parameters in eq. (1) are given as follows: z_{max} is the redshift beyond which neutrino contributions to the observed EBL become negligible, which we have numerically determined to be ~ 2. In Sciama's theory [2], each decay produces a 14.4 ± 0.5 eV photon, implying a peak decay wavelength $\lambda_\gamma = 861$ Å. To this we associate a 3σ uncertainty of 30 Å (OW). The constant I_0 is given by $I_0 = I_{bound} + I_{free}$, where I_{free} is the neutrino intensity due to free-streaming neutrinos and I_{bound} is the intensity due to neutrinos bound in galactic haloes. A detailed invesitgation (OW) leads to the expressions $I_{bound} = (1.0 \times 10^{-8}$ erg cm^{-2} s^{-1} sr^{-1} Å$^{-1})h_0^2 f_h f_\epsilon f_\tau^{-1}$ and $I_{free} = (3.7 \times 10^{-7}$ erg cm^{-2} s^{-1} sr^{-1} Å$^{-1})h_0^{-1} f_f f_\tau^{-1}$ respectively, where the f_i are factors reflecting various uncertainties in the model. In particular, $f_h = 1.00 \pm 0.25$, $f_\epsilon = 1.00 \pm 0.04$, $f_\tau = 1.0 \pm 0.5$ and $f_f = 1.00 \pm 0.08$ refer to the mass of galactic dark matter halos, the extent to which decay photons given off by bound neutrinos are absorbed by neutral hydrogen in their host galaxies, the neutrino decay lifetime, and the neutrino rest energy, respectively. We take the Hubble parameter h_0 to be 0.57 ± 0.02 [2].

The total extinction $\tau(\lambda_0, z)$ also arises from two distinct sources: hydrogen gas (τ_{gas}) and dust (τ_{dust}) along the line of sight. The gas contribution is fairly well-understood [3,5]. We adopt a recent model for intergalactic extinction in which dust is clumped into damped Lyα absorption systems whose numbers and density profiles are sufficient to obscure a portion of the light reaching us from $z \sim 3$, but not enough to excessively redden quasar spectra (Fall & Pei [6]; hereafter "FP"). The *mean* opacity at observed wavelength λ_0 due to these dusty clumps (out to redshift z) is:

$$\bar{\tau}_{dust}(\lambda_0, z) = \int_0^z \frac{\tau_*(z')(1 + z')}{(1 + \Omega_0 z')^{1/2}} \xi\left(\frac{\lambda_0}{1 + z'}\right) dz'. \tag{2}$$

Here $\tau_*(z)$ is the comoving dust density in units of optical depth per Hubble radius and $\xi(\lambda)$ is the ratio of extinction at wavelength λ to that in the B-band (4400 Å). We follow the approach of FP, in which dust opacity depends on redshift:

$$\tau_*(z) = \tau_*(0)(1 + z)^\delta, \tag{3}$$

where $\tau_*(0)$ and δ are fixed by observational data on extinction due to hydrogen at $z = 0$ and (for damped Lyα systems) at $z = 2.4$, together with estimates of dust-to-gas ratios (see FP for discussion). The data are consistent with *lower limits* of $\tau_*(0) = 0.005$, $\delta = 0.275$ (model A), *best-fit values* of $\tau_*(0) = 0.016$, $\delta = 1.240$ (model B), or *upper limits* of $\tau_*(0) = 0.050$, $\delta = 2.063$ (model C), assuming a critical density $\Omega_0 = 1$. To calculate the extinction $\xi(\lambda)$ in the 300 – 2000 Å range, we use a numerical Mie scattering routine in conjunction with various interstellar dust models. We find that the a model consisting of a mixture of spherical silicate particles and polycyclic aromatic hydrocarbon (PAH) nanostructures produce

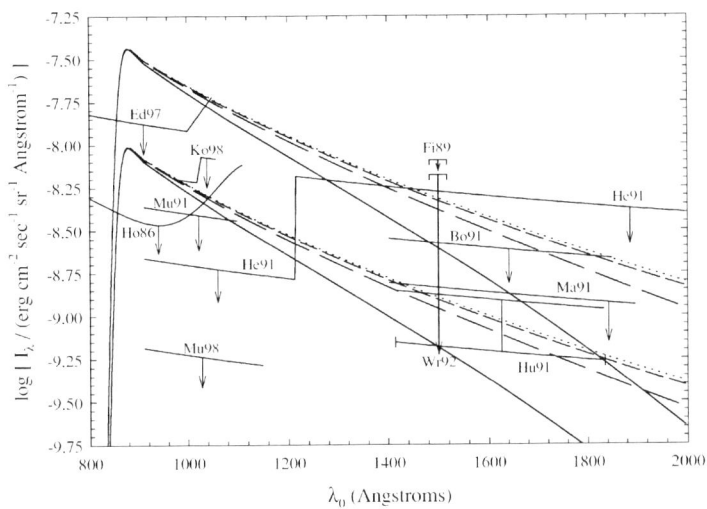

FIGURE 1. Comparison of predicted strength of a neutrino decay signal with observational upper limits on the intensity of the EBL (see text for the meaning of the various lines)

the most extinction in the waveband of interest [7], and therefore minimizes the predicted decay signal.

Inserting $\xi(\lambda) = \tau(\lambda)/\tau_B$ into eq. (2), and substituting this into the intensity integral (1) along with the gas extinction, we obtain the results shown in Fig. 1. In this plot the dashed, long-dashed and heavy solid lines are predicted intensities of the EBL due to decaying neutrinos (after absorption by gas and dust) for dust opacity models A, B and C respectively. There are two groups of lines; these have been obtained by letting the model parameters and theoretical uncertainties in section II take their maximum and minimum possible values. The faint dotted line illustrates the effects of ignoring dust (absorption by hydrogen gas only). The remaining curves are observational upper limits on EBL intensity in this waveband reported by Bowyer [8] (Bo91), Edelstein et al. [9] (Ed97), Fix et al. [10] (Fi89), Henry [11] (He91), Holberg [12] (Ho86), Hurwitz et al. [13] (Hu91), Korpela & Bowyer [14] (Ko98), Martin et al. [15] (Ma91), Murthy et al. [16] (Mu91), Murthy et al. [4] (Mu98) and Wright [17] (Wr92).

As expected, the inclusion of intergalactic dust into the model results in significant reductions in EBL intensity. At 1600 Å, we find that the strength of the neutrino decay signal is cut by 4%, 17% and 56% for dust models A,B and C respectively compared to the gas-only case. At these wavelengths, we therefore approach the factor-two reduction in intensity attributed to dust extinction by Sciama – *if* one adopts the upper limits on dust density consistent with quasar obscuration

(FP, model C). Shortward of 1216 Å, the neutrino decay signal is cut (at 1000 Å) by only 1%, 3% and 10% respectively. This is not enough to alter the findings in OW. As before, *our ability to rule out Sciama's hypothesis stands or falls on the validity of the Voyager limits.* The new limit derived by Murthy et al. ("Mu98"), in particular, is crucial – if valid, it is more than an order of magnitude below the minimum intensity consistent with the theory.

III CONCLUSIONS

Motivated by recent results from the Super-Kamiokande experiment, which indicate that neutrinos may have finite masses, we have re-examined the decaying neutrino hypothesis of Sciama, including for the first time the effects of absorption by intergalactic dust. The strongest constraints come from a new analysis of Voyager data by Murthy et al. [4], which appear to rule out the theory, but ideally should be confirmed by further experiments. At best, Sciama's theory suggests that the density of dust grains is at the upper end of the range suggested by reddening and obscuration of quasars. At worst, the theory with more conservative density estimates is incompatible with observation.

REFERENCES

1. Fukuda Y. et al., *Physical Review Letters* **81**, 1562 (1998).
2. Sciama D. W., *Monthly Notices of the Royal Astronomical Society* **289**, 945 (1997).
3. Overduin J. M., Wesson P. S., *The Astrophysical Journal* **483**, 77 (1997).
4. Murthy J., Hall D., Earl M., Henry R. C., in press in *The Astrophysical Journal* (1998).
5. Zuo L., Phinney E. S., *The Astrophysical Journal* **418**, 28 (1993).
6. Fall S. M., Pei Y. C., *The Astrophysical Journal* **402**, 479 (1993).
7. Overduin, J. M., Seahra, S. S., Duley, W. W., Wesson, P. S., in press in *Astronomy & Astrophysics* (1999).
8. Bowyer S., 1991, *Annual Review of Astronomy & Astrophyics* **29**, 59
9. Edelstein J., Bowyer S., Lampton M., *The Astrophysical Journal* **485**, 523 (1997).
10. Fix J. D., Craven J. D., Frank L. A., *The Astrophysical Journal* **345**, 203 (1989).
11. Henry R. C., *Annual Review of Astronomy & Astrophyics* **29**, 89 (1991).
12. Holberg J. B., *The Astrophysical Journal* **311**, 969 (1986).
13. Hurwitz M., Bowyer S., Martin C., *The Astrophysical Journal* **372**, 167 (1991).
14. Korpela E. J., Bowyer S., Edelstein J., *The Astrophysical Journal* **495**, 317 (1998).
15. Martin C., Hurwitz M., Bowyer S., *The Astrophysical Journal* **379**, 549 (1991).
16. Murthy J., Henry R. C., Holberg J. B., *The Astrophysical Journal* **383**, 198 (1991).
17. Wright E. L., *The Astrophysical Journal* **391**, 34 (1992).

IV. QUANTUM GRAVITY

Can Quantum Cosmology Give Observational Consequences of Many-Worlds Quantum Theory?

Don N. Page

CIAR Cosmology Program, Institute for Theoretical Physics
Department of Physics, University of Alberta
Edmonton, Alberta, Canada T6G 2J1

Abstract.

Although many people have thought that the difference between the Copenhagen and many-worlds versions of quantum theory was merely metaphysical, quantum cosmology may allow us to make a physical test to distinguish between them empirically. The difference between the two versions shows up when the various components of the wavefunction have different numbers of observers and observations. In the Copenhagen version, a random observation is selected from the sample within the component that is selected by wavefunction collapse, but in the many-worlds version, a random observation is selected from those in all components. Because of the difference in the samples, probable observations in one version can be very improbable in the other version.

INTRODUCTION

Ever since Hugh Everett III formulated his many-worlds alternative [1,2] to the Copenhagen version of quantum theory, there has been considerable discussion of its merits. Many people, including some of the original supporters of the many-worlds version, have expressed the opinion that the many-worlds version is empirically indistinguishable from the Copenhagen version, so that the difference is merely metaphysical.

For example, in the first wide popularization of the many-worlds or Everett-Wheeler-Graham (EWG) version of quantum theory, Everett's bulldog Bryce De-Witt stated [3], "Clearly the EWG view of quantum mechanics leads to experimental predictions identical with those of the Copenhagen view."

Everett's Ph.D. supervisor John Wheeler, who initially supported the many-worlds version [4], has recently summarized it as follows [5]: "Does it offer any new insights? Does it predict outcomes of experiments that differ from outcomes predicted in conventional quantum theory? The answer to the first question is emphatically yes. The answer to the second question is emphatically no."

CP493, *General Relativity and Relativistic Astrophysics,* edited by C. P. Burgess and R. C. Myers
© 1999 American Institute of Physics 1-56396-905-X/99/$15.00

Roland Omnès, though never a supporter of the many-worlds version to my knowledge, has thought about it deeply and concluded [6], "If quantum mechanics were absolutely true and Everett were right, no experiment would be able to confirm or reject it. ...It is not science because no experiment can show it to be wrong."

However, David Deutsch has argued [7] that the many-worlds version of quantum theory would be confirmed if an observer could "split" into two copies which make different observations, remember that they observed but not what they observed, and then are rejoined coherently. Although there are conceptual loopholes (such as claiming after the experiment that the observer's memory of having made a definite observation is merely a false memory), I believe this argument is fairly strong evidence that the difference between the Copenhagen and many-worlds versions of quantum theory is, in principle at least, a matter that could be experimentally tested. Nevertheless, this proposed test appears to be technically very difficult.

Because of the difficulties of Deutsch's proposed experiment, here I wish to raise the possibility that quantum cosmology might in principle lead to empirical distinctions between the Copenhagen and the many-worlds versions of quantum theory.

By the Copenhagen version, I essentially mean what I might more accurately call a single-history version, in which quantum theory gives probabilities for various alternative sequences of events, but only one sequence actually occurs. Each such alternative sequence might be called a "history" or a "world."

In the many-worlds version, in contrast, all of the possible histories or worlds with nonzero quantum probabilities actually occur, with the quantum probabilities being not probabilities for the histories to be actualized (since all are), but instead essentially measures for the magnitude of the existence of the various histories.

CONSEQUENCES OF DIFFERENT NUMBERS OF OBSERVERS

There can be significant differences in typical observations if the number of observers varies greatly from "world" to "world." Consider the following toy models:

Quantum Cosmology Model I

World 1: Observers; measure or probability 10^{-100}

World 2: No observers; measure or probability $1 - 10^{-100}$

In a single-history version of this Model I, World 1 is very improbable to occur at all, so any observation would be strong evidence against the single-history version. In a many-worlds version, World 1 does occur, so observations are not evidence against that theory.

Quantum Cosmology Model II

World A: 10^{10} observers during collapse; measure $1 - 10^{-30}$

World B: 10^{90} observers during expansion; measure 10^{-30}

In a single-history version of Model II, World B is very improbable, so a random observation should expect to see a collapsing universe, Hubble constant $H < 0$, and the probability that $H > 0$ is observed is only 10^{-30}.

In contrast, in a many-worlds version of Model II, all of the observations occur, with measures presumably given by something like the expectation values of positive operators each associated with a corresponding observation [8,9]. I shall assume that the observers are sufficiently similar that the total measure of a certain set of observations in a certain world (e.g., of whether the universe is expanding) is roughly proportional to the total number of observers in that world who make the observation, multiplied by the quantum measure of that world. I shall also assume that the fraction of observers who do observe whether the universe is expanding or contracting is the same in both World A and World B.

Then the total measure for World A observations of a collapsing universe is roughly proportional to the 10^{10} observers times the quantum measure of nearly unity for that world, or 10^{10}, whereas the total measure for World B observations of an expanding universe is roughly proportional to 10^{90} observers of that world times the quantum measure of 10^{-30} for that world, or 10^{60}. Thus a random observation chosen from the sample of all existing observations in the many-worlds version is about 10^{50} times more likely to be from World B, seeing $H > 0$, than it is to be from World A, seeing $H < 0$, a situation qualitatively the reverse of the relative probabilities in a single-history version, such as the Copenhagen version of quantum theory.

Now if one accepted the basic quantum measures of the two worlds in Quantum Cosmology Model II but was not very certain whether a single-history or a many-worlds version of quantum theory were correct, then if one made an observation of whether the universe were expanding or contracting, it would give strong evidence as to which version is correct.

One way to explain the difference between sampling a random observation in single-history versus many-worlds quantum theories is with lottery tickets. Suppose that we have a quantum cosmological model with the following two worlds:
World 1: N_1 observers; quantum measure or probability p_1
World 2: N_2 observers; quantum measure or probability p_2

The single-history version of quantum theory is like assigning lottery tickets to World 1 and World 2 in the ratio $p_1 : p_2$. Then a lottery ticket is chosen at random to select which world, and its observers, exist.

The many-worlds version of quantum theory is like assigning lottery tickets to each observer in World 1 and 2 with ratio $p_1 : p_2$, so that the ratio of the total number of lottery tickets in world 1 to that in world 2 is $N_1 p_1 : N_2 p_2$. All the observers exist, but with different measures for their reality, analogous to holding different numbers of lottery tickets. Choosing a measure-weighted observer (or, better, observation) at random is analogous to choosing a lottery ticket at random. The choice really is not made (since all observations really exist in the many-worlds version), but for saying which observations are typical, is is helpful to imagine their being chosen randomly.

PRELIMINARY EVIDENCE FROM HARTLE-HAWKING

We cannot yet calculate probabilities for our observations from an accepted model of the quantum cosmology quantum measures, so we cannot yet perform a definitive test of whether the single-history or the many-worlds version of quantum theory is correct. However, we can examine some highly speculative preliminary suggestions from the Hartle-Hawking 'no-boundary' proposal [10–12] applied to a $k = +1$ Friedmann-Robertson-Walker model with a minimally coupled massive scalar field (potential $\frac{1}{2}m^2\phi^2$).

In this minisuperspace model, an approximation of the stationary phase approximation for the path integral in which the scalar field starts at a value ϕ_i large compared with the Planck value (unity here) leads to the universe nucleating with initial size

$$a_i^2 = \frac{3}{4\pi m^2 \phi_i^2} = \frac{p}{\pi} \tag{1}$$

and quantum measure roughly proportional to

$$e^{-2S_E} \approx e^{\pi a_i^2} = e^p \tag{2}$$

with $p \equiv \pi a_i^2$. Observations suggest $m \sim 10^{-6}$ [13].

This is actually a measure density, and it is not clear what the prefactor should be. One simple choice is $dp = 2\pi a_i da_1$. The resulting measure would diverge if integrated to $p = \infty$ or $a_i = \infty$, but this would correspond to $\phi_i = 0$, where the approximation is invalid. To get an inflationary solution, one needs $\phi_i > \phi_{\min} \sim 1$, so $a_i < a_m = \sqrt{3/4\pi}/(m\phi_{\min}) \sim 1/m$ or $p < p_m = 3/(4m^2\phi_i^2) \sim 1/m^2$. Cut off the measure density there and normalize it, so we get the simple idealization

$$P(p < p') \approx \frac{e^{p'} - 1}{e^{p_m} - 1} \approx e^{-p_m}(e^{p'} - 1) \tag{3}$$

for $p' \equiv \pi a_i^2 < p_m \equiv \pi a_m^2 \sim 1/m^2 \sim 10^{12} \gg 1$.

After the universe nucleates, it undergoes slow-roll inflation with ϕ decreasing from ϕ_i to $\phi_e \sim 1$ and the volume increasing to

$$V_e = V_i \left(\frac{a_e}{a_i}\right)^3 \approx \frac{\sqrt{27\pi}}{4m^3\phi_i^3} e^{6\pi(\phi_i^2 - \phi_e^2)} = 2\sqrt{\pi} e^{-6\pi\phi_e^2} p^{3/2} \exp\frac{4.5\pi}{m^2 p} \sim p^{3/2} \exp\frac{4.5\pi}{m^2 p}, \tag{4}$$

which implies that for $m^3 V_e \gg 1$,

$$p \sim \frac{4.5\pi/m^2}{\ln(m^3 V_e) + 1.5 \ln\ln(m^3 V_e)}. \tag{5}$$

The entropy density after reheating is

$$s_e \sim T_e^3 \sim \rho_e^{3/4} \sim (m^2 \phi_e^2)^{3/4} \sim m^{3/2} \sim 10^{-9}. \tag{6}$$

By comparison, the entropy density of radiation today is

$$s_0 \approx \frac{86\pi^2}{165} T_0^3 \approx 1.22 \times 10^{-95}. \tag{7}$$

Assuming essentially adiabatic expansion after reheating, one gets that the volume of the universe today is

$$V_0 \approx \frac{s_e}{s_0} V_e \sim 10^{95} m^{3/2} V_e \sim 10^{95} m^{3/2} p^{3/2} \exp \frac{4.5\pi}{m^2 p} \sim 10^{86} p^{3/2} e^{1.4 \times 10^{13}/p}. \tag{8}$$

Now to get something analogous to Quantum Cosmology Model II above, we need to consider what values of p give observers mainly seeing the universe either contracting or expanding, and how many observers are produced as a function of p.

Let us make the crude assumption that observers require a universe of an age at least of the order of 10^{60}, a tenth of the age of our actual universe, and hence a volume of the order of 10^{181}, in order for suitable habitats to have evolved (e.g., planets around stars). This would give a lower limit on the volume at the end of inflation of about

$$V_e \gtrsim 10^{86} m^{-3/2} \sim 10^{95}. \tag{9}$$

Inserting this back into the approximate relation between V_e and p gives

$$p < p_{\max} \sim \frac{4.5\pi/m^2}{\ln(10^{77}) + 1.5 \ln \ln(10^{77})} \sim \frac{4.5\pi/m^2}{185} \sim 7.6 \times 10^{10} \tag{10}$$

as the crude condition for the existence of observers.

However, if p is sufficiently near this upper limit $p_{\max}$ for the existence of observers, then the universe will just barely last long enough for them, and they will mostly exist near the end of the lifetime of the universe, when it is collapsing. For most observers to see the universe expanding, V_e must be sufficiently larger that the lifetime of the universe is long enough for most observers to exist while the universe is still expanding. If the present age of the universe is a typical time for observers, then one might estimate that the universe must still be expanding at an age of roughly 10^{61} for most observers to see the universe expanding, and hence for it to have a volume of at least of the order of 10^{184} then. This leads us to $V_e \gtrsim 10^{89} m^{-3/2} \sim 10^{98}$ and

$$p < p_{\exp} \sim \frac{4.5\pi/m^2}{\ln(10^{80}) + 1.5 \ln \ln(10^{80})} \sim \frac{4.5\pi/m^2}{192} \sim 7.4 \times 10^{10} \tag{11}$$

as the crude condition for most observers to see the universe expanding.

In other words, in the Hartle-Hawking minisuperspace model under consideration, if $0 < p < p_{exp} \sim 7.4 \times 10^{10}$, observers will exist and will mostly see the universe expanding; if $p_{exp} < p < p_{max} \sim 7.6 \times 10^{10}$, observers will exist but will mostly see the universe contracting; and if $p_{max} < p$, essentially no observers will exist.

First, consider a Copenhagen or other single-history version of this quantum minisuperspace model, in which the wavefunction collapses to give a single macroscopic history or world, a classical Friedmann-Robertson-Walker universe characterized by ϕ_i, a_i, or p.

Using the results above, the probability for the wavefunction to collapse to classical universe that lasts long enough for observers is

$$P(\text{observers}) \approx \frac{e^{p_{max}} - 1}{e^{p_m} - 1} \approx e^{p_{max} - p_m} \sim 10^{-401000000000}, \tag{12}$$

and the probability for it to have observers that mostly see the universe expanding is

$$P(\text{observers seeing expansion}) \approx e^{p_{exp} - p_m} \sim 10^{-402000000000}, \tag{13}$$

both of which are utterly tiny.

Therefore, unless one had an uncertainty less than roughly $e^{-0.924 \times 10^{12}} \sim 10^{-40100000000}$ that this single-history model was correct, the evidence that observers exist would be overwhelming evidence against it.

Even if one somehow claimed that observers were necessary (i.e., that the wavefunction could not collapse to a world with no observers), the conditional probability of a world with observers mostly seeing the universe expand, given the condition that observers exist, is only

$$P(\text{observers seeing expansion}|\text{observers exist}) \approx e^{p_{exp} - p_{max}} \sim 10^{-1000000000}. \tag{14}$$

Thus the observation that the universe is expanding would be strong evidence against the single-history version of this model.

On the other hand, if one takes a many-worlds version of this Hartle-Hawking minisuperspace model, all components of the wavefunction exist that have positive measure, no matter how small, so observers will exist in a generalization of the model that allows sufficient structure for observers. Therefore, the existence of observers would not be evidence against a many-worlds version of a model sufficiently general to allow observers within at least some components of the wavefunction.

However, we can still ask for the relative probabilities of observations that the universe is contracting or expanding. In a many-worlds version, this will be roughly proportional to the bare quantum probability for each world multiplied by the the number of observers for that world. The unnormalized bare quantum probability was given above as $dP \approx e^p dp$ for $0 < p < p_m \sim 1/m^2 \sim 10^{12}$. For $0 < p < p_{max}$, observers exist, and it is reasonable to assume that the number of them is very roughly proportional to the volume V_0 of the universe when the entropy density is roughly the value of 10^{-95} that we observe, assuming that our observations of this quantity are typical. Then in this range of p, we get an unnormalized observational probability density roughly proportional to

$$dP_{\text{obs}} \approx V_0 dP \approx V_0 e^p dp \sim 10^{95} m^{3/2} p^{3/2} \exp\left(\frac{4.5\pi}{m^2 p} + p\right) dp. \qquad (15)$$

The integral of this over $0 < p < p_{\text{max}}$ diverges for $p \to 0$ ($V_0 \to \infty$), because of the divergence in the number of observers there, so effectively all of the observational probability occurs at that limit (infinitely large universes with infinitely many observers, presumably almost all seeing the universe expanding, since the stars would have all burned out in the infinite time it takes the infinitely large universe to recollapse.)

Therefore, in a many-worlds version of this Hartle-Hawking quantum cosmological model, one would presumably expect with very nearly unit probability that a random observation would see the universe expanding, the opposite of one's expectation for a single-history version of the same model. Thus if one accepted the basic model and allowed at least some reasonable uncertainty as to whether a single-history or a many-worlds version of the model is correct (before considering the evidence of the sign of the Hubble constant), then one's observation of whether the universe is expanding or contracting would give very strong evidence in support of either the many-worlds or the single-history version respectively. This is very similar qualitatively to the toy Quantum Cosmology Model II discussed above, except here the quantitative differences are even grossly more severe.

Of course, this preliminary evidence from a particular implementation of the Hartle-Hawking no-boundary proposal is highly speculative and is meant to be mainly illustrative, because of the many uncertainties of the model.

CONCLUSIONS

If the amount of observations (roughly, the number of observers) varies for different wavefunction components, then observation probabilities depend on whether only one component occurs in actuality (a single-history version of quantum theory, where observations truly are made only in one history, world, or component of the wavefunction), or whether many do (a many-worlds version, where observations truly are made in many histories, worlds, or components of the wavefunction). In particular, if components with relatively few observations dominate the quantum amplitude, but other components with testably different observations dominate the expectation value of the number of observations, which observations are most probable varies between single-history and many-worlds quantum theories.

The Hartle-Hawking wavefunction might allow a test from the observed expansion of the universe, but as of now it is highly speculative whether it is correct and what relative probabilities it would give for observing the universe expanding.

I acknowledge helpful discussions with Meher Antia, Jerry Finkelstein, Valeri Frolov, Jim Hartle, Jacques Mallah, and William Unruh. This research was supported in part by the Natural Sciences and Engineering Research Council of Canada.

A shorter version of this paper has been circulated [14] and has been reported on in the lay literature [15].

REFERENCES

1. Everett, H. III, *Rev. Mod. Phys.* **29**, 454 (1957).
2. DeWitt, B. S., and Graham, N., eds., *The Many-Worlds Interpretation of Quantum Mechanics*, Princeton: Princeton University Press, 1973.
3. DeWitt, B. S., *Physics Today* **23**, September 1970, p. 30.
4. Wheeler, J. A., *Rev. Mod. Phys.* **29**, 463 (1957).
5. Ford, K. W., and Wheeler, J. A., *Geons, Black Holes, and Quantum Foam: A Life in Physics*, New York: W. W. Norton, 1998, p. 270.
6. Omnès, R., *The Interpretation of Quantum Mechanics*, Princeton: Princeton University Press, 1973, pp. 327, 345.
7. Deutsch, D., *Int. J. Theor. Phys.* **24**, 1 (1985).
8. Page, D. N., "Sensible Quantum Mechanics: Are Only Perceptions Probabilistic?" (University of Alberta report Alberta-Thy-05-95, June 7, 1995), quant-ph/9506010.
9. Page, D. N., *Int. J. Mod. Phys.* **D5**, 583 (1996).
10. Hawking, S. W., in *Astrophysical Cosmology: Proceedings of the Study Week on Cosmology and Fundamental Physics*, edited by Brück, H. A., Coyne, G. V. and Longair, M. S., Vatican: Pontificiae Academiae Scientiarum Scripta Varia, 1982, pp. 563-574.
11. Hartle, J. B., and Hawking, S. W., *Phys. Rev.* **D28**, 2960 (1983).
12. Hawking, S. W., *Nucl. Phys.* **B239**, 257 (1984).
13. Linde, A., *Particle Physics and Inflationary Cosmology*, Chur: Harwood, 1990.
14. Page, D. N., "Observational Consequences of Many-Worlds Quantum Theory" (University of Alberta report Alberta-Thy-04-99, May 3, 1999), quant-ph/9904004.
15. Antia, M., *The Economist*, May 22, 1999, p. 145.

Exotic Spaces and Quantum Gravity

Kristin Schleich and Donald M. Witt

Department of Physics and Astronomy
University of British Columbia
Vancouver, BC Canada V6T 1Z1

Abstract. It is well known that in four or more dimensions, there exist exotic manifolds; manifolds that are homeomorphic but not diffeomorphic to each other. This is in contrast to the uniqueness of the differentiable structure on manifolds in one, two and three dimensions. As exotic manifolds are not diffeomorphic, one can argue that functional integrals for gravity should include a sum over not only physically distinct geometries and topologies but also inequivalent differentiable structures. But can the inclusion of exotic manifolds in such sums make a significant contribution to these quantum amplitudes? In a word, yes. Simply connected exotic Einstein manifolds with positive curvature exist in seven dimensions. Their metrics are found numerically; they are shown to have volumes of the same order of magnitude. Their contribution to the semiclassical evaluation of the partition function for Euclidean quantum gravity in seven dimensions is evaluated and found to be nontrivial. Consequently, inequivalent differentiable structures should be included in the formulation of sums over histories for quantum gravity.

The sum over histories formulation of quantum amplitudes has been particularly useful in the study of many interesting effects in quantum gravity. For example, the partition function in Euclidean quantum gravity with positive cosmological constant Λ is

$$Z = \sum_{(M^n, g)} \exp(-I[g])$$

$$I[g] = -\frac{1}{16\pi G} \int (R - 2\Lambda) d\mu(g) \tag{1}$$

where the sum is formally over physically distinct histories consisting of a manifold M^n with riemannian metric g. As detailed in [1], although a complete specification of the set of physically distinct histories is a difficult and perhaps unsolvable problem, one can readily describe the subset of these histories relevant to semiclassical approximations: two histories are physically distinct if (i) their manifolds have different topology, or (ii) their metrics are not related by a coordinate transformation or (iii) in four or more dimensions, their manifolds have the same topology

CP493, *General Relativity and Relativistic Astrophysics*, edited by C. P. Burgess and R. C. Myers
© 1999 American Institute of Physics 1-56396-905-X/99/$15.00

but inequivalent differentiable structures; i.e. they are exotic manifolds. The first two points are familiar in discussions present in the literature. The third is less so, though certainly not unmentioned; the inclusion of exotic manifolds was first proposed in the context of Kaluza-Klein theories by Freund [2] and their role in quantum gravity emphasized in [1]. However, a key issue not addressed in these previous arguments for the inclusion of inequivalent differentiable structures into functional integrals is their significance: will their presence make a difference to the results of a calculation?

In this talk we will summarize recent results showing that exotic manifolds contribute significantly in the semiclassical limit of functional integrals for quantum gravity [3], [4]. We do so by analyzing a particular set of exotic manifolds; the Wallach spaces $M_{-56788,5227}$ and $M_{-42652,61213}$. We show that these manifolds carry metrics that are solutions to the Euclidean Einstein equations with positive cosmological constant. A computation of the expectation value of the volume demonstrates that these solutions make a significant contribution. Furthermore, these exotic manifolds both admit Lorentzian solutions. Thus exotic manifolds make a significant contribution to semiclassical approximations of quantum amplitudes.

To begin, note that intuitively, an n-manifold is a space formed by sewing patches of $\mathbb{R}^n$ together. More precisely

Definition. *A metrizable space M^n is a smooth manifold if (i) every point has a neighborhood U_α homeomorphic to a subset of $\mathbb{R}^n$ via a mapping $\phi_\alpha : U_\alpha \to \mathbb{R}^n$. (ii) Given any two neighborhoods with nonempty intersection, the mapping $\phi_\beta \phi_\alpha^{-1} : \phi_\alpha(U_\alpha \cap U_\beta) \to \phi_\beta(U_\alpha \cap U_\beta)$ is a smooth mapping between subsets of $\mathbb{R}^n$.*

Although the above definition utilizes a choice of atlas $\{U_\alpha, \phi_\alpha\}$ it is clear that different choices for the atlas may be equivalent. For example, a change of coordinates corresponds to a change of the maps in the atlas. Additionally, an atlas generated by subdividing the certain neighborhoods of the original one does not produce a distinct manifold. One can precisely characterize equivalences between manifolds in terms of mappings. A *homeomorphism* is a continuous invertible map $h : M^n \to N^n$ such that its inverse $h^{-1} : N^n \to M^n$ is also continuous. A *diffeomorphism* is a homeomorphism $h : M^n \to N^n$ such that h is a differentiable invertible map whose inverse $h^{-1} : N^n \to M^n$ is also differentiable.

So what do these different notions of equivalence mean for physics? Two manifolds are topologically equivalent, that is one manifold can be continuously deformed into the other, if they are homeomorphic. However, physical theories are formulated in terms of fields satisfying differential equations. Essential properties of the solutions of such equations are preserved under diffeomorphisms but not under homeomorphisms. Thus the correct notion of equivalence of manifolds for physics is that of a diffeomorphism; two manifolds are physically equivalent if they are diffeomorphic. If M^n is homeomorphic to N^n but not diffeomorphic, then M^n and N^n have *inequivalent differentiable structures*; they are also termed *exotic manifolds*. The existence of inequivalent differentiable structures on the same topological manifold is reflected in the basic properties of the spaces. As there is no differentiable mapping between the exotic spaces, quantities such as its geometry and the spectra of differential operators such as the Laplacian would have different values. In short, exotic manifolds would have the same topology but exhibit different physics.

The possibility of exotic manifolds occurs in four or more dimensions; homeomorphic manifolds are not necessarily diffeomorphic.[1] The existence of such exotic manifolds was first established by Milnor with the construction of exotic 7-spheres [5]. Much is presently known about exotic manifolds (see the synopsis and references in [3]), including many explicit examples of such manifolds.

In particular the Wallach spaces, a set of seven dimensional simply connected manifolds exhibit members with inequivalent differentiable structures. The Wallach spaces $M_{k,l}$ are coset spaces of the form $SU(3)/i_{k,l}(S^1)$ where $i_{k,l}$ is the embedding of S^1 in $SU(3)$ [6]. The generator of $i_{k,l}$ can be written in terms of the generators of the standard maximal torus as $\mathfrak{h}_{k,l} = i/\sqrt{2\Gamma_{k,l}} \operatorname{diag}(k, l, -k-l)$ where $\Gamma_{k,l} = k^2 + kl + l^2$. The connection between the integers k, l and the topology of these spaces can be intuitively understood by recognizing that the action of the exponential of this generator on the subgroup of the maximal torus identifies points along a circle winding around it; $e^{2\pi i \theta} \rightarrow \operatorname{diag}(e^{2\pi i k\theta}, e^{2\pi i l\theta}, e^{-2\pi i(k+l)\theta})$. Different choices of the integers k, l result in a circle that winds around the maximal torus a different number of times; that is the homotopy of this circle depends on k, l. Similarly, the action of this exponential on the full $SU(3)$ group manifold will produce a manifold with homotopy depending on k, l.

A characterization of the topology of this family is given by Kreck and Stolz [7]. This characterization implies, through a difficult calculation in number theory, that there are there Wallach spaces that are homeomorphic but not diffeomorphic:

Corollary. *If $M_{k,l}$ and $M_{\bar{k},\bar{l}}$ are homeomorphic Wallach spaces with $\Gamma_{k,l} < 2955367597$, then $M_{k,l}$ is diffeomorphic to $M_{\bar{k},\bar{l}}$. On the other hand, the Wallach spaces $M_{-56788,5227}$ and $M_{-42652,61213}$ are homeomorphic but not diffeomorphic.*

A metric for the Wallach spaces can be written in terms of the left invariant covectors σ_i associated with the lie algebra elements of the coset space, $ds^2 = \kappa \sigma_0^2 + a(\sigma_1^2 + \sigma_2^2) + b(\sigma_3^2 + \sigma_4^2) + c(\sigma_5^2 + \sigma_6^2)$. This metric is homogeneous for all choices of the parameters κ, a, b, c.

An Einstein metric with positive scalar curvature satisfies $R_{ab} = E g_{ab}$ for some positive constant E. Fixing the overall scale of the metric by the condition $\kappa = 24\pi^2\Gamma$ where $\Gamma = \Gamma_{k,l}$, one finds that Einstein metrics satisfy

$$
\begin{aligned}
E &= \frac{3}{\Gamma}\left(\frac{(k+l)^2}{a^2} + \frac{l^2}{b^2} + \frac{k^2}{c^2}\right) & E &= -\frac{3(k+l)^2}{\Gamma a^2} + \frac{4}{a} + \frac{a^2 - c^2 - b^2 + 2bc}{abc} \\
E &= -\frac{3l^2}{\Gamma b^2} + \frac{4}{b} + \frac{b^2 - a^2 - c^2 + 2ac}{abc} & E &= -\frac{3k^2}{\Gamma c^2} + \frac{4}{c} + \frac{c^2 - a^2 - b^2 + 2ab}{abc}
\end{aligned} \tag{2}
$$

An abstract proof that the Wallach spaces admit Einstein metrics was given by Wang [8]. An explicit proof that this set of equations has precisely two solutions is given in [3]. From this approach, numerical values for the metric parameters can

---

[1] This is in contrast to the situation in one, two and three dimensions, in which any two manifolds that are homeomorphic are diffeomorphic.

TABLE 1. Metric parameters and volumes for Euclidean and Lorentzian Einstein metrics on the diffeomorphic spaces $M_{-56788,5227}$ and $M_{-42652,61213}$

Wallach Space k	l	Solution Type	Metric Parameters a	b	c	β
-56788	5227	Euclidean	1.78458	3.71019	4.66230	12.2181
			5.00997	4.30654	1.91680	15.6404
		Lorentzian	2.2839	0.963429	2.68894	4.39432
			1.87143	0.834748	4.15477	4.73945
			4.72684	0.853535	1.97277	5.6113
			2.38286	8.86525	2.4986	25.9964
-42652	61213	Euclidean	2.74990	4.01931	1.60339	7.6311
			4.87457	2.03556	5.23786	18.9611
		Lorentzian	1.35519	3.25177	2.02606	6.06546
			1.22819	4.35505	1.80371	6.44032
			1.30972	2.25295	6.28415	11.0707
			8.46397	2.54221	2.13119	23.1629

be obtained using the Newton-Raphson method. The resulting metric parameters and volume for the homeomorphic exotic Wallach spaces are given in Table 1. β is the dimensionless volume $\beta = V/V_0$ where V_0 is the volume of the metric on the Wallach space induced by the biinvariant metric on $SU(3)$. Perusal of the table reveals that the volumes of the Einstein metrics on $M_{-56788,5227}$ and $M_{-42652,61213}$ are comparable. Comparison of the volumes of the Einstein metrics on the exotic Wallach spaces with those on Wallach spaces in the same homotopy class shows that those of $M_{-56788,5227}$ and $M_{-42652,61213}$ fall near the median values for the homotopy class. Therefore, there is nothing particularly distinguishing about the Einstein metrics of these exotic Wallach spaces, either from each other or Wallach spaces of the same homotopy type.

The Wallach spaces also admit homogeneous Lorentzian metrics. These metrics describe homogeneous anisotropic cosmologies with closed timelike curves. Again one can solve the Einstein equations with positive cosmological constant for these metrics [4]. In contrast to the Euclidean case, there are precisely four Lorentzian Einstein metrics on each Wallach space. A comparison of the volumes of the four Lorentzian Einstein metrics, given in Table 1, yields similar results to those seen in the Euclidean case. It is amusing to note that the minimal volume metric in this case now occurs on $M_{-56788,5227}$, in contrast to the Euclidean case.

The consequences of including exotic spaces in functional integrals for gravity can be seen in the semiclassical evaluation of the logarithmic derivative of the partition function (1); this quantity is formally the expectation value of the volume over all 7-manifolds for fixed cosmological constant Λ. The graphs of the four values of β for four different sets of Wallach spaces as a function of Λ are given in Figure 1. Clearly the results exhibit a dependence on the set of spaces used.

The examples constructed in this paper show several key features: First inequivalent differentiable structures will contribute on an equal footing with topology in

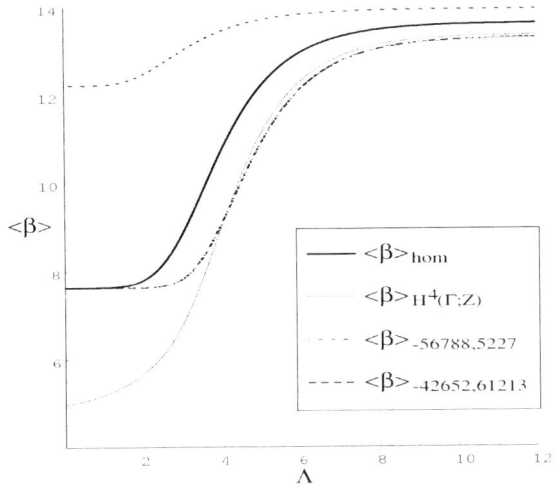

FIGURE 1. The expectation value $< \beta >$ as a function of cosmological constant for different sets of Wallach spaces. The curves $< \beta >_{-56788,5227}$ and $< \beta >_{-42652,61213}$ are for the respective Wallach spaces; the curve $< \beta >_{hom}$ is computed for the topological manifold. The curve $< \beta >_{H^4(\Gamma;Z)}$ is computed for all Wallach spaces of the same fourth cohomology as the exotic manifolds.

functional integrals for gravity. Secondly, there is no way to predict *a priori* which differentiable structure will dominate the functional integral. Therefore the contribution of all exotic manifolds to functional integrals for gravity must be considered. Clearly, an important question is the relevance of these results to four dimensional Euclidean functional integrals. Unfortunately, coset spaces will not yield exotic 4-manifolds. However, the lack of an easy construction by no means implies that Einstein metrics do not exist on exotic 4-manifolds! Furthermore, as exotic manifolds are in some sense most numerous in four dimensions, it could be argued that their contribution would be most important in this dimension.

REFERENCES

1. Schleich, K. and Witt, D., *Nucl. Phys. B* **402**, 411 (1993).
2. Freund, P., *Physica* **15D**, 263 (1985).
3. Schleich, K. and Witt, D., *Class. Quant. Grav.* **16**, 2447 (1999).
4. Schleich, K. and Witt, D., *Exotic spaces in quantum gravity II: Lorentzian quantum gravity* (in preparation).
5. Milnor, J., *Ann. Math.* **64**, 399 (1956).
6. Aloff, S. and Wallach, N., *Bull. of the Amer. Math. Soc.* **81**, 93 (1975).
7. Kreck, M. and Stolz, S., *J. Diff. Geom.* **33**, 465 (1991).
8. Wang, M.,*Duke Math. J.* **49**, 23 (1982).

Phase Transition in Quantum Gravity

Viqar Husain and Sebastian Jaimungal

Department of Physics and Astronomy,
University of British Columbia,
6224 Agricultural Road, Vancouver, BC V6T 1Z1, Canada

Abstract. A fundamental problem with attempting to quantize general relativity is its perturbative non-renormalizability. However, this fact does not rule out the possibility that non-perturbative effects can be computed, at least in some approximation. We outline a quantum field theory calculation, based on general relativity as the classical theory, which implies a phase transition in quantum gravity. The order parameters are composite fields derived from spacetime metric functions. These are massless below a critical energy scale and become massive above it. There is a corresponding breaking of classical symmetry.

A quantum theory of gravity should have the property that its long distance regime is classical general relativity (GR). On the other hand, the only criteria for the short distance regime, in the absence of experimental guides, is self-consistency and finiteness. At first sight this appears to bestow a large degree of freedom on attempts at formulating a theory of quantum gravity. However, as is well known, there have been many unsuccessful attempts over the years, with partial successes coming mainly from string theory [1], and the loop quantum gravity program [2].

From a conventional standpoint, a first question might be what classical theory is to be quantized to obtain quantum gravity. An obvious starting point, classical GR, leads to the non-renormalizability problem as far as perturbative quantum field theory (QFT) approaches are concerned. This problem in itself, however, does not imply that GR cannot be quantized. For example, the quantum theories of certain mini-superspace sectors of GR are known. Unfortunately, since these sectors correspond to quantum mechanical systems, they do not reveal much about what might be interesting properties of quantum gravity. Indeed, it may be argued that quantum mechanical reductions can reveal nothing significant about the full underlying quantum theory of gravity.

A next attempt at a starting point, short of full GR or supergravity, might be midi-superspace models – reductions of GR which are still field theories. One such reduction is obtained by imposing two Killing field symmetries. This reduction has been extensively studied with a view to quantization [3]. The most recent work is a claim that this dimensionally reduced theory can be completely quantized by

CP493, *General Relativity and Relativistic Astrophysics*, edited by C. P. Burgess and R. C. Myers
© 1999 American Institute of Physics 1-56396-905-X/99/$15.00

finding a representation of the complete classical observable algebra [4]. Although apparently complete mathematically, this quantization has so far given little physical insight into the underlying quantum theory. It is therefore important to probe such models a bit further.

In this work we attempt to extract physical consequences from a path integral quantization of the two-Killing field reduction of GR. The model is a two dimensional field theory containing two local degrees of freedom which interact non-linearly. Non-perturbative considerations reveal that the quantum theory contains a phase transition. The order parameters for the transition are composite fields made from spacetime metric functions. We find that the composite fields are massless when the ultraviolet (UV) cutoff is below a critical energy scale, and are massive when it is above this scale.

Our starting point is the spacetime metric

$$ds^2 = e^{2A} \left(-dt^2 + dz^2\right) + g_{ab} \, dx^a dx^b , \tag{1}$$

where $A = A(t,z)$, $x^a = (x^1, x^2)$, and the 2×2 metric

$$g_{ab} = R \begin{pmatrix} \cosh W + \cos\Phi \, \sinh W & \sin\Phi \, \sinh W \\ \sin\Phi \, \sinh W & \cosh W - \cos\Phi \, \sinh W \end{pmatrix}$$

is parameterized by the three functions $R(t,z)$, $\Phi(t,z)$ and $W(t,z)$. The metric (1) has two commuting space-like Killing vector fields ∂_{x^1} and ∂_{x^2}, whose orbits have the topology of T^2. The t, z coordinates have ranges $0 < t < \infty$ and $-\infty \leq z \leq \infty$. With these conventions, the metric is that of the Schmidt spacetime [5]. The vacuum Einstein equations in the gauge $R(t,\theta) = t$, give the two coupled two-dimensional evolution equations

$$\ddot{W} + \frac{1}{t}\dot{W} - W'' + \sinh W \cosh W \, (\Phi'^2 - \dot{\Phi}^2) = 0 , \tag{2}$$

$$\ddot{\Phi} + \frac{1}{t}\dot{\Phi} - \Phi'' + 2 \frac{\cosh W}{\sinh W} (\dot{\Phi}\dot{W} - \Phi'W') = 0 , \tag{3}$$

for $W(t,\theta)$ and $\Phi(t,\theta)$, and the two "constraint" equations

$$\dot{A} + \frac{1}{4t} - \frac{t}{4} \left[\dot{W}^2 + W'^2 + \sinh^2 W \, (\dot{\Phi}^2 + \Phi'^2) \right] = 0 , \tag{4}$$

$$A' - \frac{t}{2} (\dot{W}W' + \sinh^2 W \, \dot{\Phi}\Phi') = 0 . \tag{5}$$

The evolution equations involve $W(t,\theta)$ and $\Phi(t,\theta)$ and, given a solution to these equations, $A(t,\theta)$ is obtained by integration of the constraint equations (4-5).

The evolution equations (2-3) may be derived from a two-dimensional $\sigma-$model-like action

$$S_2(W,\Phi) = \frac{1}{2} \int dt d\theta \, t \, \sqrt{-\eta} \, \eta^{ab} G_{AB}(Y) \, \partial_a Y^A \partial_b Y^B , \tag{6}$$

239

where $Y^1 = W$, $Y^2 = \Phi$, $\eta^{ab} = \text{diag}(-, +)$, $a, b, \cdots = t, z$, and $G_{AB}(Y)\, dY^A dY^B = dW^2 + \sinh^2 W\, d\Phi^2$ is the unit hyperboloid metric. The t factor in the integrand cannot be absorbed by a rescaling of fields, or by introducing a curved two-dimensional metric. However this can be done if the model is embedded in three dimensions (see below).

Equations (2-3) can also be derived from the standard three-dimensional $SL(2, \mathbb{R})$ non-linear σ-model [6],

$$S_3[X^i, \lambda, \mu^i] = \int_M dt\, dx\, dz\ \sqrt{-\eta} \left\{ \eta^{ab} g_{ij} \partial_a X^i \partial_b X^j + \lambda \left(g_{ij} X^i X^j + 1 \right) + \mu^i g_{ij} D X^j \right\},$$

where $g_{ij} = \text{diag}(+, +, -)$, $\eta_{ab} dx^a dx^b = -dt^2 + dx^2 + dz^2$, $\lambda(t, x, z)$ and $\mu^i(t, x, z)$ are Lagrange multiplier fields, $D = t\partial_x + x\partial_t$, and $i = 1, \cdots, 3$. The Lagrange multiplier μ^i enforces the appropriate reduction to two dimensions, and the multiplier λ enforces the constraints which give $X^i = X^i(W, \Phi)$.

For quantization there is the option of using the two or the three-dimensional action given above. Using the latter allows inclusion of quantum fluctuations in all three dimensions, whereas the former restricts these to two dimensions. Thus we expect on intuitive grounds that the quantum theories obtained via these two routes will not be the same. Because of the richer range for quantum fluctuations and the standard $\sigma-$model approach it allows, we use the three-dimensional action S_3.

A quantum theory may be defined by the path integral

$$Z = \int [dX][d\lambda][d\mu]\ e^{-iM_P S_3[X^i, \lambda, \mu^i]} \tag{7}$$

where M_P is the Planck mass. In this path integral, it is possible to perform the Gaussian integral over all the dynamical fields X^i. This leads to the quantum effective action for the fields λ and μ^i.

The saddle point evaluation of the remaining path integral, in its regime of validity, gives a first non-perturbative approximation for the quantum dynamics of X^i. This requires solutions of the Euler-Lagrange (EL) equations for λ and μ^i derived from the quantum effective action. It is here that we find evidence of a phase transition.

The quantum effective action obtained after integrating over the fields X^i is

$$S_3^{eff} = -\frac{3}{2}\, \text{Tr} \ln\left(\Box + \lambda\right) - iM_p \int dt\, dx\, dz\ \left[\lambda - \frac{1}{4}\, D\mu^i \left(\Box + \lambda\right)^{-1} D\mu^i \right], \tag{8}$$

where $\Box = -\eta^{ab}\partial_a \partial_b$ is the three-dimensional flat space Laplacian. The EL equations for λ and μ^i are well defined only in the presence of an UV cutoff, Λ, due to a trace term which appears in one of the equations. At this stage therefore, a second energy scale, Λ, enters our quantum model; all solutions of the EL equations depend on the cutoff. The simplest solutions are

$$\lambda = \lambda^* = \text{const.} \quad \text{and} \quad \mu^i = \mu^{i*} = \text{const.}. \tag{9}$$

240

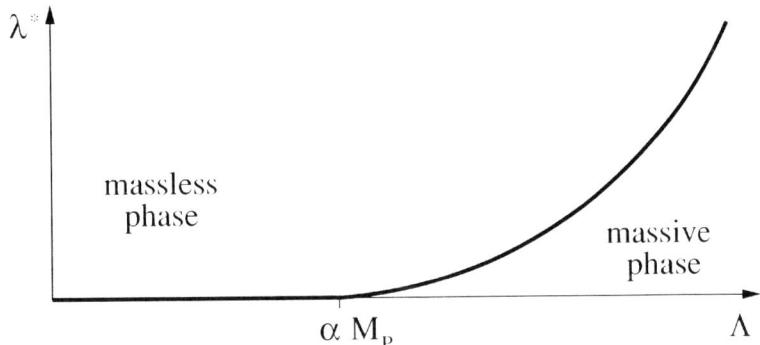

FIGURE 1. Spontaneous symmetry breaking occurs for $\Lambda < \alpha M_P$ and gravitational excitations become massless.

These particular solutions exist *only if* $\Lambda \geq \alpha M_P$ [7], where α is dimensionless constant of order unity.

What do these solutions imply for the phase of the theory? In the saddle point approximation, the two-point function is

$$\langle X^i(x)X^j(y)\rangle \sim g^{ij}\int^{\Lambda}\frac{d^3k}{(2\pi)^3}\frac{e^{ik(x-y)}}{k^2-\lambda^*},$$

This simple result demonstrates that the X^i fields are massive with mass λ^* on scales $\Lambda > \alpha M_P$, and so we can speak of a "massive phase" in this regime. A careful treatment reveals that

$$\lambda^*(\Lambda, M_P) = \text{const.}\,\frac{\Lambda^4}{\alpha^2}\left(\frac{1}{M_P}-\frac{\alpha}{\Lambda}\right)^2, \tag{10}$$

so that the mass is zero at $\Lambda = \alpha M_P$.

It is surprising that integrating out all the X^i fields in eqn. (7) leads to these solutions of the effective action only if $\Lambda \geq \alpha M_P$. Is it possible to extend this solution to the region $\Lambda < \alpha M_P$? Indeed, it is. One must go back to the original path integral and assume that one of the X^i fields is a constant, that is, it has a non-zero vacuum expectation value (vev). This modifies the EL equations for λ and μ^i, such that now $\lambda = 0$ and $\mu^i = \mu^{i*} = $ constant are indeed solutions, (while the vev is found to depend on Λ and M_P). Thus, in the $\Lambda < \alpha M_P$ parameter regime, we are in a "massless phase," and there is spontaneous breaking of the $SL(2,\mathbb{R})$ symmetry (due to the direction picked out by the constant component of the X^i field). These results are summarized in Figure 1.

The steps we have outlined are standard for exploring the possibility of phase transitions in QFT. What is surprising, as we have seen, is that an analogous treatment applies to the path integral (7) for reduced Einstein gravity. *This appears to be the first concrete indication from GR itself that a phase transition occurs at very short distances.*

What do these results imply for quantum gravity? Perhaps the main insight is that gravitational excitations are massive above the Planck energy scale, and massless, with spontaneous symmetry breaking, below this scale. The symmetry breaking below M_P in this reduction of GR is relatively slight — it is a breaking of global $SL(2, \mathbb{R})$. Nevertheless, it is significant in that it occurs at all.

It is important to emphasize that this phenomena is essentially due to the two non-linearly interacting degrees of freedom in this model; it is missed if only one field is present. The implication of our result for full quantum gravity is the possibility that the graviton is a massless Goldstone boson associated with spontaneous symmetry breaking at the Planck scale.

Several questions remain for further research: Is the dimensionally reduced model considered here renormalizable? If so what is the β–function and its fixed points (if any)? What is the spacetime metric at scales below the Planck length?

With respect to the first question, it is well known that the regular σ–model in three dimensions is renormalizable. Therefore it is clear that this must be the case also for the present model, even with the dimensional reduction constraint it contains. Surprisingly, however, this result is not straightforward to establish if the constraint is first solved classically and an action in two dimensions is used as the starting point. This is because the resulting σ–model like action has an explicit time factor in the integrand, which results in the kinetic operator not being Sturm–Lioville. This crucially affects the standard evaluation of the Gaussian integral. Thus, establishing renormalizability and calculating the β–function is straightforward from the three-dimensional perspective, but not from the two-dimensional one. Indeed the model may not even be renormalizable from the latter perspective, where the σ model constraint is explicitly time dependent. If so, this would be yet another indication of the important differences between dimensionally reducing classically and then quantizing, versus quantizing by including the constraint quantum mechanically as we have done here.

REFERENCES

1. M. Green. J. Schwarz, and E. Witten, *Superstring Theory* (University of Cambridge Press (New York,1990)); J. Polchinski, *String Theory: Superstring Theory and Beyond*, (Cambridge University Press (New York, 1998)).

2. For a recent review see: C. Rovelli, "Loop Quantum Gravity," gr-qc/9710008.

3. An incomplete selection of papers is: K. Kuchař, Phys. Rev. D4, 955 (1971); B. Berger, Ann. Phys. **83**, 458 (1974); V. Husain, Class. Quantum Grav. **4** 1587 (1987); D. Korotkin and H. Nicolai, Phys. Rev. Lett. **74**, 1272 (1995); G. Mena-Marugan Phys. Rev. D**56**, 908 (1997).

4. D. Korotkin and M. Samptleben, Phys. Rev. Lett. **80**, 14 (1998).

5. B. G. Schmidt, Class. Quantum Grav., **13** (1996) 2811.

6. A. Ashtekar and V. Husain, Int. J. Mod. Phys. D**7**, 549 (1998).

7. A. M. Polyakov, *Gauge Fields and Strings*, (Harwood, NY, (1988)).

Schrödinger's Equation
in General Relativity

Chris Vuille*

*Department of Physical Sciences, Embry-Riddle Aeronautical University
Daytona Beach, FL 32114

Abstract. A process of operator replacement leads to a generalization of Schrodinger's equation to curved space-time, linking general relativity and quantum mechanics in a natural way. This fourth-order equation, which reduces and specializes to the Klein-Gordon equation in the flat space limit, must be solved in tandem with the Einstein field equations with suitable stress-energy. The flat space propagator, for large momenta, varies as $1/p^4$, meaning that divergences should be less problematic than in standard theories. Approximate solutions can be obtained for static, spherically-symmetric scalar fields.

DERIVATION OF THE FIELD EQUATION

Schrodinger's equation represents one of the great triumphs of modern physics. In this paper, it will be generalized to an arbitrary four-manifold by replacing the differential operators with their relativistic counterparts, followed by standard operator replacement as necessary.

The free-space Schrodinger equation reads

$$i\hbar \frac{\partial \Psi}{\partial t} = -\frac{\hbar^2}{2m} \nabla^2 \Psi \tag{1}$$

This equation should also be expected to hold in a frame which is nearly comoving with the particle. The idea, then, is to transform the equation into coordinates that are adapted to the particle's worldline. In this way, the differential operators in equation 1 can be properly transcribed to operators on an arbitrary four-manifold.

First, the mass energy, which has essentially been scaled out of the Schrodinger equation, must be reinserted:

$$i\hbar \frac{\partial \Psi}{\partial t} = -\frac{\hbar^2}{2m} \nabla^2 \Psi + mc^2 \Psi \tag{2}$$

Now let p^a be the relativistic four-momentum of the particle described by Ψ. Then the proper time derivative is given by:

CP493, *General Relativity and Relativistic Astrophysics*, edited by C. P. Burgess and R. C. Myers
© 1999 American Institute of Physics 1-56396-905-X/99/$15.00

$$\frac{d}{d\tau} = \frac{p^a}{m} \bigtriangledown_a \tag{3}$$

The metric of the space-time can then be written as

$$g_{ab} = \frac{p_a p_b}{m^2 c^2} - \frac{\eta_a \eta_b}{c^2} - Q_{ab} \tag{4}$$

where η_a is a covector orthogonal to the four-velocity of the particle and Q_{ab} is the metric of the two-space perpendicular to the direction of motion. It is always possible to rewrite the metric in this fashion: furthermore, rearranging the above expression, a projection operator can be created that projects four-vectors into the three-space orthogonal to the particle momentum. With these tools in hand, the following generalizations of the Schrodinger differential operators can be made:

$$\frac{\partial}{\partial t} \longrightarrow \frac{d}{d\tau} = \frac{p^a}{m} \bigtriangledown_a \tag{5}$$

$$\bigtriangledown^2 \longrightarrow \left(\frac{\eta^a \eta^b}{c^2} + Q^{ab} \right) \bigtriangledown_a \bigtriangledown_b = \left(\frac{p^a p^b}{m^2 c^2} - g^{ab} \right) \bigtriangledown_a \bigtriangledown_b \tag{6}$$

In addition, we have

$$mc^2 = (1/m) \, g^{ab} p_a p_b \tag{7}$$

Equation 7 leads to an alternative to the usual interpretation of the mass energy as a multiplicative operator; however, it will result in a highly esthetic form for the wave equation. The differential operators in equations 5, 6, and 7 shall replace the operators in Schrodinger's equation, resulting in a direct generalization of that equation to a four manifold:

$$i\hbar \frac{p^a}{m} \bigtriangledown_a \Psi = \frac{\hbar^2}{2m} \left(g^{ab} - \frac{p^a p^b}{m^2 c^2} \right) \bigtriangledown_a \bigtriangledown_b \Psi + \left(\frac{1}{m} \right) g^{ab} p_a p_b \Psi \tag{8}$$

Equation 8 is relativistically invariant, however 4-momentum vectors have reappeared in the equation, which must now be replaced by the appropriate quantum operators given by

$$p_a = i\hbar \bigtriangledown_a \tag{9}$$

and

$$p^a = i\hbar g^{ab} \bigtriangledown_b \tag{10}$$

Inserting equations 9 and 10 into equation 8 results in

$$-\frac{\hbar^2}{2m} \bigtriangledown^a \bigtriangledown_a \Psi = \frac{\hbar^4}{2m^3 c^2} \bigtriangledown^a \bigtriangledown^b \bigtriangledown_a \bigtriangledown_b \Psi \tag{11}$$

Some rearrangement of this fourth-order equation is desirable in order to better understand its properties. Note that the definition of the curvature tensor (conventions follow Carmeli [1]) gives

$$\nabla^b \nabla_a \nabla_b \Psi = \nabla_a \nabla^b \nabla_b \Psi + R^c{}_a \nabla_c \Psi \tag{12}$$

where $R^c{}_a$ is the Ricci curvature. Substituting this expression into equation 11 and rearranging the terms results in

$$\nabla^a \nabla_a \left(\nabla^b \nabla_b \Psi + \frac{m^2 c^2}{\hbar^2} \Psi \right) = -\nabla^a \left(R^c{}_a \nabla_c \Psi \right) \tag{13}$$

This, then, is a fully covariant equation for a quantum wave in general relativity which reduces to the D'Alembertian of the Klein-Gordon equation when the spacetime is flat. The equation can be trivially integrated to produce a third order equation with an arbitrary divergence-free vector field. For higher spin, the scalar wave function may be replaced by the appropriate spinor expression.

GENERAL REMARKS

Following the method in [2], it is easy to verify that in flat space the propagator is

$$S(p) = \frac{\hbar^4}{\left(p^2 - m^2 c^2 \right) p^2} \tag{14}$$

By contrast, the Dirac propagator varies as $1/p$ for large p, while the Klein-Gordon propagator varies as $1/p^2$. The propagator in equation 14 is therefore less likely to lead to divergences, and may result in a finite quantum field theory.

A Lagrangian for equation 11 exists, though a higher-order variational theory is needed. Following Lovelock and Rund [3], this principle can be written as

$$-\nabla_a \nabla_b \frac{\partial \pounds}{\partial \nabla_a \nabla_b \Psi} + \nabla_a \frac{\partial \pounds}{\partial \nabla_a \Psi} - \frac{\partial \pounds}{\partial \Psi} = 0 \tag{15}$$

By inspection, the correct Lagrangian for this theory is given by

$$\pounds_\Psi = -\frac{1}{2} \frac{m^2 c^2}{\hbar^2} g^{ab} \nabla_a \Psi \nabla_b \Psi + \frac{1}{2} g^{ac} g^{bd} \nabla_a \nabla_b \Psi \nabla_c \nabla_d \Psi \tag{16}$$

With equation 16, a canonical stress-energy tensor could be derived, and hence Einstein's equations for the gravity field. Alternatively, varying the Langrangian density with respect to the metric would also give a stress-energy, a method generally preferred by relativists. The resulting equations, Einstein's field equations coupled to the generalized Schrodinger's equation, can be solved numerically under a number of simplifying assumptions. Preliminary work on the spherically-symmetric

245

static case indicates that, unlike the Dirac and Klein-Gordon equations, the mass appears to play a role similar to the energy eigenvalues of the original Schrodinger equation, constrained in this case by the gravity field rather than the classic potentials. One significant difficulty is determining appropriate initial values for Ψ and its derivatives. Further development of the equations is in progress.

ACKNOWLEDGEMENT

I would like to thank Fred Elston for a number of valuable conversations.

REFERENCES

1. Carmeli, C., *Classical Fields: An Introduction to General Relativity and Gauge Theory*, New York, John Wiley and Sons, 1984.
2. Halzen, F., and Martin, A. *Quarks and Leptons*, New York, John Wiley and Sons, 1984.
3. Lovelock, D. and Rund, H., *Tensors, Differential Forms, and Variational Principles*, New York, John Wiley and Sons, 1975.

Quantum Time

Jonathan Oppenheim[1]

University of British Columbia
6224 Agricultural Road
Vancouver B.C., V6T 1Z1

Abstract. After discussing the role of time in both General Relativity and Quantum Gravity, it is argued that even in ordinary Quantum Mechanics, the role of time is ambiguous. A rather surprising result is that measurements of the time-of-arrival of a free particle must always be more inaccurate than $1/E$ where E is the kinetic energy of the particle. This inaccuracy is not related to the Heisenberg uncertainty relations. It is a dynamical relation which applies to individual measurements of a single quantity.

It is widely believed that one of the difficulties of constructing a Quantum Theory of Gravity, is that time plays an incompatible role in Quantum Mechanics and General Relativity. In Quantum Mechanics, time is an external parameter. Other observables, like position and momentum, are described by operators. These operators evolve in time, while time itself cannot be represented by an operator. On the other hand, in General Relativity, time is much more a part of the theory. Both time and space, bend and twist in the presence of massive objects, and both space and time are represented by coordinates.

These coordinates are, of course, subject to coordinate transformations, and in particular, the theory is invariant under reparametrization of the time coordinate. One consequence of this, is that if one tries to naively quantize gravity, one finds that the wave-function must satisfy the Wheeler-DeWitt equation

$$\mathbf{H}\Psi(g_{ab}, \pi_{ab}) = 0 \tag{1}$$

where the wavefunction depends on the 3-metric and conjugate momenta and $\mathbf{H}$ is known as the Hamiltonian constraint and is the generator of time-translations when the equations of motion are satisfied. Because the Hamiltonian constraint must always be satisfied, the only possible observables are those which commute with the constraint. However, observables which commute with the constraints, don't evolve in time, making the physics of the system rather hard to describe.

---

[1] This talk is based on work done in collaboration with Y. Aharonov, S. Popescu, B. Reznik and W.G. Unruh. Some of the ideas presented here, first appeared in Phys. Rev. A, **57** 4130 (1998).

The situation is somewhat analogous to being inside a box, and having some external observer weigh the box with high accuracy. In order to keep the box at this fixed weight, the external experimenter cannot measure observables which evolve in time. Quantum Mechanics also dictates that the observer will see people inside the box in a superposition of many different ages. This is because time is (in a sense) conjugate to energy. It has been suggested by Benni Reznik that this gives us a rather interesting way to put animals in superpositions of being alive and dead. Take an animal (Schrödinger's poodle, for example), stick her in a box, and weigh the box accurately. The poor poodle will be in a superposition of herself at different stages of her life. If we weigh the box very accurately, and later look at the age of the poodle, we will sometimes find that the poodle is so old that she died (or so young that she was just matter waiting to be born). If we measure the weight of the box with infinite accuracy, then essentially any time we look inside the box, we will find nothing but poodle dust.

In General Relativity, which describes the entire universe, all of us, observers and the observed, are in some sense living inside a box of fixed energy. To describe such a system, we may need to re-examine the way we think of time.

For experiments which occur at a fixed time, Quantum Mechanics provides us with a useful formalism. Observables are represented by operators, and in the Heisenberg representation, they evolve in time. The possible results of any measurement at any instant of time t can be found by acting these operators on the wave function of the system at the time t.

However, in General Relativity (and even in many laboratory experiments), we are often interested in performing experiments which are not fixed in time. For example, if we wish to measure space-time distances, then we will probably want to know how long it takes for a photon to travel between two points. This is a continuous measurement which does not occur at any particular time. We may also want to know whether one event is in the past or future of another event. Both these measurements are, in some sense, a measurement of time itself. These types of measurements are necessary in order to determine the components of the metric tensor.

Another physical property, which appears in the context of Quantum Cosmology, is the maximum size the universe will attain. This is not a property of the universe at a fixed time, but rather, a property of the universe over all time. In classical physics, one could make measurements on a system at a fixed time in order to predict the evolution of the system for all time. However, in Quantum Mechanics, this is not always the case.

These examples, which arise out of trying to understand Quantum Gravity, have led us to examine the role that time plays in ordinary Quantum Mechanics.

Within standard Quantum Mechanics, we are used to asking the question "where is the particle at time t?" However, it is also perfectly natural to ask "at what time is the particle at a certain location." This question is often posed in the laboratory, yet Quantum Mechanics does not seem to provide an unambiguous answer.

It is often stated that the time of an event is not a standard observable in

Quantum Mechanics because it cannot be represented by an operator. While this is certainly true, it is important to realize that the difference between time, and other observables is not merely formal.

For example, if at time t a particle is detected at location X, then we can say with certainty that at the same time t, the particle was not at any other location X'. However, if we turn on a detector located at position x, and detect a particle at time T, then it is quite possible that this particle might also have been detected at any number of other times T'. One can also find that the particle never arrives at the location x, or that it is always at x.

This leads us to consider the first occurrence of an event, such as the arrival-time of a particle, since a particle can only arrive once to a particular location. In order to measure the arrival time, one needs to detect the particle at time t_A, and also know that the particle was not there at any previous time. In other words, one must continuously monitor the location of arrival in order to find out when the particle arrives.

However, the probability to find a particle at $t = T$ is generally *not* independent of the probability to find the particle at some other time $t = T'$. Measurements made at different times do not commute and will disturb one another. Each successive measurement disturbs the system, which can make measurements of the time of an event problematic.

We will now attempt to make a measurement of the time-of-arrival. In order to do so, we will need a clock. This ideal clock can be represented by the Hamiltonian

$$\mathbf{H_{clock}} = \mathbf{P_y} \ . \tag{2}$$

The Hamiltonian for this clock is unbounded from above and below, nonetheless, using a sufficiently massive particle, we can approximate the ideal situation to arbitrary accuracy. To read the time of the clock, we measure the coordinate $\mathbf{y}$ conjugate to $\mathbf{P_y}$ and using the Heisenberg equations of motion it is not hard to see that $y = t$. Quantum mechanics puts no limitation on how accurate this clock can be measured, and there is no difficulty in measuring the time – one simply measures the coordinate y to any degree of accuracy one wishes. However, we will see that if we want to use this clock to measure the time of an event, then Quantum Mechanics will place a limitation on how accurately this can be done. For the time-of-arrival of a free particle, this inaccuracy is given by

$$\delta t > 1/E \tag{3}$$

where E is the kinetic energy of the particle. This inaccuracy is not equivalent to the so-called "Heisenberg energy-time uncertainty principle". It refers to individual measurements of a single quantity. Quantum Mechanics places no limitation on how accurately we can make a single measurement of position or momentum (although an accurate measurement of position will disturb the momentum and visa-versa). For measurements of time-of-arrival however, we cannot make a single measurement arbitrarily accurate. If we do so, we will find that the particle never arrives.

The reason for this is rather simple. Because $\mathbf{y}$ is conjugate to $\mathbf{P_y}$, accurate clocks have a large spread in momentum. This means that in general the momentum of an accurate clock can take on fairly large values. For an infinitely accurate clock, the momentum will almost always be infinite. Accurate clocks therefore, are very massive and energetic, and this makes them very hard to use to measure the time of an event. To measure the time-of-arrival of a particle, the particle itself will have to turn off the clock when it arrives. If the clock is very energetic, then the particle will not be able to turn off the clock and no record will exist that the particle arrived. In fact, the particle will get reflected and will never arrive. The situation is very different from usual measurements, where an external observer supplies the energy to make the measurement. If we were to measure the time of the clock, we would have to supply a large amount of energy to make the measurement accurate. However, in order to measure the time-of-arrival, it is the particle who must supply the energy to stop the clock.

To see this, let us use the clock to measure the time-of-arrival. This could involve a coupling such as

$$\mathbf{H} = \frac{1}{2m}\mathbf{P_x}^2 + \theta(-\mathbf{x})\mathbf{P_y}. \tag{4}$$

Here, the particle's motion is confined to one spatial dimension, x, and $\theta(x)$ is a step function.

We imagine that initially, the particle is localised in a region $x < 0$ and we wish to know when the particle arrives to the origin. While the particle is travelling along the negative x-axis, the clock runs. When the particle crosses the origin, the clock stops. We can see immediately what the problem will be. The above Hamiltonian is equivalent to a particle which must cross a step-potential. If the step is very high, then the particle will be reflected, and the clock will not turn off. The more accurate we wish the clock to be, the greater the possible values of p_y will be, and this will make it very hard for the particle to cross the step potential and turn off the clock.

At $t \rightarrow \infty$ the clock shows the time of arrival:

$$\mathbf{y}_\infty = \mathbf{y}(t_0) + \int_{t_0}^\infty \theta(-\mathbf{x}(t))dt \tag{5}$$

A crucial difference between the classical and the quantum case is that in the classical case the back-reaction can be made negligibly small by choosing $P_y \rightarrow 0$. In this case, the particle follows the undisturbed solution, $x(t) = x(t_0) + \frac{p_x}{m}(t - t_0)$. If initially we set $y(t_0) = t_0$ and $x(t_0) < 0$ the clock finally reads:

$$y_\infty = y(t_0) + \int_{t_0}^\infty \theta[-x(t_0) - \frac{p_x}{m}(t - t_0)]dt = -\frac{mx(t_0)}{p_x}. \tag{6}$$

The classical time-of-arrival is $t_A = y_\infty = -mx(t_0)/p_x$. The same result would have been obtained by measuring the classical variable $-mx_0/p_x = -mx(t)/p_x + (t - t_0)$, at arbitrary time t. Classically, the continuous measurement procedure using a clock, and the indirect measurement of x_o/p_x give the same result.

On the other hand, in quantum mechanics the uncertainty relation dictates a strong back-reaction, i.e. in the limit of $\Delta y = \Delta t_A \to 0$, p_y must have a large uncertainty, and the state of the particle will be strongly affected by the act of measuring. Therefore, the two classically equivalent measurements become inequivalent in quantum mechanics.

If k is the momentum of the particle, then it is not hard to see that if p_y is small, then the clock reading will be peaked around the classical time of arrival

$$y = \frac{mx_o}{k_0}. \tag{7}$$

However, the transition probability is given by

$$T = \sqrt{\frac{E + p_y}{E}} \left[\frac{2\sqrt{E}}{\sqrt{E} + \sqrt{E + p_y}} \right]^2. \tag{8}$$

Since the possible values obtained by p_y are of the order $1/\Delta_y \equiv 1/\Delta t_A$, the probability to trigger the clock remains of order one only if

$$\bar{E}\Delta t_A > 1. \tag{9}$$

Here Δt_A stands for the initial uncertainty in position of the dial $\mathbf{y}$ of the clock, and is interpreted as the accuracy of the clock. $\bar{E}$ can be taken as the typical initial kinetic energy of the particle.

In measurements with accuracy better then $1/\bar{E}$ the probability to succeed drops to zero like $\sqrt{E\Delta t_A}$, and the time-of-arrival of most of the particles cannot be detected. Furthermore, the probability distribution of the fraction which has been detected depends on the accuracy Δt_A and can become distorted with increased accuracy.

While in this talk, I have only demonstrated the inaccuracy relation in this rather simple model, the relationship seems to hold no matter how we try to measure the time-of-arrival of the particle. However, as of yet, no general proof of this relation has been found.

This new inaccuracy relation $\delta t > 1/E$ suggests that at some level, the time of an event is ambiguously defined. One of the lessons of Quantum Mechanics is that only those objects which are measurable have physical meaning. What then, is the relationship between time observables and the parameter t in the Schrödinger equation? Certainly Quantum Mechanics is entirely self-consistent, and yet, questions remain about the role of time in the theory. Not only do we not understand the role of time in Quantum Gravity, but it appears that we do not quite understand the role of time in ordinary Quantum Mechanics. It is hoped that by understanding *quantum time*, some new insight can be gained into Quantum Gravity.

Gravitational Mass in Asymptotically de Sitter Space-Times with Compactified Dimensions[1]

Tetsuya Shiromizu

Department of Applied Mathematics and Theoretical Physics, University of Cambridge, Silver Street, Cambridge CB3 9EW, Untited Kingdom

Abstract. We define gravitational mass in asymptotically de Sitter space-times with compactified dimension. We give simple examples with negative energy in higher dimensions. They do not have the lower bound on the mass. We also give a positive energy argument in higher dimensions and realise that elementary fermion cannot exist in our examples.

INTRODUCTION

Superstring or M-Theory may offer the proper theory of gravity [2]. Such kind of theories are formulated in higher dimensions and it is believed that the extra dimension space will be compactified to be less than Planck length.

The stability of such space-times is important and has been discussed. It was shown that there is the instanton which may indicate the decay of Kaluza-Klein vacuum [3] [4] [5]. As Witten pointed out, the decay mode is excluded by the existence of (massless) elementary fermion related to supersymmetry [6].

On the other hand, the stability of the asymptotically anti-de Sitter(AdS) space-time with compactified dimension has been focused recently by Horowitz & Myers [7] because AdS/CFT correspondence [8] links the stability of super-Yang-Mills theory to that of AdS space-time. So they suggested a positive energy conjecture in locally asymptotically AdS space-times.

In the actual cosmological context, same argument of stability is important. In this presentation, for simplicity, we consider D-dimensional Einstein gravity with a positive cosmological constant. A positive cosmological constant is essential for inflation universe.

In asymptotically flat space-times with compactified extra dimensions, there are examples of momentarily static initial slices such that the energy can be negative regardless of the size of the compactified dimensions [5]. A parallel argument helps

CP493, *General Relativity and Relativistic Astrophysics,* edited by C. P. Burgess and R. C. Myers
© 1999 American Institute of Physics 1-56396-905-X/99/$15.00

us to discuss about the lower bound of the energy in asymptotically de Sitter space-times. Since a positive cosmological constant prompts the rapid expansion of the universe, we might be able to obtain an implication about the effect to the energy from the dynamics of the compactified dimensions.

The definition of the energy in higher dimensions is straightforward extension of work by Nakao et al [9] based on [10] [11]. We consider D-dimensional space-times which satisfy the Einstein equation with a positive cosmological constant Λ,

$$R_{IJ} - \frac{1}{2}g_{IJ}R + \Lambda g_{IJ} = 8\pi G_D T_{IJ}, \tag{1}$$

where suffices I, J runs over $0, 1, ..., D-1$, T_{IJ} is the energy-momentum tensor and G_D is D-dimensional Newton's constant. We decompose the space-time metric into the D-dimensional de Sitter metric $\bar{g}_{IJ}$ and the rest h_{IJ};

$$g_{IJ} = \bar{g}_{IJ} + h_{IJ}. \tag{2}$$

Hereafter the notation 'over-bar' indicates the quantities of the background de Sitter space-time. We remember that h_{IJ} is not necessarily small, but we impose that it vanishes at infinity.

Repeating Nakao et al's argument, we obtained so-called Abbott-Deser energy in higher dimensions,

$$\begin{aligned}
E_{AD} &= a(t)\Big(E_{\text{ADM}} + \Delta P_{\text{ADM}}(-\bar{\xi})\Big) \\
&= \frac{a(t)}{16\pi G_D}\int d\bar{S}_i(\partial_j h^{ij} - \bar{g}^{ij}\partial_j h^k_k) \\
&\quad - \frac{a(t)}{8\pi G_D}\int d\bar{S}_i\Big[K^i_j - K\delta^i_j + (D-2)H\delta^i_j\Big]\bar{\xi}^j, \tag{3}
\end{aligned}$$

where K_{ij} is the extrinsic curvature of $t =$ constant $(D-1)$-dimensional hypersurface. By using the momentum $\pi_{ij} = K_{ij} - q_{ij}K$, where q_{ij} is the metric of $t -$ constant hypersurface, the second term in the second line of eq. (3) can be written as

$$\Delta P_{\text{ADM}}(-\bar{\xi}) = P_{\text{ADM}}(-\bar{\xi}) - \bar{P}_{\text{ADM}}(-\bar{\xi}) = -\frac{1}{8\pi G_D}\int d\bar{S}_i(\pi^i_j - \bar{\pi}^i_j)\bar{\xi}^j. \tag{4}$$

As a result, E_{AD} is written in terms of the sum of the ADM energy and net momentum. We note that the momentum of background de Sitter space-time is subtracted automatically.

In four dimension the AD mass can be negative when system has a large outward momentum [9]. On the other hand, the positivity of the AD or ADM mass is guaranteed for systems without the net momentum, $\Delta P_{\text{ADM}} = 0$, in four dimension [13].

EXAMPLES IN FIVE DIMENSIONS

Let us consider an initial slice with $K_{ij} = \pm H g_{ij}$ and $H = \sqrt{\Lambda/6}$. In this slice the Hamiltonian constraint becomes $^{(4)}R = 0$. Thus one can use the argument on the momentarily static slices in asymptotically flat cases because the Hamiltonian constraint is just same one.

One can see easily that the euclidian Reissner-Nordstrom metric with imaginary 'charge' ie satisfies the Hamiltonian constraint [5]. This metric of the hypersurface is given by

$$^{(4)}g = V(r)d\chi^2 + \frac{dr^2}{V(r)} + r^2 d\Omega_2^2 \tag{5}$$

where $V(r) = 1 - 2m/r - e^2/r^2$ and $r \geq r_+ := m + \sqrt{m^2 + e^2}$. To avoid a conical singularity at $r = r_+$, we assume the period $\chi_p = 4\pi/V'(r_+) = 2\pi r_+^2/(r_+ - m)$ along the χ-direction.

On the present slice, the mass is constructed by only the ADM energy component, $E_{AD} = E_{ADM} = m\chi_p/2G_5 = m/2$. By the same argument as Brill & Horowitz [5], it is shown that the mass becomes negative and does not have the lower bound. The mass can be set to be arbitrary negative regardless of the radius of the compactified space.

As a second example, we take a dynamical solution [14]. The metric is

$$ds^2 = -dt^2 + a(t)^2\Big[\frac{d\chi^2}{\Delta} + \Delta dr^2 + r^2\Delta^2 d\Omega_2^2\Big], \tag{6}$$

where $\Delta(r) = 1 - m/r$, $a(t) = e^{\pm Ht}$ and $H = \sqrt{\Lambda/6}$. The $t =$ constant hypersurface has the extrinsic curvature $K_j^i = \pm H\delta_j^i$. However, this space-time has timelike naked singularity at $r = m$. To avoid the naked singularity, we change the sign of the mass parameter, $m \to -m$. After that the radial coordinate r can run up to $r = 0$ and the conical singularity occurs at $r = 0$ in general cases. Near $r = 0$, the metric is written as

$$ds^2 \simeq -dt^2 + a(t)^2\Big[\frac{r}{m}d\chi^2 + \frac{m}{r}dr^2 + m^2 d\Omega_2^2\Big]. \tag{7}$$

Here we introduce a new coordinate $R = (rm)^{1/2}$ and the metric is

$$ds^2 \simeq -dt^2 + a^2\Big[4\Big\{dR^2 + R^2 d\Big(\frac{\chi}{2m}\Big)^2\Big\} + m^2 d\Omega_2^2\Big]. \tag{8}$$

Hence, the metric is regular everywhere except for $a = 0$ singularity if one assumes the period $\chi_p = 4\pi m$ along the χ-direction. The physical size of the compactified dimensions is given by $4\pi a(t)m$. Thus, it decreases/increases if one takes the collapsing/expanding chart. A cosmological constant make the compactified space dynamical as well as the four dimensional part. The AD mass is

$$E = E_{\text{ADM}} = -\frac{a(t)^3 m \chi_P}{G_5} = -a^3(t)m. \tag{9}$$

This is not conserved due to the expansion of the universe! This comes from the non-vanishing boundary term $\int d^{D-2} S_i T^i$ which vanishes for asymptotically flat cases.

POSITIVE ENERGY THEOREM, ELEMENTARY FERMION AND STABILITY

In locally asymptotically flat cases with compactified dimensions, the break down of positive energy theorem means non-existence of the Witten spinor [6]. In supergravity side, that means there are not supersymmetry because the spinor is related to infinitesimal generator of local supersymmetry [15].

In asymptotically de Sitter cases, the situation is different from the above. As we stated, the energy can be negative even if the extra dimension is not compactified and the topology is trivial. In this section, we discuss the positive energy theorem, based on [13] [16], in cases whose extra dimension is not compactified. Conversely, we can see easily that the Witten spinor cannot exist for examples given in the previous section.

Following Kastor&Traschen [16], we define the cosmological supercovariant derivative operator on a spinor ϵ as

$$\hat{\nabla}_I \epsilon = \left(\nabla_I + \frac{i}{2} H \gamma_I \right) \epsilon = \left(\partial_I \mid \Gamma_I + \frac{i}{2} H \gamma_I \right) \epsilon =: \left(\partial_I + \Gamma'_I \right) \epsilon. \tag{10}$$

The cosmological Witten equation is define by

$$\gamma^i \hat{\nabla}_i \epsilon = 0. \tag{11}$$

The solution is given by a constant spinor ϵ_0 satisfying $\gamma^0 \epsilon_0 = -i\epsilon_0$ in the expanding flat slice of de Sitter space-time.

By using the Bianchi identity, the Hamiltonian and momentum constraints, we obtain an identity

$$\int dS^i \epsilon^\dagger \hat{\nabla}_i \epsilon = \int dV \left[|\hat{\nabla}_i \epsilon|^2 + 4\pi G_D (\epsilon^\dagger T_{\hat{0}\hat{0}} \epsilon + \epsilon^\dagger T_{i\hat{0}} \gamma^i \gamma^0 \epsilon) \right] \tag{12}$$

Using the cosmological Witten equation, the left-hand side becomes

$$\int dS^i \epsilon^\dagger \hat{\nabla}_i \epsilon = \frac{1}{4} \int dS^i \epsilon_0^\dagger (\partial_j h_i^j - \partial_i h_j^j) \epsilon_0 - \frac{1}{2} \int dS_i \tilde{K}_j^i \bar{\epsilon}_0 \gamma^j \epsilon_0$$
$$= 4\pi G_D \left(E_{\text{ADM}} |\epsilon_0|^2 + \Delta P_{\text{ADM}} (\bar{\epsilon}_0 \gamma \epsilon_0) \right), \tag{13}$$

where $\tilde{K}_j^i$ is the traceless part of K_j^i. Here we note that $\bar{\epsilon}_0 \gamma^i \epsilon_0 = -\epsilon_0^i |\epsilon_0|^2 = O(1/r^{D-3})$ due to $\gamma^0 \epsilon_0 = -i\epsilon_0$. This and $\tilde{K}_j^i = O(1/r^{D-2})$ lead us that the net momentum term $\Delta P_{\text{ADM}} (\bar{\epsilon}_0 \gamma \epsilon_0)$ vanishes. Finally we obtain an inequality

255

$$E = \frac{a(t)}{4\pi G_D} \int dS^i \epsilon^\dagger \hat{\nabla}_i \epsilon = a(t) E_{\text{ADM}} |\epsilon_0|^2 \geq 0 \qquad (14)$$

under the dominant energy condition on the energy-momentum tensor T_{IJ}.

We can see that E is not equal to the AD mass, E_{AD}. When $\tilde{K}^i_j = O(1/r^{D-1})$ holds and the momentum term of the AD mass vanishes, the AD mass equals to the ADM energy and the positivity is guaranteed. For our purpose, it is worth imposing $\tilde{K}^i_j = O(1/r^{D-1})$ because we are interested in the stability of space-time, not its dynamics. As eq. (14) this implies the positivity of the AD mass. The apparent contradiction with examples given in the previous section indicates that the Witten spinor does not exist in such examples.

SUMMARY

In this presentation, we defined the gravitational mass in asymptotically de Sitter space-time with extra dimensions and obtained a refined expression related to the ADM energy and momentum associated to the timelike Killing vector of the background de Sitter space-time. Furthermore, we gave one dynamical solution and one initial data with the negative energy in five dimensions. Does these solution indicate the quantum decay of de Sitter space-time? We cannot reply to the question instantly because we do not know whether the instanton exists or not. Naively speaking, we can guess from the previous section that the decay occurs unless one imposes the existence of elementary fermion or supersymmetry.

REFERENCES

1. T. Shiromizu, hep-th/9902049, to be published in Phys. Rev. **D**
2. J. Polchinski, *String Theory* I & II(Cambridge Univ. Press, 1998)
3. E. Witten, Nucl. Phys. **B195**, 481(1982)
4. D. Brill and H. Pfister, Phys. Lett. **B228**, 359(1989)
5. D. Brill and G. T. Horowitz, Phys. Lett. **262**, 437(1991)
6. M. M. Taylor-Robinson, Phys. Rev. **D55**, 4822(1997)
7. G. T. Horowitz and R. C. Myers, Phys. Rev. **D59**, 026005(1999)
8. J. M. Maldacena, Adv. Theor. Math. Phys. **2**, 231(1998);
   E. Witten, Adv. Theor. Math. Phys. **2**, 253(1998)
9. K. Nakao, T. Shiromizu and K. Maeda, Class. Quant. Grav. **11**, 2059(1994)
10. L. F. Abbott and S. Deser, Nucl. Phys. **195**, 76(1982)
11. S. Deser and M. Soldate, Nucl. Phys. **B311**, 739(1988)
12. K. Nakao, K. Maeda, T. Nakamura and K. Oohara, Phys. Rev. **D44**, 1326(1991)
13. T. Shiromizu, Phys. Rev. **D49**, 5026(1994)
14. T. Maki and K. Shiraishi, Class. Qaunt. Grav. **12**, 159(1995)
15. G. T. Horowitz and A. Strominger, Phys. Rev. **D27**, 2793(1983);
    S. Deser, Phys. Rev. **D27**, 2805(1983)
16. D. Kastor and J. Traschen, Class. Quant. Grav. **13**, 2753(1996)

Dimensional Reduction of the Effective Action and the Multiplicative Anomaly

Patrick J. Sutton[1]

Theoretical Physics Institute, Department of Physics,
University of Alberta, Edmonton, AB, Canada T6G 2J1

Abstract. In a wide class of four-dimensional spacetimes which are direct or semidirect products of a homogeneous n-dimensional space and a $(4-n)$-dimensional space, a field can be decomposed into modes. As a result of this mode decomposition, the main objects which characterize the free quantum field, such as Green functions and heat kernels, can effectively be reduced to objects in a $(4-n)$-dimensional spacetime with an external dilaton field. We study the problem of the dimensional reduction of the effective action for such spacetimes. While before renormalization the original four-dimensional object can be presented as a "sum over modes" of $(4-n)$-dimensional objects, this property is violated after renormalization. We calculate the corresponding anomalous terms and relate their origin with the effect of the multiplicative anomaly. This effect is demonstrated with some simple examples.

INTRODUCTION

Many spacetimes of physical interest can be expressed as a semidirect product of a homogeneous space with another space, where the line element takes the form

$$ds^2 = \gamma_{ab}(x)dx^a dx^b + e^{-2\phi(x)}\omega_{ij}(y)dy^i dy^j \ . \tag{1}$$

Here ω_{ij} is the metric of an n-dimensional homogeneous space, and γ_{ab} is arbitrary. The scalar ϕ is known as the dilaton. Let us consider a massless, minimally-coupled scalar field Φ propagating on such a spacetime. For the line element (1), the d'Alembertian operator $\Box$ decomposes as

$$\Box = \Delta_\gamma - n\nabla\phi \cdot \nabla + e^{2\phi}\Delta_\omega \ , \tag{2}$$

where Δ_γ, Δ_ω are the d'Alembertians corresponding to the metrics γ_{ab}, ω_{ij} respectively, and ∇ denotes the covariant derivative with respect to the metric γ_{ab}. Meanwhile, the field $\Phi(x, y)$ may be decomposed in terms of the eigenmodes $Y_\lambda(y)$ of Δ_ω on the homogeneous space as [1]

---

[1] e-mail: psutton@phys.ualberta.ca

CP493, *General Relativity and Relativistic Astrophysics*, edited by C. P. Burgess and R. C. Myers
© 1999 American Institute of Physics 1-56396-905-X/99/$15.00

$$\Phi(x, y) \longrightarrow e^{\frac{n}{2}\phi} U(x) Y_\lambda(y) . \tag{3}$$

Combining (2,3) shows that the bare quantum field theory in four dimensions described by

$$\Box \Phi = 0 \tag{4}$$

reduces to a collection of $(4 - n)$-dimensional theories

$$(\Delta_\gamma - V(\lambda)) U(x) = 0 , \tag{5}$$

one for each eigenmode Y_λ of Δ_ω. Here the induced potential V is

$$V(\lambda) = -\Delta_\gamma \phi + (\nabla \phi)^2 + \lambda e^{2\phi} . \tag{6}$$

It is not difficult to show that the non-renormalized quantum field theories (4,5) are equivalent in the sense that four-dimensional quantities like the effective action $W^\Box$ and the heat kernel $K^\Box$ can be constructed as the sum over modes λ of the corresponding $(4 - n)$-dimensional quantities:

$$W^\Box = \sum_\lambda W^{\Delta_\gamma - V(\lambda)} , \tag{7a}$$

$$K^\Box(x, y; x', y'|s) = e^{\frac{n}{2}(\phi(x) + \phi(x'))} \sum_\lambda K^{\Delta_\gamma - V(\lambda)}(x, x'|s) Y_\lambda(y) \bar{Y}_\lambda(y') . \tag{7b}$$

As we shall see from some simple examples, relations like (7) break down for renormalized quantum field theories, for the simple reason that the divergent quantities which must be subtracted to obtain finite results do not obey decompositions of this form. We call this effect the **dimensional reduction anomaly**. It is related to the multiplicative anomaly [2,3], in which for operators D_1, D_2 with divergent determinants

$$\det(D_1 D_2) \neq \det(D_1) \times \det(D_2) \tag{8}$$

after renormalization.

FLAT SPACE EXAMPLE

The simplest example of the dimensional reduction anomaly comes from the decomposition of a flat-space quantum field theory into spherical modes.

The divergences in flat-space quantum field theory in d dimensions are eliminated by subtracting off the contribution of the first term in the Schwinger-DeWitt expansion of the heat kernel [4]:

$$K_{div}^{\square}(x, x'|s) = \frac{1}{(4\pi s)^{\frac{d}{2}}} \exp\left\{-m^2 s - \frac{\sigma}{2s}\right\} . \tag{9}$$

Here $\sigma(x, x')$ is one-half the square of the geodesic distance between x and x', which for four-dimensional flat spacetime in spherical coordinates is

$$2\sigma = \Delta t^2 + \Delta r^2 + 2rr'(1 - \cos\theta) \tag{10}$$

where θ is the angle between x and x'. The part of $K_{div}^{\square}$ with angular momentum number l is extracted by integrating over the two-sphere with the Legendre polynomial $P_l(\cos\theta)$:

$$K_{div|l}^{\square}(t, r; t', r'|s) = 2\pi \int_0^\pi d\theta \sin\theta \, P_l(\cos\theta) \, K_{div}^{\square}(x, x'|s) . \tag{11}$$

For example, for the $l = 0$ mode we have in the coincidence limit $(t', r') = (t, r)$,

$$r^2 K_{div|l=0}^{\square} = \frac{e^{-m^2 s}}{4\pi s}\left[1 - \exp\left(-\frac{r^2}{s}\right)\right] . \tag{12}$$

Meanwhile, for the effective two-dimensional theory $V = 0$ for $l = 0$, so the divergences in two dimensions are removed by subtracting just the free heat kernel,

$$K_{div}^{\Delta_\gamma - V(0)} = \frac{e^{-m^2 s}}{4\pi s} . \tag{13}$$

We see that the standard heat kernel (12) obtained by renormalizing before performing the mode decomposition has a term proportional to $e^{-r^2/s}$ which is missed by naïve renormalization in two dimensions [2]. It is not difficult to show that this anomalous term is precisely equivalent to imposing Dirichlet boundary conditions at $r = 0$ in the two-dimensional theory, and is a natural consequence of the fact that $r \geq 0$ for the four-dimensional spacetime. This shows that for the two-dimensional theory to correctly reproduce results for renormalized quantities in four dimensions, an additional boundary condition must be imposed.

For $l \neq 0$ there are additional anomalous terms connected with the nonvanishing induced potential $V(l)$ (the centrifugal barrier).

STATIC SPACE EXAMPLE

A slightly more complicated example demonstrates the effect of spacetime curvature on the anomaly. We consider the anomaly in $\langle\Phi^2\rangle$ in a four-dimensional static spacetime that results from decomposing the field in terms of Fourier time modes $\exp\{i\omega t\}$ before rather than after renormalization.

---

[2] From (7b) we expect $r^2 K_{div|l}^{\square} = K_{div}^{\Delta - V(\lambda)}$ in the absence of the anomaly, with $e^{-2\phi} = r^2$.

The line element for a static spacetime may be written as

$$ds^2 = e^{-4\phi(x)}dt^2 + h_{ab}(x)dx^a dx^b \ . \tag{14}$$

One may obtain $\langle \Phi^2 \rangle$ as the coincidence limit of the Green function, which is divergent and hence requires renormalization for a finite $\langle \Phi^2 \rangle$. The quantity to be subtracted in four dimensions may be shown to be

$$G^{\square}_{(div)}(t, x; t', x') = \int_0^\infty ds \frac{\Delta^{\frac{1}{2}}}{(4\pi s)^2} \exp\left\{ -m^2 s - \frac{2\sigma + \epsilon^2}{4s} \right\} \left[1 - \frac{1}{6}Rs \right] \ , \tag{15}$$

where R is the scalar curvature and Δ is a two-point function of the geometry [4]. Here ϵ is a regularization parameter to be taken to zero at the end of the calculation. The Fourier decomposition of this divergent quantity can be evaluated approximately using an expansion of (15) for small curvatures, with the result

$$\int_{-\infty}^\infty dt\, e^{i\omega t} G^{\square}_{(div)}(t, x; 0, x) \simeq \frac{e^{2\phi}}{4\pi} \left[\frac{1}{\epsilon} - \mu + \frac{1}{12}\frac{R}{\mu} + \frac{m^2}{12}\frac{R_t^t}{\mu^3} + \frac{m^4}{2}\frac{(\nabla\phi)^2}{\mu^5} + O(\epsilon) \right] \tag{16}$$

where $\mu \equiv (m^2 + \omega^2 e^{4\phi})^{\frac{1}{2}}$.

Meanwhile, for the three-dimensional theory on the space with metric h,

$$G'^{\Delta_h - V}_{(div)}(x, x) = \int_0^\infty ds \frac{1}{(4\pi s)^{\frac{3}{2}}} \exp\left\{ -m^2 s - \frac{\epsilon^2}{s} \right\}$$

$$= \frac{1}{4\pi} \left[\frac{1}{\epsilon} - m + O(\epsilon) \right] \ . \tag{17}$$

The anomaly for each ω is the difference[3] between (16,17), which is the difference in the renormalized values of $\langle \Phi^2 \rangle$ as computed with renormalization occuring before and after the decomposition:

$$\Delta\langle\Phi^2(x)\rangle_\omega = \frac{1}{4\pi} \left[\mu - m - \frac{1}{12}\frac{R}{\mu} - \frac{m^2}{12}\frac{R_t^t}{\mu} - \frac{m^4}{2}\frac{(\nabla\phi)^2}{\mu^5} \right] \ . \tag{18}$$

These results are easily extended to include arbitrary coupling to the scalar curvature.

A particularly interesting feature of this anomaly is that from it one may obtain an approximation for $\langle \Phi^2 \rangle$ in the original four-dimensional static space which reproduces the Killing [5] and Page-Brown-Ottewill [6] approximations in the conformally-coupled case, and generalizes these approximations to nonzero mass and arbitrary coupling. It also reproduces the analytic approximation of Anderson et. al. [7] for static spherically symmetric spacetimes, but is valid for arbitrary static spacetimes. Details of this procedure may be found in a future publication.

---

[3] As in (12), the factor $e^{2\phi}$ in (16) is expected from (7b) and is not part of the anomaly.

DISCUSSION

Quantum field theories residing on a spacetime which is the semidirect product of a homogeneous spacetime may be decomposed in terms of field modes on the homogeneous space. This effectively reduces the system from a D-dimensional theory to a collection of $(D - n)$-dimensional theories, one for each field mode on the homogeneous space. However, as has been demonstrated, while the bare theories in D and $(D - n)$ dimensions are equivalent under this decomposition, the renormalized theories are not. As a result, a D-dimensional renormalized quantity is not equal to the sum over all modes of the corresponding renormalized quantity for each $(D - n)$-dimensional theory. This effect, termed the dimensional reduction anomaly, could be of significance in problems such as the calculation of the s-mode contribution to Hawking radiation for four-dimensional black holes using two-dimensional models [1,8,9]. In addition, the anomaly may be of some use in obtaining approximations for field-theoretic quantities in curved spacetimes.

This work was done in collaboration with Valeri Frolov and Andrei Zelnikov of the University of Alberta.

REFERENCES

1. Mukhanov, V., Wipf, A., and Zelnikov, A., *Phys. Lett* **B 332**, 283 (1994).
2. Evans, T. S., *Phys. Lett.* **B 457**, 127 (1999).
3. Elizalde, E., Filippi, C., Vanzo, L., and Zerbini, S., hep-th/9804071.
4. Birrell, N. D., and Davies, P. C. W., *Quantum fields in curved space*, Cambridge University Press, 1982.
5. Frolov, V. P., and Zel'nikov, A. I., *Phys. Rev.* **D 35**, 3031 (1987).
6. Brown, M. R., Ottewill, A. C., and Page, D. N., *Phys. Rev.* **D 33**, 2840 (1986).
7. Anderson, P. R., Hiscock, W. A., and Samuel, D. A., *Phys. Rev.* **D 51**, 4337 (1995).
8. Kummer, W., and Vassilevich, D. V., hep-th/9811092.
9. Balbinot, R., and Fabbri, A., *Phys.Rev.* **D 59** 044031 (1999).

Definitions of Particle Mass
in Kaluza-Klein Theory

P. S. Wesson and W. N. Sajko

Department of Physics, University of Waterloo, Waterloo, Ontario, N2L 3G1, Canada;
wesson@astro.uwaterloo.ca

Abstract. The definition of particle rest mass in a 5D space, like the definition of mass in 4D general relativity, is difficult. We review 6 alternatives. It is plausible that the mass of a particle derives from the geometry of a 4D space embedded in a 5D space, and that the Klein-Gordon equation of 4D quantum theory is a wave equation in a conformal factor of 5D Kaluza-Klein theory.

I   INTRODUCTION

The definition of mass or energy in general relativity is difficult, insofar as what appears to be a unique physical object has several different and inequivalent definitions in 4D Einstein theory [1]. Some clarification of this problem has in recent years been gained by the consideration of 5D Kaluza-Klein theory [2]. Algebraically, it is always possible to embed a 4D space with field equations $G_{\alpha\beta} = 8\pi T_{\alpha\beta}$ in a 5D space with field equations $R_{AB} = 0$, provided the geometry is left unrestricted and dependent on the extra coordinate $x^4 = \ell$. (See ref. 3. Greek indices run 0-3, Latin indices run 0-4; and G, R, T label the Einstein, Ricci and energy-momentum tensors respectively.) This approach identifies the *density* of matter $\rho = \rho(x^4)$ as a property of the extra dimension in a 5D manifold. There are now several treatments available of this, including numerous exact solutions of the 5D field equations which reproduce known 4D physics [2]. They share the property that the form of $\rho = \rho(x^4)$ is changed by a 5D coordinate transformation $\bar{x}^A = \bar{x}^A(x^B)$, in contrast to the situation in general relativity where $\rho(x^\alpha)$ is unchanged by a coordinate transformation $\bar{x}^\alpha = \bar{x}^\alpha(x^\beta)$. This is simply because ρ is a 4D object and the theory is 5D in nature.

Following from this, if one considers that a fluid of density ρ consists of particles of rest mass m, one should on physical grounds be able to define $m = m(x^4)$. There have in fact been several attempts to do this, involving conformal invariance in 4D geometry [4], Mach's principle as a 4D manifestation of 5D geometry [5], and a redefinition of the 4D energy- momentum tensor as derived from 5D geometry [6],

CP493, *General Relativity and Relativistic Astrophysics*, edited by C. P. Burgess and R. C. Myers
© 1999 American Institute of Physics 1-56396-905-X/99/$15.00

[7]. However, these approaches are all from classical theory, and meet a counter-approach derived from quantum theory: in the latter, the rest mass m of a particle is effectively defined from the Klein-Gordon equation $\Box^2 \psi + m^2 \psi = 0$. [Here $\Box^2 = \eta^{\alpha\beta}(\partial/\partial x^\alpha)(\partial/\partial x^\beta)$ is the wave operator in a flat 4D space which in the diagonal case has signature $(+, -, -, -)$, and $\psi = \psi(x^\alpha)$ is the wave function which effectively spreads the wave corresponding to a particle of mass m throughout 4D Minkowski space.] In what follows, definitions of m will be considered from classical theory with a view to their compatibility with the definition from quantum theory.

As in general relativity [1], it will be found that there are several overlapping but in general inequivalent definitions of m. However, judgement is aided by certain technical results in the literature [8]- [11]. It appears that the mass of a particle may be related to the geometry of a 4D space embedded in a 5D space.

II   DEFINITIONS OF PARTICLE MASS IN KALUZA-KLEIN SPACE

In this section, some alternative ways will be listed of defining the rest mass m of a particle as it appears in 4D Einstein theory in terms of the geometry of 5D Kaluza-Klein theory. Apart from the success of the induced-matter version of Kaluza-Klein theory applied to fluids [2], a motivation for the following is to replace the fact that m is just a *given* in 4D theory by a rationale to *derive* $m = m(x^4)$ from 5D theory.

(a) Wave equations in Kaluza-Klein theory could in principle yield in an appropriate limit the Klein-Gordon equation of quantum theory. In the latter, the energy, momentum and rest mass of a particle are related by $E^2 - p^2 = m^2$, so for 4D mass vectors k^α that represent E, p one might expect to find a relation like $k^\alpha k_\alpha = m^2$. Alternatively, if the mass is itself represented by a component of a wave vector (see below), one might expect to find a relation like $k^A k_A = 0$. Now one can certainly obtain wave equations in 5D, most readily by following the procedure used in 4D. That is, consider a nearly-flat metric ($g_{AB} = \eta_{AB} + h_{AB}, \eta_{AB} >> h_{AB}$), linearize the Ricci tensor using the harmonic coordinate condition, and obtain $\Box^2 h_{AB} = 0$ [8], [9]. However, two problems arise. First, the harmonic coordinate condition is not covariant, so the physics is not preserved between hypersurfaces $x^4 = $ constants. Second, the particles associated with the waves are massless [11], describing gravitational, electromagnetic and scalar interactions of infinite range.

(b) The field equations in Kaluza-Klein theory number 15, and as noted above in the form $R_{AB} = 0$ contain the 10 Einstein equations $G_{\alpha\beta} = 8\pi T_{\alpha\beta}$ with an effective or induced energy-momentum tensor [2]. As is well known, the Einstein equations applied to a gravitating fluid of density ρ can be reduced to Poisson's equation $\nabla^2 \phi = 4\pi\rho$ where ϕ is a potential. (Actually, the standard procedure is ambiguous, but the important thing is that the Einstein equations can be contracted to $G = 8\pi T$ and reduced to 1 field equation.) By analogy, it is possible that the Kaluza-Klein equations could be applied to a particle and reduced to the Klein-

Gordon equation $\Box^2\psi + m^2\psi = 0$ (this would presumably be equivalent to $R = 0$ where R is the 5D Ricci scalar). However, it is not clear how to pick out a particle-like potential, and the latter would have to be complex in order to match the properties of the wave function.

(c) The geodesic equation in 5D minimizes an interval given by $dS^2 = g_{AB}dx^A dx^B$, and yields equations of motion for the 5-velocities $U^A \equiv dx^A/dS$ which are subject to the metric constraint $U^A U_A = 0$ or 1 depending on whether the path is null or not. The same kind of constraint exists in 4D, and in fact the weak-field limit for a massive particle is equivalent to the relation $E^2 - p^2 = m^2$ noted above. The latter is in turn equivalent to the Klein-Gordon equation $\Box^2\psi + m^2\psi = 0$ with $\psi = \psi_o e^{i(Et-px)}$ as the wave function for a free particle. The point is that while the Klein-Gordon equation of flat-space quantum theory resembles a field equation in $\psi = \psi(x^\alpha)$, and while $\psi\psi^*$ may be interpreted as a probability density, that equation is connected to dynamics. This conclusion is not altered by generalizations of the ordinary Klein-Gordon equation to curved 4D and 5D spaces (where the partial derivative is replaced by a covariant derivative). Thus if one wishes to define particle mass m from the Klein-Gordon equation, the latter should be derivable from or at least compatible with the geodesic equation.

(d) The extra coordinate $x^4 = \ell$ of Kaluza-Klein theory can under certain circumstances play the role of particle rest mass m [2], [10]. To be specific, if the 5 available coordinate degrees of freedom for the metric are used to set $g_{4\alpha} = 0$ and $g_{44} = -1$, thereby suppressing the electromagnetic and scalar fields, the "mechanical" part of the metric can be written $dS^2 = \ell^2 \, g_{\alpha\beta}(x^\alpha, \ell) \, dx^\alpha dx^\beta - d\ell^2$. This so-called canonical form leads to great simplification in dynamical problems, and if $\partial g_{\alpha\beta}/\partial\,\ell = 0$ the spacetime part of the geodesic in Kaluza-Klein theory is identical to the geodesic in Einstein theory. On a hypersurface $\ell = $ constant, $dS = \ell\,ds$ where $ds^2 = g_{\alpha\beta}\,dx^\alpha dx^\beta$, so already it is clear that ℓ plays the part of m in the action principle $\delta\,[\int m\,ds] = 0$ of particle physics. It transpires that under appropriate conditions the constant of the motion in the 5D problem becomes just the energy in the 4D problem if one makes the identification $\ell = m$. This is important; but it should be appreciated that the correspondence is only exact in one system of coordinates on a classical manifold, and that no contact is apparent with the Klein-Gordon equation of quantum theory.

(e) Redefining the energy-momentum tensor in terms of particle properties rather than fluid properties opens a new vista for explaining m [6], [7]. In one such approach, a curved 4D space embedded in a 5D space has principal curvatures related to the de Broglie wavelengths of a particle [7]. The 4D scalar curvature, as measured by either G or R, is related to the Compton wavelength or mass of the particle. A specific model can be constructed wherein the 4D Ricci tensor is related to the product of the 4-momenta via $R_{\alpha\beta} = k\,p_\alpha p_\beta$ where k is a coupling constant. Then $R = km^2$ measures the mass and is compatible with the relation $E^2 - p^2 = m^2$ of particle physics. The 4D field equations are $G_{\alpha\beta} = k(p_\alpha p_\beta - m^2 g_{\alpha\beta}/2)$, wherein the right-hand side is now a dynamical object relevant to a particle. (Here and elsewhere units are chosen to make Newton's constant, the speed of light and

264

Planck's constant equal to unity, and the last in particular would appear in the last equation if conventional units were restored.) This approach has great philosophical appeal, saying that a particle is a 4D object which viewed as a wave is defined by the geometry of the 4D space it inhabits. However, it only really works in 5D, since the 4D space is embedded in a (possibly flat) 5D space which induces the new $T_{\alpha\beta}$. There are also two technical objections to this approach. First, it is not always possible to write the Ricci tensor as a product of two 4-vectors, and this needs to be elucidated, presumably using the Segré classification scheme [9]. Second, while the Klein-Gordon equation can be formally recovered from the extra field equation $R_{44} = 0$, there is an ambiguity involved that concerns a conformal factor which appears in the 4D and extra parts of the 5D metric.

(f) Conformal factors have been implicated in defining the mass of a particle both in 4D [4] and 5D [7], and to end this list it may be relevant to make some observations about such factors. A conformal function of coordinates applied to a known metric effectively introduces a new scalar field which can in principle be related to m. Also, while classical physics in the old frame may be couched in real quantities, the conformal function may be complex, providing a link to quantum physics. However, a conformal transformation is not in general a coordinate transformation, so in effect conformal factors introduce new solutions. Consider the following 4D algebra [8]. A new metric g_{ij} is obtained from an old metric $\bar{g}_{ij}$ by the application of a conformal factor $e^{\Psi(x)}$, where $g_{ij} = e^{\Psi}\bar{g}_{ij}$. The Christoffel symbols are related by $\Gamma^i_{jk} = \bar{\Gamma}^i_{jk} + A^i_{jk}$ where $A^i_{jk} \equiv \left(\delta^i_j\Psi_k + \delta^i_k\Psi_j - \bar{g}_{jk}\bar{g}^{ia}\Psi_a\right)/2$. Here $\Psi_k \equiv \partial\Psi/\partial x^k$ and A^i_{jk} (since it is obtained from derivatives of a scalar) is a tensor. Define the scalar quantities $\chi \equiv \bar{g}^{ab}\Psi_a\Psi_b$ and $\Box\Psi = \bar{g}^{ab}\Psi_{;ab}$ where the semicolon denotes the covariant derivative. Then the Ricci tensors for the new and old metrics are related by

$$R_{jk} = \bar{R}_{jk} + \Psi_{;jk} - \frac{1}{2}\Psi_j\Psi_k + \frac{1}{2}\bar{g}_{jk}(\Box\Psi + \chi), \qquad (2.1)$$

and the Ricci scalars are related by

$$R = g^{jk}R_{jk} = e^{-\Psi}\bar{g}^{jk}R_{jk} = e^{-\Psi}\left(\bar{R} + 3\Box\Psi + \frac{3}{2}\chi\right). \qquad (2.2)$$

Now in 4D, these relations are not very useful. Firstly, generating a new R_{jk} by a conformal transformation generates a new G_{jk} which can be matched by a new T_{jk} which is independent (alternatively: a conformal transformation changes the geometry, and there is always a corresponding energy-momentum tensor, even if it is unphysical). Secondly, while a scalar wave equation $\Box\Psi + \chi/2 = 0$ would result from the condition $R = \bar{R} = 0$, there is no reason in 4D to impose such a condition. In 5D, the situation is different. The field equations have to hold, so $R_{AB} = \bar{R}_{AB} = 0$ become an alternative set of 15 field equations in the conformal factor Ψ, and $R = \bar{R} = 0$ becomes a wave equation with a (non-linear χ) source. In the appropriate limit, and with the 5D Ψ a separable function of a 4D ψ and an extra factor with wave number m, we obtain $\Box^2\psi + m^2\psi = 0$, or the regular Klein-Gordon equation. Of course this equation (or equivalently $R = \bar{R} = 0$) is derived in effect by contracting a set of 15 relations, and the latter may in practice be unduly restrictive. (Also, it may or may not be the case that the 15 wave relations when contracted with Ψ^k give 5 relations compatible with the 5 components of the geodesic equation mentioned above.) However, in principle it is possible that a classical wave equation in a (complex) conformal factor will yield the quantum Klein-Gordon equation.

III CONCLUSION

The 6 alternative ways reviewed above of defining the rest mass m of a particle overlap somewhat, most notably (a), (b) and (f). There is also a commonality involving a conformal factor in (d), (e) and (f). It would be pedantic to choose one or another of the alternatives (a)-(f), and it is in fact clear that the definition of m is still some way from resolution. However, a plausible view is that m is derived from the geometry of a 4D space embedded in a 5D space, and that the Klein-Gordon equation of 4D quantum theory is a wave equation in a (complex) conformal factor of 5D Kaluza-Klein theory. The details need to be worked out.

IV ACKNOWLEDGEMENTS

Thanks for comments go to H. Liu, and for financial support to NSERC and OGS.

REFERENCES

1. Hayward, S. A., *Phys. Rev.* **D49**, 831 (1994).
2. Overduin, J. M., Wesson, P. S., *Phys. Rep.* **283**, 303 (1997).
3. Wesson, P. S., Ponce de Leon, J., *Jour. Math. Phys.* **33**, 3883 (1992).
4. Hoyle, F., Narlikar, J. V., *Action at a Distance in Physics and Cosmology* (Freeman, San Francisco, 1974).
5. Wesson, P. S., Ponce de Leon, J., *Gen. Rel. Grav.* **26**, 555 (1994).
6. Matute, E. A., *Class Quant. Grav.* **14**, 2771 (1997).
7. Liu, H., Wesson, P. S., Int. Jour. Mod. Phys. **D5**, 737 (1998).
8. Synge, J. L., *Relativity: The General Theory* (N. Holland, Amsterdam, 1960).
9. Kramer, D., Stephani, H., Herlt, E., MacCallum, M., Schmutzer, E., *Exact Solutions of Einstein's Field Equations* (Cambridge U.P., Cambridge, 1980).
10. Wesson, P. S., Liu, H., *Int. Jour. Theor. Phys.* **36**, 1865 (1997).
11. Sajko, W. N., Wesson, P. S., Liu, H., *J. Math. Phys.* **39**, 2193 (1998).

Energy and Motion
in Kaluza-Klein Gravity

W. N. Sajko[1] and P. S. Wesson[2]

Department of Physics, University of Waterloo, Waterloo, Ontario, N2L 3G1, Canada

Abstract. We show that the assumption of 5D null particle trajectories in five-dimensional Kaluza-Klein gravity can correspond to either four-dimensional massive or null trajectories when the parameterization is chosen properly. We retain the extra-coordinate dependence in the metric and show the possibility of a cosmological variation in the rest masses of particles, and a consequent departure from 4D geodesic motion by a fifth force.

I   INTRODUCTION

The modern version of non-compactified 5D Kaluza-Klein gravity, in which the cylinder condition has been eliminated in favour of retaining the extra coordinate dependence in the metric components, has had great success in describing 4D general relativity with an induced energy-momentum tensor (see [1] for a recent review). The 5D spacetime can be viewed as a foliation of 4D sheets on which general relativity holds and a stress-energy tensor is induced through the metric dependence on the extra coordinate [2]. This procedure is always mathematically possible due to local embedding theorems which state that a 4D Riemannian manifold can be locally embedded in a 5D Ricci-flat Riemannian manifold [3], [4]. In what follows we first derive the 4D induced matter from a 5D vacuum using a 4+1 approach analogous to the 1+3 lapse-shift decomposition used for 4D GR. We then show that 5D null geodesics reproduce null 4D geodesics, but not 4D geodesics for massive particles. The acceleration of massive particles is due to a fifth force which depends on a scalar field and has an explicit dependence on the extra dimension. We interpret the velocity in the fifth dimension and we relate this to a mass variation for massive particles. We close off by examining a 5D cosmological solution and make some final comments.

---

[1] email : wnsajko@astro.uwaterloo.ca
[2] email : wesson@astro.uwaterloo.ca

CP493, *General Relativity and Relativistic Astrophysics*, edited by C. P. Burgess and R. C. Myers
© 1999 American Institute of Physics 1-56396-905-X/99/$15.00

II 4D INDUCED MATTER FROM 5D VACUUM

We wish to derive the induced matter resulting from the reduction of a 5D vacuum to a 4D hypersurface. Consider the following ansatz for the 5D metric which explicitly depends on the extra coordinate $x^4 \equiv l$. We factor out a conformal dependence on the 4D metric and include a scalar field so that the 5D metric can be written as

$$\hat{g}_{AB} = \begin{pmatrix} \frac{l^2}{L^2} g_{\alpha\beta}(x^\Sigma, l) & 0 \\ 0 & -\phi^2(x^\Sigma, l) \end{pmatrix}, \tag{1}$$

where a length scale L has been introduced. Here and in what follows we use hats to designate 5D quantities and unhatted for 4D quantities. Also, uppercase Latin letters are used for the 5D manifold, and lowercase Greek indices are used for the 4D manifold. To determine the induced matter on the 4D hypersurfaces ($l = l_o = const.$) we decompose the 5D metric using a 4+1 decomposition. This procedure was initially carried out in [2], and for the above metric (1) the components of the 5D vacuum field equations $\hat{R}_{AB} = 0$ are:

$$\hat{R}_{\alpha\beta} = 0 \quad \Rightarrow \quad R_{\alpha\beta} = \frac{1}{\phi}\nabla_\alpha \nabla_\beta \phi + \frac{1}{\phi}\partial_l K_{\alpha\beta} - (K K_{\alpha\beta} - 2K_{\alpha\gamma} K^\gamma{}_\beta), \tag{2}$$

$$\hat{R}_{l\alpha} = 0 \quad \Rightarrow \quad \nabla_\alpha (K^\alpha{}_\beta - \delta^\alpha{}_\beta K) = 0, \tag{3}$$

$$\hat{R}_{ll} = 0 \quad \Rightarrow \quad \Box\phi = -\partial_l K + K^{\alpha\beta} K_{\alpha\beta}\phi. \tag{4}$$

Here the extrinsic curvature of the embedded 4D hypersurfaces is defined as

$$K_{\alpha\beta} = -\frac{1}{2\phi}\partial_l \left(\frac{l^2}{L^2} g_{\alpha\beta}(x^\Sigma, l) \right). \tag{5}$$

It is evident that the extra coordinate dependence in the 4D metric plays a crucial role in inducing matter in 4D. However, if the cylinder condition ($\partial_l g_{\alpha\beta} = 0$) is enforced the only consistent solution to the above equations is

$$\partial_l g_{\alpha\beta} = 0 \quad \Rightarrow \quad R_{\alpha\beta} = -\frac{3}{L^2} g_{\alpha\beta}, \quad \phi = 1. \tag{6}$$

This can be identified as deSitter vacuum matter ($\rho c^2 = -p = \Lambda c^4/8\pi G$), provided the constant L is identified via

$$\frac{3}{L^2} \equiv \Lambda. \tag{7}$$

When the cylinder condition is lifted ($\partial_l g_{\alpha\beta} \neq 0$) the extra terms generated by the derivatives wrt the extra coordinate and the scalar field terms can be viewed as the matter contribution to the stress-energy, whereas terms containing the constant L are related to the vacuum stress-energy. We now investigate how the induced matter influences particle dynamics.

III MOTION, MASS VARIATION, AND THE FIFTH FORCE

To study particle dynamics in 5D Kaluza-Klein gravity with the metric (1) we begin by making a choice of path parameterization. We choose the 5D null parameterization based on mathematical simplicity and the appealing feature that it naturally sets the velocity in the extra dimension. To see this we write the null condition using the 5D metric,

$$\text{5D null}\quad d\hat{s}^2 = 0 \quad \Rightarrow \quad \frac{l^2}{L^2}\, g_{\alpha\beta}\frac{dx^\alpha}{d\lambda}\frac{dx^\beta}{d\lambda} = -\epsilon\,\phi^2\frac{dl^2}{d\lambda^2}\,, \tag{1}$$

where a path parameter λ has been inserted. Now choosing $\lambda = s$ to measure 4D proper distances we obtain

$$\left(\frac{\dot{l}}{l}\right)^2 = -\frac{\epsilon L^2}{\phi^2}\, g_{\alpha\beta}\, u^\alpha u^\beta = -\frac{\epsilon L^2}{\phi^2}\Sigma\,, \tag{2}$$

where $\Sigma = u_\alpha u^\alpha = 1, 0$ for massive particles and photons respectively (here and below $\dot{l} \equiv dl/ds$ and $u^\alpha \equiv dx^\alpha/ds$). This relation constrains the velocity $\dot{l}$ but does not give it physical meaning. For this we turn to the space-time-mass approach of Kaluza-Klein gravity [5], in which mass is geometrized via $l = Gm/c^2$. Thus it is evident that the variation of rest mass is a function of 4D path parametrization. Since in 4D we have $u_\alpha u^\alpha = 0$ for photons, this implies $\dot{l}/l = 0$ so there is no variation in a photon's rest mass and the mass of a photon may consistently be set to zero. However, for 4D paths which have $u_\alpha u^\alpha = 1$ we have a variation in rest mass given by

$$\frac{\dot{m}}{m} = \pm\frac{1}{\phi L}\,. \tag{3}$$

Thus the scalar field ϕ behaves as a Higgs-type field which generates a rest mass variation. We now turn our attention to the acceleration equation.

Some algebra is required to show that the 4D acceleration equation implied by the 5D null geodesic can be reduced to the form

$$u^\beta \nabla_\beta u^\alpha = -h^{\alpha\gamma}\left(\Sigma\frac{\phi_\gamma}{\phi} + \partial_l g_{\gamma\beta}\, u^\beta\, \dot{l}\right) \equiv f^\alpha\,, \tag{4}$$

where f^α is a force per unit rest mass and $h^{\alpha\beta} = g^{\alpha\beta} - u^\alpha u^\beta$ is the usual projection tensor. When the cylinder condition is imposed, $f^\alpha = 0$, which gives geodesic motion for both photons and massive particles in the 4D deSitter vacuum. However, when the cylinder condition is lifted, photons will still travel along null 4D geodesics since they obey $\Sigma = 0$ and $\dot{l} = 0$, but massive particles will experience a type of fifth force since $\Sigma = 1$ and $\dot{l} \neq 0$. This in principle could be used to test for the existence of the scalar field. In particular, (4) can be related to the weak equivalence principle and tests of it, though the departure from the classical results are tiny [7], [8]. We now consider some examples to elucidate these ideas.

IV   A COSMOLOGICAL EXAMPLE

In this section we consider a simple example of how the rest masses of particles may vary in a cosmological frame which employs a comoving coordinate system, and make some comments about observability. The 5D cosmological solution we investigate is a one-parameter class found by Ponce de Leon [9]

$$d\hat{s}^2 = \frac{l^2}{L^2} \left(c^2 dt^2 - \left(\frac{ct}{L}\right)^{2/\alpha} \left(\frac{l}{L}\right)^{2\alpha/(1-a)} d\vec{x} \cdot d\vec{x} \right) - \left(\frac{\alpha}{1-\alpha}\right)^2 \left(\frac{ct}{L}\right)^2 dl^2 . \quad (1)$$

Since the 4D metric components have extra coordinate dependence, the induced field equations are not that of a pure vacuum, but of more general matter. If we assume that the cosmological matter can be modeled as a perfect fluid, the energy density and the equation of state are [10]:

$$\rho c^2 = \frac{3}{8\pi G} \left(\frac{Lc}{\alpha l t}\right)^2 \quad \text{and} \quad p = \left(\frac{2}{3}\alpha - 1\right) \rho c^2 . \quad (2)$$

Thus we see that the parameter α characterizes the equation of state. In fact, using the definition of the cosmological constant for the deSitter vacuum (7) we can define an effective cosmological constant

$$\Lambda_{eff} = \Lambda_{vac} \left(\frac{L^2}{\alpha l c t}\right)^2 , \quad (3)$$

which varies both in time and in the extra dimension. Three of the usual choices for α are: $\alpha \in (0,1)$ for inflation, $\alpha = 2$ for radiation, and $\alpha = 3/2$ for dust ($\alpha \neq 0, 1$ since these choices introduce singularities into the 5D metric). A simple calculation of (3) leads to the rest mass variation

$$\frac{\dot{m}}{m} = \pm \left(\frac{1-\alpha}{\alpha}\right) \frac{1}{ct} . \quad (4)$$

For the present universe, which is well approximated by dust, the variation of rest masses is only borderline detectable [1], [8], and subject to significant astrophysical uncertainties.

The acceleration equation for the Ponce deLeon metric is simplified by the comoving coordinate system. In general, the assumption that the spatial velocities are constant, implies that the scalar field can only depend on time. Thus we can conclude that any 5D metric in the form of (1), which has the 4D section $g_{\alpha\beta}$ written in comoving coordinates with a time-dependent scalar field, will not impart a fifth force and the motion will be 4D geodesic. Hence, we need to look at other models for examples of a non-zero fifth force.

V FINAL COMMENTS

By retaining the extra coordinate $x^4 = l$ in 5D Kaluza-Klein gravity we have seen that a 5D vacuum may induce non-trivial matter on 4D hypersurfaces $l = l_o$. The same can be said of particle dynamics. We used the assumption of 5D null geodesics to induce particle motion in the 4D subspace with the added feature of variable rest mass for massive particles, once we interpreted the extra dimension with mass. The acceleration for null particles remained the same as in 4D, but the motion for massive particles was augmented by a type of fifth force. This force has a contribution from a scalar field and crucially depends on the existence of the extra dimension. This motion was investigated for a cosmological metric, which induced a time-varying cosmological constant. For this metric, the resulting fifth force is undetectable due to the nature of the comoving coordinate system. It seems that we should turn to l-dependent analogues of the Schwarzschild metric to observe and test any deviations from the classical tests of GR due to the fifth force. Work on this is underway, and we expect to relate 5D dynamics to the upcoming Space Test of the Equivalence Principle.

VI ACKNOWLEDGEMENTS

The authors would like to thank T. Y. Jung for comments, and OGS and NSERC for financial support.

REFERENCES

1. Overduin, J. M., Wesson, P. S., *Phys. Rep.* **283**, 303 (1997).
2. Sajko, W. N., Wesson, P. S., Liu, H., *Jour. Math. Phys.* **39**, 2193 (1998).
3. Rippl, S., Romero, C., Tavakol, R., *Class. Quant. Grav.* **12**, 2411 (1995).
4. Romero, C., Tavakol, R., Zalaletdinov, R., *Gen. Rel. Grav.* **28**, 365 (1996).
5. Wesson, P. S., Ponce de Leon, J., Liu, H., Mashhoon, B., Kalligas, D., Everitt, C. W . F., Billyard, A., Lim, P., Overduin, J., *Int. Jour. Mod. Phys.* **A11**, 3247 (1996).
6. Mashhoon, B., Wesson , P. S., Liu, H., *Gen. Rel. Grav.* **30**, 555 (1998).
7. Kalligas, D., Wesson, P. S., Everitt, C. W. F., *Astrophys. Jour.* **439**, 548 (1995).
8. Will, C. M., *Int. Jour. Mod. Phys.* **D1**, 13 (1992).
9. Ponce de Leon, J., *Gen. Rel. Grav.* **20**, 1115 (1988).
10. Wesson, P. S., *Astrophys. Jour.* **394**, 19 (1992).

V. CLASSICAL GENERAL RELATIVITY

Transformation of Vacuum Solutions of Conformal Gravity to Flat Space

A. Edery[1] and M. B. Paranjape[2]

Abstract. We show how to transform the spherically symmetric vacuum solution to conformal gravity to the flat metric via coordinate and conformal transformations.

Conformal gravity is a metric theory of gravity whose action is given by

$$I = \alpha \int \sqrt{-g}\, C_{\lambda\mu\sigma\tau} C^{\lambda\mu\sigma\tau} d^4x \tag{1}$$

where α is a dimensionless constant. It is the simplest action that can be constructed which is conformally invariant i.e. invariant under the conformal transformation $g_{\mu\nu}(x) \rightarrow \Omega^2(x) g_{\mu\nu}(x)$ where $\Omega^2(x)$ is a non-singular, non-vanishing, function. Interest in conformal gravity was rekindled in the early 90's after the metric exterior to a static spherically symmetric source was obtained [4]. Taking the standard form

$$d\tau^2 = B(r)\, dt^2 - A(r)\, dr^2 - r^2\left(d\theta^2 + \sin^2\theta\, d\varphi^2\right) \tag{2}$$

the static spherically symmetric vacuum solutions are [1,4]

$$B(r) = A^{-1}(r) = 1 - \frac{(2-3\gamma\beta)\beta}{r} - 3\beta\gamma + \gamma\, r - k r^2 \tag{3}$$

where β, γ and k are constants. The above solution is valid up to a conformal factor. Constraints from phenomenology imply that γ, k and $\gamma\beta << 1$ (see [4–6]). The constant $\gamma\beta$ is taken to be negligible. β has to do with the Schwarzschild mass parameter, and breaks the conformal invariance. The remaining parameters, γ, k do not violate conformal invariance, and describe the solution at large distance. We set $\beta = 0$ since we are interested with the import of the new parameters implied by conformal gravity.

We wish to find the coordinate transformation from the r, t coordinates to a new set of coordinates ρ, τ where the metric (2) with (3) is written in a form which is manifestly conformal to flat. We write

[1] present address: Department of Physics, McGill University, 3600 University St., Montréal, Québec, Canada, H3A 2T8. E-mail: edery@hep.physics.mcgill.ca
[2] Groupe de Physique des Particules, Département de Physique, Université de Montréal, C.P. 6128, succ. centreville, Montréal, Québec, Canada, H3C 3J7.

CP493, *General Relativity and Relativistic Astrophysics*, edited by C. P. Burgess and R. C. Myers
© 1999 American Institute of Physics 1-56396-905-X/99/$15.00

$$ds^2 = (1 + \gamma r - kr^2)dt^2 - 1/(1 + \gamma r - kr^2)dr^2 - r^2(d\theta^2 + \sin^2\theta \, d\varphi^2)$$
$$= \Omega^2(\rho, \tau)\left[d\tau^2 - d\rho^2 - \rho^2(d\theta^2 + \sin^2\theta \, d\varphi^2)\right] \tag{1}$$

where τ and ρ are the new coordinates and $\Omega(\rho, \tau)$ is the conformal factor. We immediately have the following relations:

$$r = \rho \, \Omega \tag{5}$$

$$\Omega^2(d\tau^2 - d\rho^2) = (1 + \gamma r - kr^2)dt^2 - 1/(1 + \gamma r - kr^2)dr^2. \tag{6}$$

The coordinates r and t are now functions of both ρ and τ so that $dr = r'd\rho + \dot{r}d\tau$ and $dt = t'd\rho + \dot{t}d\tau$ where a prime and dot on r and t represent partial derivatives with respect to ρ and τ respectively.

Equations (5) and (6) lead to three partial differential equations for r and t, which can be reduced to two partial differential equations for r upon eliminating t and integrating once,

$$\frac{r^2}{\rho^2(\dot{r} + r')} = f(\tau - \rho) \tag{7}$$

$$r'^2 - \dot{r}^2 = \frac{r^2(1 + \gamma r - kr^2)}{\rho^2} \tag{8}$$

where $f(\tau - \rho)$ is an arbitrary function of $\tau - \rho$. Evidently the new coordinates u and v

$$u = \tau - \rho \quad ; \quad v = \tau + \rho \tag{9}$$

are useful, and (7) reduce to

$$\frac{2r^2}{(v - u)^2 \, \partial r/\partial v} = f(u) \tag{10}$$

with solution

$$r = \frac{f(u)(v - u)}{2 + h(u)(v - u)} \tag{11}$$

where $h(u)$ is an arbitrary function of u. Substituting (10) into (8) one obtains

$$-2\int \frac{dr}{1 + \gamma r - kr^2} = \int f(u)du = g(u) + p(v) \tag{12}$$

where $dg(u)/du = f(u)$ and $p(v)$ is an arbitrary function of v. The solution to the above equation depends on whether the polynomial $1 + \gamma r - kr^2$ has roots or not.

We will analyze only the case with roots and refer the reader to our more detailed [7] article for the full analysis. Then the integral of $1/(1 + \gamma r - kr^2)$ is given by

$$\frac{-1}{k(r_+ - r_-)} \ln \left| \frac{r - r_+}{r - r_-} \right| \; ; \quad k > -\frac{\gamma^2}{4} \tag{13}$$

where the two roots r_+ and r_-, which can have negative values, are given by

$$r_\pm = \frac{\gamma}{2k} \pm \sqrt{\frac{\gamma^2}{4k^2} + \frac{1}{k}}. \tag{14}$$

Substituting (13) for the integral in (12) one obtains

$$r = \frac{r_+ \pm r_- e^{k(r_+ - r_-)(g(u)+p(v))/2}}{1 \pm e^{k(r_+ - r_-)(g(u)+p(v))/2}} \tag{15}$$

where the negative sign corresponds to the region where $\infty > r > r_+$ and $0 < r < r_-$ whereas the plus sign corresponds to the region where $r_+ > r > r_-$. We now equate r in (11) to r in (15). Note that $f(u)$ in (11) is $g'(u) \equiv dg(u)/du$. One obtains the following equality

$$\frac{g'(u)e^{-k(r_+ - r_-)(g(u)+p(v))/2}}{r_+ e^{-k(r_+ - r_-)(g(u)+p(v))/2} \pm r_-} \pm \frac{g'(u)e^{k(r_+ - r_-)(g(u)+p(v))/2}}{r_+ \pm r_- e^{k(r_+ - r_-)(g(u)+p(v))/2}} = \frac{2}{v - u} + h(u). \tag{16}$$

After integrating the above equation and some algebra we obtain

$$\ln \left(r_+ e^{-k(r_+ - r_-)(g(u)+p(v))/4} \pm r_- e^{k(r_+ - r_-)(g(u)+p(v))/4} \right) = \ln(v - u) + S(u) + T(v) \tag{17}$$

where $S(u)$(related to $h(u)$) and $T(v)$ are arbitrary functions of u and v repectively. Exponentiating both sides (17) reduces to

$$r_+ e^{-k(r_+ - r_-)P(v)/2} \pm r_- e^{k(r_+ - r_-)g(u)/2} = (v - u)A(u)B(v). \tag{18}$$

The functions $A(u), B(v), g(u)$ and $p(v)$ are arbitrary functions of u and v and we can therefore write the above equation as

$$(v - u)A(u)B(v) = N(v) + M(u) \tag{19}$$

where all the functions above are arbitrary functions of u and v. The coordinate r, given by (15), can be expressed in terms of the functions $M(u)$ and $N(v)$ i.e.

$$r = \frac{r_+ r_- (M(u) + N(v))}{r_- N(v) + r_+ M(u)} \tag{20}$$

where the above is valid for the entire region $\infty > r > 0$. Fortunately, equation (19) can be solved algebraically. We actually arrive at the same equation in case

277

where the polynomial has no roots. To solve (19) we note first that the right hand side of the equation does not contain any mixed terms of u and v and therefore the mixed terms on the left hand side must vanish. We write $A(u)$ as

$$A(u) = A_0 + a(u') \tag{21}$$

where $A_0 = A(u_0)$ is a constant and $a(u')$ is a function of $u' \equiv u - u_0$ which vanishes at $u' = 0$. Similarly

$$B(v) = B_0 + b(v'). \tag{22}$$

With $A(u)$ and $B(v)$ given above, the left hand side of (19) yields

$$(v' - u' + C_0)(A_0 B_0 + A_0 b(v') + B_0 a(u') + a(u')b(v')) \tag{23}$$

where $C_0 = v_0 - u_0$ is a constant. The mixed terms must vanish and we obtain the following equation

$$v'a(u')B_0 + v'a(u')b(v') - u'b(v')A_0 - u'a(u')b(v') + C_0 a(u')b(v') = 0. \tag{24}$$

The solution to (24) is obtained by separating the variables i.e.

$$b(v') = \frac{-v'B_0}{C_0 + v' - u'(1 + A_0/a(u'))}. \tag{25}$$

The function $b(v')$ is a function of v' only and therefore the term $u'(1 + A_0/a(u'))$ must be a constant (call it D). We therefore obtain the following solutions

$$a(u') = \frac{-u'A_0}{D + u'}; \quad b(v') = \frac{-v'B_0}{C_0 + D + v'}. \tag{26}$$

The solution to $(v - u)A(u)B(v) = M(u) + N(v)$ is therefore

$$A(u) = \frac{A}{B + u}, \ B(v) = \frac{C}{B + v}, \ M(u) = \frac{-AC\,u}{B(B + u)}, \ N(v) = \frac{AC\,v}{B(B + v)} \tag{27}$$

where the solution (26) was substituted into equations (21) and (22) and the quantities A, B and C are constants related to the constants A_0, B_0, C_0 and D. With the above solution we can finally obtain the coordinate r, which is then given by (20) and yields

$$r = \frac{r_+ r_- (v - u)}{vr_-(1 + u/B) - ur_+(1 + v/B)}. \tag{28}$$

Thus we find that the spherically symmetric solution to conformal gravity can be explicitly brought to a flat Minkovski metric, in the appropriate system of coordinates and change of conformal gauge.

ACKNOWLEDGEMENTS

We thank Robert Mann for useful discussions and NSERC of Canada and FCAR du Québec for financial support.

REFERENCES

1. R. Riegert, Phys. Rev. Lett. **53**, 315 (1984).
2. R. Riegert, Phys. Lett. **105A,** 110 (1984).
3. D. G. Boulware, G. T. Horowitz and A. Strominger, Phys. Rev. Lett. **22**, 1726 (1983).
4. D. Kazanas and P.D. Mannheim, Astrophys. J. **342,** 635 (1989).
5. P.D. Mannheim, Astrophys. J. **479,** 659 (1997), and references therein.
6. A. Edery and M. B. Paranjape, Phys. Rev. D **58** 024011(1998).
7. A. Edery and M. B. Paranjape, "Causal Structure of Vacuum Solutions to Conformal(Weyl) Gravity", astro-ph/9808345, Gen. Rel. Grav., to be published (1998).
8. M. A. Walker, Astrophys. J. **430,** 463 (1994).

Conformal Symmetry

W.R. Wood

Faculty of Natural and Applied Sciences, Trinity Western University
7600 Glover Road, Langley, British Columbia, V2Y 1Y1

Abstract. In recent years, conformal transformations have been used in a variety of gravitational applications, as well as in other fields. The precise form that these transformations take typically depends on the problem at hand. A brief introduction to the topic of conformal symmetry will lead to a framework from which these applications, such as the mapping of Riemannian spaces and self-similarity, can be put into a broader context. The systematic approach presented suggests a new perspective from which conformal symmetry can be studied.

INTRODUCTION

The concept of conformal symmetry has proven to be useful in a variety of contexts over the years. However, a coherent framework within which different applications of conformal symmetry can be related to one another is lacking. This has led to a confusing and sometimes inconsistent use of terminology. In what follows it is shown that the theory of G-Structures provides a natural means by which the various applications of conformal transformations can be related to one another. The clarification of the relationship between different types of conformal transformations will provide a foundation to seek a better understanding of the physical meaning of conformal symmetry in nature.

At its most fundamental level, conformal symmetry refers to changes under which angles and localized shapes are preserved, but size may be scaled. By way of example, the Mercator projection of the 2-sphere onto a plane preserves angles, but areas, especially near the poles, are distorted. Another classic application of conformal symmetry occurs in the conformal mapping used in complex analysis. In the context of differential geometry, the metric plays a central role in discussions of conformal symmetry since angles are defined in terms of the metric. It is this latter context that will be of particular interest here.

One may question whether conformal transformations should be viewed strictly as a mathematical tool that is useful in recasting a problem into a more convenient form, or whether some intrinsic physical meaning should be associated with conformal symmetry. Physicists have long held the view that the description of nature, at a fundamental level, is characterized by "mathematical beauty" or "simplicity"

CP493, *General Relativity and Relativistic Astrophysics,* edited by C. P. Burgess and R. C. Myers
© 1999 American Institute of Physics 1-56396-905-X/99/$15.00

[1]. From this perspective, one holds the prejudice that structures in mathematics that are "pretty" may well find application in the physical domain. The number of successful examples of this in the past, what Wigner [2] has called "the unreasonable effectiveness of mathematics in the natural sciences", encourages one to maintain this prejudice. The view adopted here is that we have not fully recognized the physical significance of conformal symmetry, and if this is to be remedied, a systematic approach to the problem is the one most likely to end in success.

From a physical point of view, there are several reasons to believe that conformal symmetry is significant. For one, theories that apply at the high energy domain, such as string theories, are often required to be conformally invariant. The existence of absolute standards of length provided by particles can then be understood to be the result of a breaking of the conformal invariance. Along these lines, it has been shown [3] that it is possible for conformal symmetry to be broken locally by forming regions of Riemannian geometry in an ambient Weyl geometry. Even in the broken symmetry state, one generally anticipates the manifestation of some form of the symmetry under certain conditions. It was from this perspective that the talks given at the Seventh Canadian Conference on General Relativity and Relativistic Astrophysics on self-similarity were of particular interest to the present author. While the significance of scale invariance has been recognized in areas such as renormalization group theory and fractals, critical phenomena in general relativity provides a link between these applications of scale invariance and the geometric analysis presented here.

The conformal symmetry associated with self-similar solutions is not the same as the conformal invariance present in Weyl geometry. Different categories of conformal symmetry are discussed in the following section. Then a framework based on G-structures and their isomorphisms is described that allows a comprehensive description of each category of conformal transformation. Some concluding remarks are given in the last section.

CATEGORIES OF CONFORMAL SYMMETRY

For present purposes, it will be sufficient to identify two categories of conformal transformations. The first type involves transformations of the metric of the form

$$\tilde{g}_{\mu\nu} = \Omega^2(x)g_{\mu\nu}, \tag{1}$$

where $\Omega(x)$ is an arbitrary function. Transformations of the type (1), together with corresponding transformations of other fields, combine to form the essence of Weyl's conformally invariant geometric theory of electromagnetism and gravitation [4]. Other applications of transformations of the type (1) occur in general relativity. For example, conformal transformations are used in defining the notion of asymptotic flatness. As well, one can map a known solution to new solution in a different "conformal frame". In this latter application, one needs to remember that Einstein's theory is not invariant under conformal transformations, so the two solutions will not, in general, be physically equivalent to each other [5]. The terms "physical" and "unphysical" frames are frequently used.

The second type of transformations are the conformal diffeomorphisms

$$\mathcal{L}_\xi g_{\mu\nu} = \rho(x)g_{\mu\nu}. \tag{2}$$

When $\rho(x)$ is an arbitrary function, ξ is called a conformal Killing vector field. These transformations have been useful in the study of the initial value problem in general relativity. They also yield the well-known 15-parameter Lie group in Minkowski space [6]. When ρ is a constant, ξ is called a homothetic vector. These are the transformations that are associated with continuous self-similar solutions in general relativity. Finally, ξ is called a Killing vector when $\rho = 0$. This case represents the absence of any level of scale invariance.

Along with the above categories of transformations, it will also be useful to distinguish the following levels of structure on a manifold M. A *conformal structure* is an equivalence class of metrics, where any two metrics in a class are related to one another by (1). In the case of Lorentz metrics, a conformal structure is equivalent to a distribution of null cones on M. A *projective structure* is an equivalence class of symmetric affine connections such that they have the same autoparallels. Physically, these correspond to the world lines of freely falling particles. A *Weyl structure* is the pair of an equivalence class of metrics together with a symmetric connection that preserves the equivalence class. Ehlers, Pirani and Schild [7] have shown that one arrives at Weyl's geometry when the particle world lines are required to lie within the future part of the null cones. A *Riemannian structure* consists of a unique symmetric metric and a connection that is metric compatible.

G-STRUCTURES: PROLONGATIONS AND MORPHISMS

The theory of G-Structures [8] is rooted in fiber bundle theory [6]. A principal bundle (P, M, π, G) is a fiber bundle in which the typical fiber F is also the structure group G. The bundle P is locally trivial: $\pi^{-1}(U_i) \to U_i \times G$, where U_i is an open neighborhood in the base space M. The group G is said to have a free right action on P that preserves the fibers. An example of a principal bundle is the linear frame bundle $L(M)$. A frame $(e_j, j = 1 \ldots n)$ consists of $n = dim(M)$ linearly independent vectors in the tangent space $T_x(M)$. The transformation between frames is given by $e_i = a^j{}_i \bar{e}_j$, where the $(a^j{}_i)$ are elements of $GL(n, R)$, the group of all $n \times n$ nonsingular matrices. The linear frame bundle is the collection of all linear frames at all points of M.

The relevance of the frame bundle to the task at hand can easily be seen by considering two special frames. The set of frames whose vectors are mutually orthogonal and all of the same length (with respect to one, and hence any, metric in the equivalence class), define a conformal structure. The corresponding matrices preserve the scalar product of signature $(r, n - r)$ up to a non-zero factor and form a subgroup of $GL(n, R)$, denoted by $CO(r, n - r)$. A Riemannian structure corresponds to the set of orthonormal frames and the subgroup $O(r, n - r)$. While $L(M)$ provides the appropriate context to describe the structures of interest, one must consider the morphisms of bundles before conformal diffeomorphisms can be incorporated into the scheme.

Here the notation of Trautman [9] will be adopted. Given a smooth map $f : M_1 \to M_2$, one may ask under what conditions the bundles over these manifolds will have a natural correspondence to one another. In the case of a principal bundle, this correspondence is achieved if and only if the diagram consisting of the three maps f, $h : P_1 \to P_2$, and $k : G_1 \to G_2$, together with their corresponding projections $\psi_i : P_i \times G_i \to P_i$ and $\pi_i : P_i \to M_i$ $(i = 1, 2)$, commutes. Trautman considers the case where $f = id$ so that $M_1 = M_2$. In this case, if both h and k are injective (one-to-one), P_1 is said to be the restriction of P_2. Examples of a restriction are the two cases considered above where $G_2 = GL(n, R)$ and $G_1 = CO(r, n - r)$ or $G_1 = O(r, n - r)$. If both h and k are surjective (onto), P_1 is said to be the reduction of P_2. In what follows, two generalizations of Trautman's analysis will be necessary. First, it is necessary to define the prolongations of G-Structures. It is also necessary to generalize the map f from the identity map.

A *first order G-Structure* (G-Structure) on M is a restriction of $L(M)$ to a G-principal bundle, i.e., a subbundle P whose structure group G is a closed Lie subgroup of $GL(n, R)$. A *second order G-Structure* on M is a restriction of the quadratic frame bundle $L^2(M)$ to a G-principal bundle on M. A second order G-Structure is said to be a prolongation of a first order G-Structure. The prolongation of a G-Structure is analogous to the prolongation of the tangent bundle $T(M)$. In mechanics, one works with accelerations which are elements of $T^2(M)$, the tangent bundle to $T(M)$. In terms of a Taylor series expansion, $T(M)$ provides first-order derivative terms of coordinates in M and $T^2(M)$ provides second-order terms. In like fashion, a second-order frame of M is defined in terms of its second order partial derivatives. More precisely, if two functions f and g map open neighborhoods about the origin in $\mathbf{R}^n$ into M such that $f(0) = g(0)$ and their partial derivatives also agree up to order k, they are said to be equivalent to order k. Each such equivalence class is called a k-jet of M and we write $j^k_0(f) = j^k_0(g)$. Finally, if f is a diffeomorphism, then $j^k_0(f)$ is a k-th order frame of M.

A map f from the manifold (M_1, g_1), where g_1 is the metric on M_1, to the manifold (M_2, g_2) induces the field $f\hat{\ }g_2$ on M_1. If this induced field is equal to $\Omega^2 g_1$, then f is called a conformal mapping. If the diagram consisting of f, its lift $\hat{f} : L(M_1) \to L(M_2)$, and the projections $\pi_i : L(M_i) \to M_i$ commutes, then it is a bundle morphism. Recall that the conformal diffeomorphism (2) is defined in terms of the Lie derivative. This requires the existence of a one-parameter family of diffeomorphisms $f_t : M \to M$, where a C^∞ map f is a diffeomorphism if it is one-to-one, onto and f^{-1} is C^∞. When f is a diffeomorphism, its lift $\hat{f}$ is called an isomorphism [10]. If $L(M_1) = L(M_2)$, $\hat{f}$ is called an automorphism. A conformal isometry is a diffeomorphism that is a conformal mapping.

CONFORMAL SYMMETRY WITHIN A
G-STRUCTURE FRAMEWORK

G-Structures over spacetime, their prolongations and morphisms provide a very convenient structure in which the different categories of conformal symmetry can be described. A conformal structure is a restriction of the linear frame bundle: $L(M) \to CO(r, n-r)$. A conformal isometry in M induces an automorphism of a CO-Structure. A projective structure is a reduction of the quadratic frame bundle: $L^2(M) \to PGL^2(r, n-r)$, where PGL is the general projective group. The projective structure cannot be reduced to a first-order structure. It has been shown [11] that the analysis of Elher, Pirani and Schild [7] can be recast into the language of G-Structures. These authors showed that a Weyl structure can be obtained by first prolonging the first-order conformal structure, followed by a restriction: $CO(r, n-r) \to CO^2(r, n-r) \to W^2(r, n-r)$. Finally, a Riemannian structure is a restriction of the linear frame bundle: $L(M) \to O(r, n-r)$. An isometry in M induces an automorphism of an O-Structure.

This analysis has demonstrated that the various categories of conformal symmetry fit very nicely into the framework afforded by G-Structures. Historically, our understanding of nature has progressed when disparate elements have been brought together into a coherent model. Hopefully the framework provided here will be useful in gleaning a better understanding of the physical properties of conformal symmetry.

REFERENCES

1. P. A. M. Dirac, *Proc. Roy. Soc.* (Edinburgh, Part II) **59**, 122 (1938-39).
2. E. P. Wigner, in *The World of Physics*, ed. J. H. Weaver, New York: Simon and Schuster, 1987, pp. 80-96.
3. W. R. Wood and G. Papini, *Phys. Rev.* **D45**, 3617 (1992).
4. H. Weyl, *Sitzungsber. Preuss. Akad. Wiss.* 465, (1918).
5. V. Faraoni, E. Gunzig, and P. Nardone, to appear in *Foundations of Cosmic Physics* (preprint gr-qc/9811047).
6. Y. Choquet-Bruhat, C. deWitt-Morette and M. Dillard-Bleick, *Analysis, Manifolds and Physics*, revised ed., New York: North-Holland, 1982.
7. J. Ehlers, F. A. E. Pirani and A. Schild, in *General Relativity, Papers in Honour of J. L. Synge*, ed. L. O'Raifeartaigh, New York: Oxford, 1972, pp. 63-84.
8. See, e.g., S. Sternberg, *Lectures on Differential Geometry*, 2nd ed., New York: Chelsea, 1983; K. Yang, *Exterior Differential Systems and Equivalence Problems*, Dordrecht: Kluwer Academic, 1992.
9. A. Trautman, *Differential Geometry for Physicists*, Napoli: Bibliopolis, 1984.
10. K. Yang, *Exterior Differential Systems and Equivalence Problems*, Dordrecht: Kluwer Academic, 1992.
11. R. A. Coleman and H. Korte, *J. Math. Phys.* **22**, 2598 (1981).

Monopole and Dyon Solutions in the Einstein-Yang-Mills Theory in Asymptotically AdS Space

Yutaka Hosotani and Jefferson Bjoraker

School of Physics and Astronomy, University of Minnesota
Minneapolis, MN 55455, U.S.A.

Abstract. Regular monopole and dyon solutions to the $SU(2)$ Einstein Yang-Mills equations in asymptotically anti-de Sitter space are discussed. A class of monopole solutions are shown to be stable against spherically symmetric linear perturbations.

I INTRODUCTION

Static solutions to the Einstein Yang-Mills (EYM) equations differ considerably depending on the value of the cosmological constant. The solutions can be separated into two families; $\Lambda \geq 0$, and $\Lambda < 0$. The solutions where $\Lambda = 0$ were discovered by Bartnik and McKinnon (BK) [1] and their asymptotically de Sitter (dS) analogs ($\Lambda > 0$) were discovered independently by Volkov et. al. and Torii et. al. [2]. The BK solutions and the cosmological extensions to them all share similar behavior and have been studied in detail (see Ref. [3] for a review). Recently, asymptotically anti-de Sitter ($\Lambda < 0$) black hole solutions [4] and soliton solutions [5] were found which are strikingly different than the BK type solutions. In particular, there exist asymptotically anti-de Sitter (AdS) solitons with no nodes in the field strength which are stable. Furthermore, in asymptotically AdS space dyon solutions are allowed.

CP493, *General Relativity and Relativistic Astrophysics*, edited by C. P. Burgess and R. C. Myers
© 1999 American Institute of Physics 1-56396-905-X/99/$15.00

II GENERAL FORMALISM

Given the spherically symmetric metric in the Schwarzschild gauge

$$ds^2 = -\frac{H\,dt^2}{p^2} + \frac{dr^2}{H} + r^2(d\theta^2 + \sin^2\theta\,d\phi^2),\tag{1}$$

where H, p are functions of t and r, the coupled static EYM equations of motion are

$$\left(\frac{H}{p}w'\right)' = -\frac{p}{H}u^2 w - \frac{w(1-w^2)}{p}\frac{}{r^2}\tag{2}$$

$$\left(r^2 p u'\right)' = \frac{2p}{H}w^2 u\tag{3}$$

$$p' = -\frac{2v}{r}p\left[(w')^2 + \frac{u^2 w^2 p^2}{H^2}\right]\tag{4}$$

$$m' = v\left[\frac{(w^2-1)^2}{2r^2} + \frac{1}{2}r^2 p^2(u')^2 + H(w')^2 + \frac{u^2 w^2 p^2}{H}\right],\tag{5}$$

where $w(r)$ and $u(r)$ are the magnetic and electric components of the $SU(2)$ Yang-Mills fields [6], $H(r) = 1 - 2m(r)/r - \Lambda r^2/3$, $v = G/4\pi e^2$ and Λ is the cosmological constant. We require that solutions to Eq.'s (2) to (5) are regular everywhere and have finite ADM mass $M = m(\infty)$. The electric and magnetic charges of solutions are given by

$$\begin{pmatrix} Q_E \\ Q_M \end{pmatrix} = \frac{e}{4\pi}\int dS_k\,\sqrt{-g}\begin{pmatrix} F^{k0} \\ \tilde{F}^{k0} \end{pmatrix} = \begin{pmatrix} u_1 p_0 \\ 1 - w_0^2 \end{pmatrix}\frac{\tau_3}{2}\tag{6}$$

where $w = w_0 + w_1/r + O(r^{-2})$ etc.

For solutions in asymptotically flat or dS space, $w(r)$ has at least one node [2]. The situation is quite different in asymptotically AdS space ($\Lambda < 0$). $H(r)$ is positive everywhere and there are solutions where w has no node.

III MONOPOLE SOLUTIONS

The solutions to Eq.'s (2) to (5), for $\Lambda < 0$, were evaluated numerically [5] using the shooting method. In the shooting method one solves Eqs. (2) to (5) at $r = 0$ in terms of two parameters, a and b, and 'shoots' for solutions with the desired asymptotic behavior. a and b are adjustable parameters which together specify the boundary conditions at the origin for w, u, H, and p: $u(r) = ar + \cdots$ and $w(r) = 1 - br^2 + \cdots$ near $r = 0$.

Purely magnetic solutions (monopoles) are found by setting $a = 0$, corresponding to $u(r) = 0$. A continuum of monopole solutions were obtained by varying the parameter b. The solutions are similar to the black hole solutions found by Ref. [4], but are also regular for all r. The number of times w crosses the axis depends on the value of the adjustable shooting parameter b.

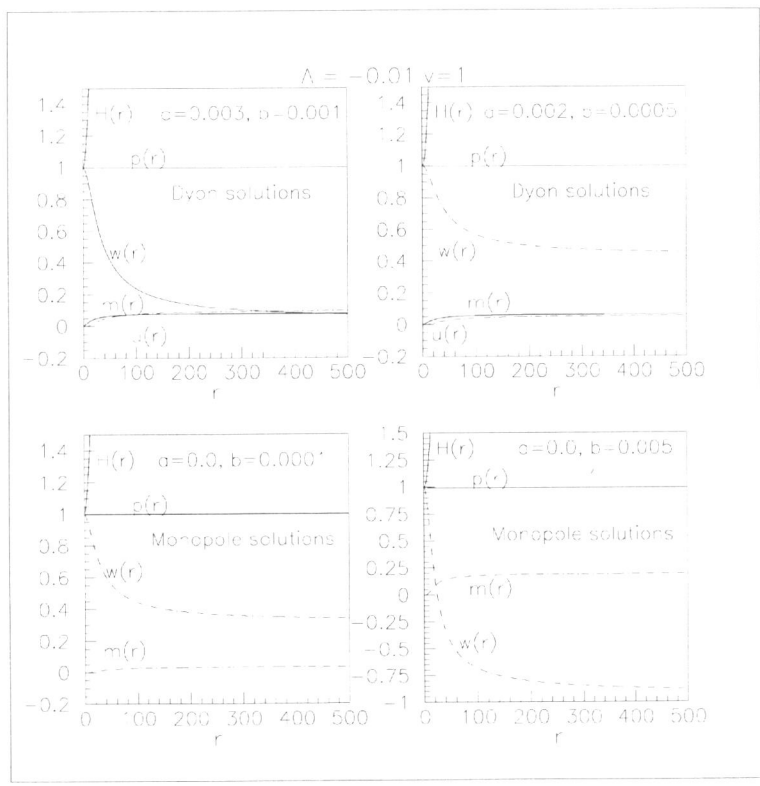

FIGURE 1. Monopole and dyon solutions for $\Lambda = -0.01$ and $v = 1$. $(a,b) = (0,0.001)$ and $(0,0.005)$ for the monopole solutions and $(a,b) = (0.003,0.001)$ and $(0.002,0.0005)$ for the dyon solutions.

As shown in fig. 1, the behavior of m and p is similar to that of the asymptotically dS solutions previously considered [2]. In contrast, there exist solutions where w has no nodes, which are not seen in the asymptotically dS or Minkowski cases.

IV DYON SOLUTIONS

Dyon solutions to the EYM equations for a given negative Λ are found if we choose the adjustable shooting parameter a to be non-zero. Fig. 1 shows how the electric component, u, of the EYM equations starts at zero and monotonically increases to some finite value. The behavior of w, m, H, and p is similar to the monopole solutions.

Again we find a continuum of solutions for a continuous set of parameters a and b, where w crosses the axis an arbitrary number of times depending these parameters. Also similar to the monopole solutions is the existence of solutions where w does not cross the axis. This is in sharp contrast to the $\Lambda \geq 0$ case where the dyon solutions are forbidden.

V STABILITY OF THE MONOPOLE SOLUTIONS

The BK solutions and the dS-EYM solutions are unstable [7–9]. In contrast, the AdS black hole solutions [4] with $u = 0$ and the monopole solutions without nodes [5] are stable against spherically symmetric linear fluctuations.

In order to derive the time dependent EYM equations, we use the most general expression for the spherically symmetric SU(2) gauge fields in the singular gauge:

$$A = \frac{1}{2e}\left\{u\tau_3 dt + \nu\tau_3 dr + (w\tau_1 + \tilde{w}\tau_2)d\theta + (\cot\theta\tau_3 + w\tau_2 - \tilde{w}\tau_1)\sin\theta d\phi\right\}, \quad (7)$$

where τ_i $(i = 1,2,3)$ are the usual Pauli matrices and u, ν, w and $\tilde{w}$ depend on r and t. The boundary conditions $u = \nu = 0$ and $w^2 + \tilde{w}^2 = 1$ at $r = 0$ ensure regularity at the origin. Linearized equations when $u(t, r) = 0$, for the gauge fields $\delta w(t, r)$, $\delta\tilde{w}(t, r)$, $\delta\nu(t, r)$, $\delta p(t, r)$, and $\delta H(t, r)$ have been derived in the literature [3]. Fluctuations decouple in terms of $\delta w(t, r)$, $\delta p(t, r)$, and $\delta H(t, r)$ which form even-parity perturbations, and in terms of $\delta\tilde{w}(t, r)$ and $\delta\nu(t, r)$ which form odd-parity perturbations.

The equation for parity-odd perturbations in $\beta = r^2 p\delta\nu/w$, where $\beta(t, r) = e^{-i\omega t}\beta(r)$, is [5]

$$\left\{-\frac{d^2}{d\rho^2} + U_\beta(\rho)\right\}\beta = \omega^2\beta \quad , \quad U_\beta = \frac{H}{r^2 p^2}(1 + w^2) + \frac{2}{w^2}\left(\frac{dw}{d\rho}\right)^2 \quad (8)$$

where $d\rho/dr = p/H$. Volkov et al. showed that for the BK solutions there appear exactly n negative eigenmodes ($\omega^2 < 0$) if w has n nodes [8]. Their argument applies to the asymptotically AdS case without modification. One concludes that the solutions with nodes in w are unstable against parity-odd perturbations.

For parity-even perturbations, where $\delta w(t, r) = e^{-i\omega t} \delta w(r)$, we find the equation

$$\left\{ -\frac{d^2}{d\rho^2} + U_w(\rho) \right\} \delta w = \omega^2 \delta w \quad , \quad U_w = \frac{H(3w^2 - 1)}{p^2 r^2} + 4v \frac{d}{d\rho} \left(\frac{Hw'^2}{pr} \right) \quad . \quad (9)$$

Although the potential $U_w(\rho)$ is not positive definite, it is regular in the entire range $0 \le \rho \le \rho_{\max}$. The first term in U_w becomes negative for $w^2 < 1/3$. The Schrödinger equation (9) was solved numerically [5]. The potential for the solutions with $(a, b) = (0, 0.001)$, which has no node in w, has the lowest eigenvalue ω^2 of 0.028 and for $(a, b) = (0, 0.005)$, which has one node, has the lowest eigenvalue of 0.023. Therefore, these solutions are stable against parity-even perturbations, even if w has one node, and differ from the solutions where $\Lambda \ge 0$ where the parity-even perturbations are always unstable.

VI   CONCLUSION

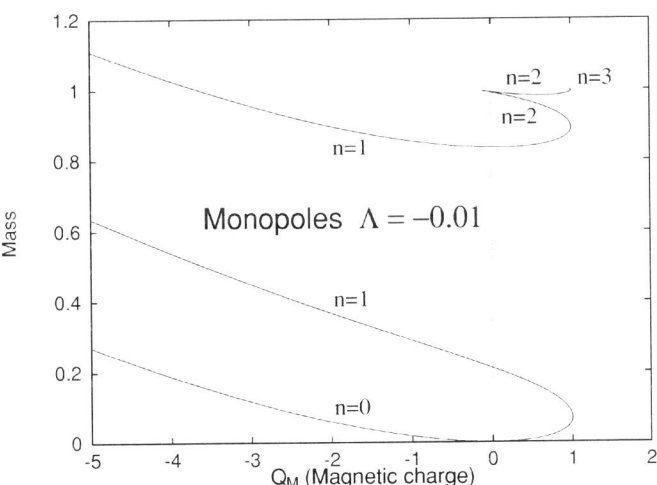

FIGURE 2. The Q_M-*Mass* plot of monopole solutions at $\Lambda = -0.01$. The number of nodes, n, in $w(r)$ is also indicated.

A continuum of new monopole and dyon solutions to the asymptotically AdS EYM equations have been found. In fig. 2 the spectrum of monopole solutions is plotted. The monopole solutions in which the magnetic component $w(r)$ of the gauge fields never vanishes, corresponding to the portion with $n = 0$ in fig. 2, were shown to be stable against linear perturbations. Solutions with $b \geq 0.7$ develop an event horizon, becoming black holes. The end point of the $n = 3$ portion of the curve in fig. 2 shows where the solutions become black holes.

ACKNOWLEDGMENTS

This work was supported in part by the U.S. Department of Energy under contracts DE-FG02-94ER-40823.

REFERENCES

1. R. Bartnik and J. McKinnon, Phys. Rev. Lett. **61** , 141 (1988).

2. M.S. Volkov, N. Straumann, G. Lavrelashvili, M. Huesler and O. Brodbeck, Phys. Rev. D **54** 7243 (1996); T. Torii, K. Maeda and T. Tachizawa, Phys. Rev. D **52** R4272 (1995).

3. M. S. Volkov and D. Gal'tsov, hep-th/9810070 (1998).

4. E. Winstanley, Class. Quant. Grav. **16**, 1963 (1999).

5. J. Bjoraker and Y. Hosotani, gr-qc/9906091 (1999).

6. E. Witten, Phys. Rev. Lett. **38** 121 (1977).

7. N. Straumann and Z. Zhou, Phys. Lett. **B237**, 353 (1990); Z. Zhou and N. Straumann, Nucl. Phys. **B360**, 180 (1991).

8. M.S. Volkov, O. Brodbeck, G. Lavrelashvili, and N. Straumann Phys. Lett. **B349** 438 (1995).

9. O. Brodbeck, M. Huesler, G. Lavrelashvili, N. Straumann and M.S. Volkov, Phys. Rev. D **54** 7338 (1996).

Perturbation Analysis of Deformed Q-balls and Primordial Magnetic Fields

Tomoko Uesugi*, Tetsuya Shiromizu[†] and Mayumi Aoki[‡]

*Institute for Cosmic Ray Research, The University of Tokyo,
Tokyo 188-8502, Japan
[†]DAMTP, University of Cambridge,
Silver Street, Cambridge, CB3 9EW, UK
[‡]Graduate School of Humanities and Sciences, Ochanomizu University,
Tokyo 112-8610, Japan

Abstract. We study the excited states of the Q-balls by performing stationary perturbation on the spherical Q-balls. We find the exact solution of the stationary perturbation of the global Q-ball. For local Q-balls we solve the equations of motion for the perturbative part approximately by using expansion about the coupling constant. Furthermore we comment on the magnetic field generated by the excited states of local Q-balls during the phase transition.

INTRODUCTION

Q-ball is non-topological soliton solution of the complex scalar field arising in the theory with an unbroken continuous symmetry [1] [2]. The generation of Q-ball [3] and the possibility of the phase transition precipitated by solitosynthesis have been investigated in Ref. [4] [5]. These studies are based on the ground state of Q-ball. If we consider the excited states of Q-ball, however, it is conceivable that Q-ball has angular momentum. Then, if it is the gauged Q-ball [6], it could also have the magnetic moment [7].

In this paper, we considered the excited states of the Q-balls with the global U(1) symmetry and with the local U(1) symmetry coupled to the U(1) gauge fields. Furthermore, we estimated the generation of the primordial magnetic field due to excited states of the Q-balls with the local U(1) symmetry during the phase transition.

CP493, *General Relativity and Relativistic Astrophysics*, edited by C. P. Burgess and R. C. Myers
© 1999 American Institute of Physics 1-56396-905-X/99/$15.00

EXCITED STATE OF GLOBAL Q-BALLS

In this section, we consider only the complex scalar field with a global $U(1)$ symmetry. The Lagrangian is $L = \int d^3x \left[\frac{1}{2}|\dot{\phi}| - \frac{1}{2}|\nabla\phi|^2 - U(|\phi|)\right]$. If we fix the total conserved charge, $Q_0 = \int d^3x \phi^2 \neq 0$, the most favorable configuration of the field is determined by variational principle to minimize the total energy of the system. The solution has stationary and spherical configuration,

$$\phi_0 = \varphi_0(r)e^{i\omega t}. \tag{1}$$

For the existence of this Q-ball solution, we consider here the potential which has two minima, that is to say, true vacuum and false vacuum.

To consider the deformation from spherical configuration, we add the stationary perturbation on the Q-balls in the ground state as follows,

$$\phi = \phi_0 + \phi_1 = \left(\varphi_0(r) + \sqrt{4\pi}\varphi_{\ell m}(r)Y_{\ell m}(\theta, \varphi)\right)e^{i\omega t}, \tag{2}$$

where $\ell \geq 1$ and $\ell \geq |m|$. By using thin wall approximation,

$$U''(\varphi_0(r)) = \mu_0^2\theta(r - R) + \mu^2\theta(R - r), \tag{3}$$
$$\text{where} \quad \mu_0^2 = U''(0), \quad \mu^2 = U''(\sigma_+), \quad (\mu_0 > \omega > \mu),$$

the equation of motion for the perturbation, $\varphi_{\ell m}$, can be solved exactly. The solution becomes

$$\varphi_{\ell m}(r) = \varphi_\ell^<(r) = A_\ell r^\ell \left(\frac{1}{r}\frac{d}{dr}\right)^\ell \left(\frac{\sin(\lambda r)}{r}\right) \quad \text{for} \quad r < R \tag{4}$$

$$= \varphi_\ell^>(r) = B_\ell r^\ell \frac{d^\ell}{d(r^2)^\ell}\left(\frac{e^{-\gamma r}}{r}\right) \quad \text{for} \quad r > R, \tag{5}$$

where $x = \lambda r$ and $\lambda = \sqrt{\omega^2 - \mu^2}$, $y = i\gamma r$ and $\gamma = \sqrt{\mu_0^2 - \omega^2}$. The relation between the prefactors, A_ℓ and B_ℓ, is determined by the matching conditions of the field on the wall.

EXCITED STATE OF LOCAL Q-BALLS

Local Q-balls in the Ground State and Phase Transition

The basic features of the local Q-balls were studied by Lee et al in Ref. [6]. We consider the theory with a complex scalar field ϕ coupled to a $U(1)$ gauge field A_μ. The Lagrangian density is

$$\mathcal{L} = \frac{1}{2}|(\partial_\mu - ieA_\mu)\phi|^2 - U(|\phi|) - \frac{1}{4}F_{\mu\nu}F^{\mu\nu}, \quad \text{where} \quad F_{\mu\nu} = \partial_\mu A_\nu - \partial_\nu A_\mu. \tag{6}$$

To find the Q-balls in the ground state, we seek for the solution such that

$$\phi = \varphi(r)e^{i\omega t} \quad \text{and} \quad A_\mu = A_0(r)\delta_{\mu 0}. \tag{7}$$

The conserved charge associated with the U(1) symmetry becomes

$$Q = \int d^3x j_0 = \int d^3x(\omega - eA_0)\varphi^2 = \int d^3x g\varphi^2, \quad \text{where} \quad g = \omega - eA_0(r). \tag{8}$$

In the case of local Q-balls, the total energy can be written in the monotonically increasing function of the charge. Since the energy for a unit of charge cannot become greater than the mass of free ϕ particles, there is a maximum value for the allowed charge for stable Q-balls, Q_{max}.

Let us consider the phase transition which takes place below the critical temperature, T_c, and above the temperature at which the standard nucleation starts. At the temperature below T_c, if the charge of the Q-ball exceeds the critical value, Q_c, pointed in Ref. [5], Q-ball can grow up into the macroscopic size. This critical value is decided by the following conditions

$$\partial E/\partial R|_{R=R_c,Q=Q_c} = \partial^2 E/\partial R^2|_{R=R_c,Q=Q_c} = 0. \tag{9}$$

To accomplish the phase transition caused by local Q-ball, Q_c must be smaller than Q_{max}. To see the details of this condition, we consider a potential

$$U(\varphi) = \frac{1}{2}\mu^2(T)\varphi^2 - \frac{1}{6}M(T)\varphi^3 + \frac{\lambda(T)}{24}\varphi^4, \tag{10}$$

Using the thin wall approximation, the condition $Q_c/Q_{max} < 1$ becomes

$$\frac{3^{3/2}e^2}{2^{5/3}} < \lambda|\epsilon|. \tag{11}$$

Stationary Perturbation on Local Q-balls

The perturbation on the fields are written by

$$\phi = (\varphi_0 + \sqrt{4\pi}\varphi_{\ell m}Y_{\ell m})e^{i\omega t} \quad \text{and} \quad A_\mu = {}^{(0)}A_\mu + {}^{(1)}A_\mu. \tag{12}$$

We assume the following ansatz on perturbation of the vector potential, ${}^{(1)}A_0 = {}^{(1)}A_r = 0$, and ${}^{(1)}A_I = \frac{\sqrt{4\pi}}{i}g_{\ell m}(r)\partial_I(Y_{\ell m} - Y_{\ell m}^*)$, where suffix I runs over θ, φ. The angular momentum induced by the modes of excitation is given by

$$J_m = \int d^3x T_{0\varphi} = m\int d^3x g_0\left[\varphi_{\ell m}^2 - eg_0\varphi_0 g_{\ell m}, \varphi_{\ell m}\right], \tag{13}$$

and the averaged magnetic field over the angular direction is

$$\overline{B^2}(r) := \frac{1}{4\pi} \int d\Omega \mathbf{B}^2 = 2\ell(\ell+1)\frac{g_{\ell m}'^2}{r^2}. \tag{14}$$

The total energy becomes

$$E = \frac{1}{2}\omega Q_0 + \int d^3x \left[\frac{1}{2}\varphi_0'^2 + U(\varphi_0)\right] + \frac{1}{2}\omega \delta Q_m + \frac{1}{2}\int d^3x \, g_0^2 \varphi_{\ell m}^2 =: E_0 + \delta E, \tag{15}$$

where $\delta E = (1/2)\omega \delta Q_m + (1/2)\int d^3x \, g_0^2 \varphi_{\ell m}^2$.

We consider only the mode of $\ell = 1$. Assumed that $e \ll 1$, the solutions for the equations of motion in the order $O(e^0)$ become

$$^{(0)}\varphi_1^< = A_1 \frac{d}{dr}\frac{\sin(\lambda r)}{r}, \qquad ^{(0)}\varphi_1^> = B_1 r \frac{d}{dr^2}\frac{e^{-\gamma r}}{r} \quad \text{and} \quad ^{(0)}g_1 = 0. \tag{16}$$

Up to the order of $O(e)$, one can obtain the relation

$$^{(1)}g_1 \simeq \varphi_0 R^{2(0)}\varphi_1. \tag{17}$$

Then the energy induced by the magnetic field becomes

$$E_M \simeq 4\pi \int_0^R dr r^2 \overline{B^2}(r) \simeq \frac{16\pi}{27}e^2 A_1^2 \sigma_+^2 R^5 \lambda^6 \simeq \frac{40\pi}{9}e^2 Q_0 \frac{\sigma_+^2}{\omega}, \tag{18}$$

where we used $\delta Q \sim Q_0$. The mean magnetic field becomes

$$B \simeq \sqrt{2E_M / \frac{4\pi}{3}R^3} \simeq eQ_0^{1/2}\left(\frac{\sigma_+^2}{\omega R^3}\right)^{1/2}, \tag{19}$$

where $\rho = \omega R$. For the potential of eq. (10), the order of the magnitude becomes

$$B \simeq 24(2\pi)^{1/2}e\frac{\mu^2}{\lambda}\rho^{-1/2}. \tag{20}$$

Finally, we estimate the mean magnetic field for each Q-balls. We assume that the gauged Q-balls can be copiously produced, and the excited states of the Q-balls obey the thermal distribution. Then the cosmological mean magnetic field generated by the excitation of the local Q-ball can be evaluated by multiplying the factor $e^{-(1/2)\beta\delta E}$, and it becomes

$$\langle B \rangle_{R_{\max}} \sim 50\frac{e}{\lambda\rho^{1/2}}\mu^2 e^{-\mathcal{E}}, \quad \text{where} \quad \mathcal{E} := (1/2)\beta\delta E = \frac{2\pi\rho}{e^2}\frac{\mu}{T}. \tag{21}$$

Following the estimate of the generation of the magnetic field in the strongly first order phase transition showed in Ref. [11], the energy of the magnetic field is equipartitioned with the energy of turbulence for the electroweak plasma. Then the field strength at the end of the phase transition becomes

$$B(R_f) \sim \rho^{1/2}v \sim g_*^{1/2}vT_f^2 \sim 10^{24}\left(\frac{T_f}{100\text{GeV}}\right)^2\text{Gauss}, \qquad (22)$$

where ρ and v are the density and the velocity of the fluid, respectively.

If the coherent scale is comoving, the number of magnetic domains inside the galaxy scale is $N \sim 10^{10}(10^{-3}/f_b)(T_f/100\text{GeV})$. Thus, averaging over the galaxy scale [10], one can obtain the present mean value

$$\langle B_{\text{now}}\rangle_{\text{galaxy}} \sim \frac{1}{N^{3/2}}\left(\frac{a_{\text{f}}}{a_{\text{now}}}\right)^2 B(R_f) \sim 10^{-21}\left(\frac{f_b}{10^{-3}}\right)^{3/2}\left(\frac{T_f}{100\text{GeV}}\right)^{-3/2}\text{Gauss}. \qquad (23)$$

The above value satisfies the onset condition of the dynamo mechanism.

SUMMARY

In this paper, we investigated the perturbation on the ground state of Q-balls. Under the thin-wall approximation, we solved the equation for the part of perturbation analytically in the global Q-balls. For the local Q-balls, we solved them by expanding in powers of coupling constant e. Furthermore we naively estimated the magnetic field generated by the excited states of the local Q-balls, but the more detailed analysis of the evolution of the magnetic field as in Ref. [12] [13] should be taken into account.

REFERENCES

1. S. Coleman, Nucl. Phys. **B262**(1985),263
2. T. D. Lee and Y. Pang, Phys. Rep. **221**, 251(1992)
3. K. Griest and E. W. Kolb, Phys. Rev. **40**(1989),3231; J. A. Frieman, A. V. Olinto, M. Gleiser and C. Alcock, Phys. Rev. **D40**, (1989),3241; K. Griest, E. W. Kolb and A. Massarotti, Phys. Rev. **40**(1989),3529
4. A. Kusenko, Phys. Lett. bf 405, 108(1997); A. Kusenko and M. Shaposhnikov, Phys. Lett. **B418**, 46(1998); G. Dvali, A. Kusenko and M. Shaposhnikov, Phys. Lett. **B417**,99(1998)
5. A. Kusenko, Phys. Lett. **B406**, 26(1997)
6. K. Lee, J. A. Stein-Schabes, R. Watkins and L. M. Widrow, Phys. Rev. **D39**, 1665(1989)
7. T. Shiromizu, Phys. Rev. **D58** 107301 (1998)
8. J. Ellis, K. Enqvist, D. V. Nanopoulos and K. A. Olive, Phys. Lett. **B225**(1989),313
9. G. Sigl, A. Olinto and K. Jedamik, Phys. Rev. **D55**(1997), 4582
10. K. Enqvist and P. Olesen, Phys. Lett. **B319**(1993),178
11. G. Baym, D. Bödeker and L. McLerran, Phys. Rev. **D53**(1996),662
12. K. Jedamik, V. Katalinic and A. V. Olinto, Phys. Rev. **D57**(1998),3264
13. P. Olesen, Phys. Lett. **B398**(1997),321; A. Brandenburg, K. Enqvist and P. Olesen, Phys. Lett. **B391**(1997),395; T. Shiromizu, Phys. Lett. **B443**(1998),127

Kinks, Energy Conditions
and Spherical Symmetry

Tina A. Harriott[1a] and J.G. Williams[2b]

[a]*Department of Mathematics and Computer Studies, Mount Saint Vincent University,
Halifax, Nova Scotia B3M 2J6, Canada*

[b]*Department of Mathematics and Computer Science, and the Winnipeg Institute for Theoretical
Physics, Brandon University, Brandon, Manitoba R7A 6A9, Canada*

Abstract. The metric for a broad class of spherically symmetric kink spacetimes is
considered and the curvature 2-forms are computed. The mass density and principal
pressures are obtained by diagonalizing the energy-momentum tensor. An example of a
one-kink spacetime with a particularly simple angular term is presented and shown to
satisfy the weak and dominant energy conditions but not the strong energy condition.

In this paper the authors extend their previous work[1] on energy conditions for
kinked spacetimes[2] by considering the general spherically symmetric spacetime,[3]

$$ds^2 \;=\; e^{-2\Lambda} \left\{ -e^{2\eta} \cos 2\alpha \; dt^2 - 2\sin 2\alpha \; dt\, dr + e^{-2\eta} \cos 2\alpha \; dr^2 \right\} \;+\; e^{-2\zeta} r^2 \, d\Omega^2,$$

with the restriction that Λ, η, ζ and α are functions only of the radial variable r.
The coefficient in front of the $d\Omega^2$ allows for a general choice of angular term. The
angle α measures the angle of tilt of the light cones. For a one-kink spacetime,
$\alpha(0) = 0$ and $\alpha(\infty) = \pi$, so that $\cos 2\alpha$ varies from 1 to -1 and back to 1 as r varies
from zero to infinity.

The computation of the Einstein tensor $G_{\mu\nu}$ begins with the introduction of an
orthonormal tetrad of basis vectors[4] $\{e_\mu\}$ and corresponding basis 1-forms $\{\omega^\mu\}$, so
that the metric can be written

$$ds^2 \;=\; -(\omega^t)^2 \;+\; (\omega^r)^2 \;+\; (\omega^\theta)^2 \;+\; (\omega^\varphi)^2.$$

---

[1] Electronic mail address: tharriot@istar.ca
[2] Electronic mail address: williams@brandonu.ca

CP493, *General Relativity and Relativistic Astrophysics*, edited by C. P. Burgess and R. C. Myers
© 1999 American Institute of Physics 1-56396-905-X/99/$15.00

The $\{\omega^\mu\}$ are

$$\omega^t = e^{-\Lambda} (e^\eta \cos \alpha \, dt + e^{-\eta} \sin \alpha \, dr),$$
$$\omega^r = e^{-\Lambda} (e^\eta \sin \alpha \, dt - e^{-\eta} \cos \alpha \, dr),$$
$$\omega^\theta = e^{-\zeta} r \, d\theta,$$
$$\omega^\varphi = e^{-\zeta} r \sin \theta \, d\varphi.$$

Introducing $Q := r^{-1} - \partial_r \zeta$, the connection 1-forms are found to be

$$\omega^t{}_r = \omega^r{}_t = e^{\Lambda+\eta} \{\sin \alpha \, \partial_r \alpha + \cos \alpha \, \partial_r(\Lambda - \eta) \} \omega^t +$$
$$e^{\Lambda+\eta} \{\cos \alpha \, \partial_r \alpha - \sin \alpha \, \partial_r(\Lambda - \eta) \} \omega^r,$$
$$\omega^t{}_\theta = \omega^\theta{}_t = e^{\Lambda+\eta} Q \sin \alpha \, \omega^\theta,$$
$$\omega^t{}_\varphi = \omega^\varphi{}_t = e^{\Lambda+\eta} Q \sin \alpha \, \omega^\varphi,$$
$$\omega^r{}_\theta = -\omega^\theta{}_r = e^{\Lambda+\eta} Q \cos \alpha \, \omega^\theta,$$
$$\omega^r{}_\varphi = -\omega^\varphi{}_r = e^{\Lambda+\eta} Q \cos \alpha \, \omega^\varphi,$$
$$\omega^\theta{}_\varphi = -\omega^\varphi{}_\theta = -r^{-1} e^\zeta \cot \theta \, \omega^\varphi.$$

Define L, M, N and P by

$$L := e^{2\eta} \partial_r \Lambda, \qquad M := -2^{-1} \partial_r(e^{2\eta} \cos 2\alpha),$$

$$N := r^{-2} e^{-2\Lambda+2\zeta} - e^{2\eta} Q^2 \cos 2\alpha, \qquad P := -(2r)^{-1} e^{2\eta+\zeta} \partial_r(rQe^{-\zeta}).$$

It follows that the curvature 2-forms can be written

$$\theta^t{}_r = \theta^r{}_t = e^{2\Lambda} \partial_r(M + L \cos 2\alpha) \, \omega^t \wedge \omega^r,$$
$$\theta^t{}_\theta = \theta^\theta{}_t = e^{2\Lambda} \{ [Q(M + L) - 2P \sin^2\alpha] \, \omega^t \wedge \omega^\theta$$
$$+ (P - QL) \sin 2\alpha \, \omega^r \wedge \omega^\theta \},$$
$$\theta^t{}_\varphi = \theta^\varphi{}_t = e^{2\Lambda} \{ [Q(M + L) - 2P \sin^2\alpha] \, \omega^t \wedge \omega^\varphi$$
$$+ (P - QL) \sin 2\alpha \, \omega^r \wedge \omega^\varphi \},$$
$$\theta^r{}_\theta = -\theta^\theta{}_r = e^{2\Lambda} \{ (QL - P) \sin 2\alpha \, \omega^t \wedge \omega^\theta$$
$$+ [Q(M - L) + 2P \cos^2\alpha] \, \omega^r \wedge \omega^\theta \},$$
$$\theta^r{}_\varphi = -\theta^\varphi{}_r = e^{2\Lambda} \{ (QL - P) \sin 2\alpha \, \omega^t \wedge \omega^\varphi$$
$$+ [Q(M - L) + 2P \cos^2\alpha] \, \omega^r \wedge \omega^\varphi \},$$
$$\theta^\theta{}_\varphi = -\theta^\varphi{}_\theta = e^{2\Lambda} N \, \omega^\theta \wedge \omega^\varphi.$$

Letting $|\alpha\beta|$ denote indices in increasing order, the equation

$$\theta^\mu{}_\nu = R^\mu{}_{\nu|\alpha\beta|} \, \omega^\alpha \wedge \omega^\beta$$

then leads to the Einstein tensor, $G_{\mu\nu}$, whose nonzero components are:

$$G_{tt} = e^{2\Lambda}\{2QM + N + 2P\cos 2\alpha - 2(QL - P)\},$$
$$G_{tr} = G_{rt} = 2e^{2\Lambda}(QL - P)\sin 2\alpha,$$
$$G_{rr} = -e^{2\Lambda}\{2QM + N + 2P\cos 2\alpha + 2(QL - P)\},$$
$$G_{\theta\theta} = G_{\varphi\varphi} = -e^{2\Lambda}\{\partial_r(M + L\cos 2\alpha) + 2QM + 2P\cos 2\alpha\}.$$

To conveniently discuss energy conditions, $G^{\mu\nu}$ must first be diagonalized (Ref. 5, p. 89). The eigenvalue equation,

$$(G_{\mu\nu} - \lambda g_{\mu\nu})\xi^\nu = 0,$$

can be used to introduce a new orthonormal tetrad of basis vectors $\{\mathbf{E}_{\hat{\alpha}}\}$ as the eigenvectors that correspond to the various choices of λ. The resulting $G_{\hat{\gamma}\hat{\kappa}} = G_{\mu\nu}E_{\hat{\gamma}}^\mu E_{\hat{\kappa}}^\nu$ will be diagonal. (Indices with a hat, $\hat{\alpha}$, will refer to the tetrad $\{\mathbf{E}_{\hat{\alpha}}\}$ and, as before, unhatted indices will refer to $\{\mathbf{e}_\mu\}$). Substituting the previous expressions for $G_{\mu\nu}$ into the eigenvalue equation, one finds a double eigenvalue, $\lambda_{\hat{2}} = \lambda_{\hat{3}}$, which can then be used to determine spacelike eigenvectors $E_{\hat{2}}$ and $E_{\hat{3}}$. There are also two *distinct* eigenvalues, $\lambda_{\hat{0}}$, $\lambda_{\hat{1}}$, given by

$$\lambda = e^{2\Lambda}\{-(2QM + N + 2P\cos 2\alpha) \pm 2[(QL - P)^2\cos^2 2\alpha]^{1/2}\}.$$

The ambiguity resulting from different interpretations of $\pm[(QL - P)^2\cos^2 2\alpha]^{1/2}$ can be resolved by the requirement that $E_{\hat{0}}$ be everywhere timelike and $E_{\hat{1}}$ be everywhere spacelike. This leads to

$$\lambda_{\hat{0}} = e^{2\Lambda}\{-(2QM + N + 2P\cos 2\alpha) + 2(QL - P)|\cos 2\alpha|\},$$
$$\lambda_{\hat{1}} = e^{2\Lambda}\{-(2QM + N + 2P\cos 2\alpha) - 2(QL - P)|\cos 2\alpha|\},$$
$$\lambda_{\hat{2}} = \lambda_{\hat{3}} = -e^{2\Lambda}\{\partial_r(M + L\cos 2\alpha) + 2QM + 2P\cos 2\alpha\},$$
$$\mathbf{E}_{\hat{0}} = (g_-, h_-, 0, 0),$$
$$\mathbf{E}_{\hat{1}} = (g_+, h_+, 0, 0),$$
$$\mathbf{E}_{\hat{2}} = (0, 0, 1, 0),$$
$$\mathbf{E}_{\hat{3}} = (0, 0, 0, 1),$$

where

$$g_\pm = \sin 2\alpha\{2|\cos 2\alpha|(1 \pm |\cos 2\alpha|)\}^{-1/2},$$
$$h_\pm = \{(1 \pm |\cos 2\alpha|)(2|\cos 2\alpha|)^{-1}\}^{1/2}.$$

Noting that $\|g^{\hat{\alpha}\hat{\beta}}\| = \|g_{\hat{\alpha}\hat{\beta}}\| = \mathrm{diag}(-1, 1, 1, 1)$ and using $G^{\hat{\alpha}\hat{\beta}} = g^{\hat{\alpha}\hat{\gamma}}g^{\hat{\beta}\hat{\kappa}}G_{\hat{\gamma}\hat{\kappa}}$ and the Einstein equations, $G^{\hat{\alpha}\hat{\beta}} = T^{\hat{\alpha}\hat{\beta}}$, the energy-momentum tensor is found to be $\|T^{\hat{\alpha}\hat{\beta}}\| = \mathrm{diag}(\rho, p_1, p_2, p_3)$, where

$$\rho = e^{2\Lambda}\{2QM + N + 2P\cos 2\alpha - 2(QL - P)|\cos 2\alpha|\},$$
$$p_1 = -e^{2\Lambda}\{2QM + N + 2P\cos 2\alpha + 2(QL - P)|\cos 2\alpha|\},$$
$$p_2 = p_3 = -e^{2\Lambda}\{\partial_r(M + L\cos 2\alpha) + 2QM + 2P\cos 2\alpha\}.$$

The weak energy condition is equivalent to $\rho \geq 0$ and $\rho + p_i \geq 0$, $i = 1,2,3$. The dominant energy condition is equivalent to $\rho \geq 0$ and $-\rho \leq p_i \leq \rho$, $i = 1,2,3$. The strong energy condition is equivalent to $\rho + p_i \geq 0$, $i = 1,2,3$, and $\rho + \sum_{i=1}^{3} p_i \geq 0$.

A simple special case is obtained by setting $e^{-2\zeta}r^2 = K^2$, where K is a positive constant. This implies $P = Q = 0$ (so that $G^{\mu\nu}$ is already diagonal), $N = e^{-2\Lambda}K^{-2}$ and

$$\rho = -p_1 = K^{-2}, \qquad p_2 = p_3 = -e^{2\Lambda}\partial_r(M + L\cos 2\alpha).$$

The weak energy condition will be satisfied if

$$K^{-2} - e^{2\Lambda}\partial_r(M + L\cos 2\alpha) \geq 0,$$

and the dominant energy condition if

$$K^{-2} \pm e^{2\Lambda}\partial_r(M + L\cos 2\alpha) \geq 0,$$

where the latter inequality must be satified for both choices of sign. For further simplification, let $e^{2\eta} = 2K^{-2}$ and $\Lambda = 0$. Then the weak energy condition will be satisfied if

$$1 + \partial_r^2(\cos 2\alpha) \geq 0,$$

and the dominant energy condition if

$$1 \pm \partial_r^2(\cos 2\alpha) \geq 0.$$

An example in which $\cos 2\alpha$ changes sufficiently slowly for the above condition to hold can be constructed by considering

$$\cos 2\alpha = 1 - 2kr\, e^{1-kr},$$

and choosing the positive constant k to be sufficiently small. Note that

$$\cos 2\alpha(0) = 1, \qquad \cos 2\alpha(1/k) = -1, \qquad \cos 2\alpha(\infty) = 1,$$

with $-1 \leq 1 - 2kr\, e^{1-kr} \leq 1$. This is consistent with α varying from 0 to π as r varies from zero to infinity, i.e. a kink is present. It is easy to check that values of k for which $0 < k \leq e^{-1/2}/2$ lead to $1 \pm \partial_r^2(\cos 2\alpha) \geq 0$, so that the corresponding spacetime satisfies the weak and dominant energy conditions. The strong energy condition requires $\partial_r^2(\cos 2\alpha) \geq 0$, which is clearly impossible for any choice of α that corresponds to having a kink present (since $\cos 2\alpha$ must vary from 1 to -1 and back to 1 again).

ACKNOWLEDGEMENTS

This work was supported by the Mount Saint Vincent University Research Committee and by the Natural Sciences and Engineering Research Council of Canada.

[1] K.A. Dunn, T.A. Harriott, and J.G. Williams, J. Math. Phys. **37**, 5637–5651 (1996), **38**, 6470–6474 (1997).

[2] D. Finkelstein and C.W. Misner, Ann. Phys. (NY) **6**, 230–243 (1959).

[3] A.Z. Wang, Mod. Phys. Lett. A **7**, 1779–1789 (1992); P.S. Letelier and A.Z. Wang, Phys. Rev. D **48**, 631–646 (1993).

[4] G.F.R. Ellis and M.A.H. MacCallum, Commun. Math. Phys. **12**, 108–141 (1969).

[5] S.W. Hawking and G.F.R. Ellis, *The Large Scale Structure of Space-time* (Cambridge U.P., Cambridge, 1973).

Perturbative Superluminal Censorship and the Null Energy Condition

Matt Visser,[†] Bruce Bassett,[¶,*] and Stefano Liberati[§,‡]

[†] *Physics Department, Washington University, Saint Louis, Missouri 63130-4899, USA*
[¶,*] *International School for Advanced Studies (SISSA), Via Beirut 2-4, 34014 Trieste, Italy*
[*] *Department of Theoretical Physics, University of Oxford, 1 Keble Road, OX1 3NP, UK*
[‡] *Istituto Nazionale di Fisica Nucleare (INFN), sezione di Trieste, Italy*

Abstract. We argue that "effective" superluminal travel, potentially caused by the tipping over of light cones in Einstein gravity, is always associated with violations of the null energy condition (NEC). This is most easily seen by working perturbatively around Minkowski spacetime, where we use linearized Einstein gravity to show that the NEC forces the light cones to contract (narrow). Given the NEC, the Shapiro time delay in any weak gravitational field is always a delay relative to the Minkowski background, and never an advance. Furthermore, any object travelling within the lightcones of the weak gravitational field is similarly delayed with respect to the minimum traversal time possible in the background Minkowski geometry.

INTRODUCTION

The relationship between the causal aspects of spacetime and the stress-energy of the matter that generates the geometry is a deep and subtle one. In this note, which is a simplified presentation based on our earlier work [1], we shall focus in somewhat more detail on the perturbative investigation of the connection between the null energy condition (NEC) and the light-cone structure. We shall demonstrate that in linearized gravity the NEC always forces the light cones to contract (narrow): Thus the validity of the NEC for ordinary matter implies that in weak gravitational fields the Shapiro time delay is always a delay rather than an advance.

This simple observation has implications for the physics of (effective) faster-than-light (FTL) travel via "warp drive". It is well established, via a number of rigorous theorems, that any possibility of effective FTL travel via traversable wormholes necessarily involves NEC violations [2–5]. On the other hand, for effective FTL travel via warp drive (for example, via the Alcubierre warp bubble [6], or the Krasnikov FTL hyper-tube [7]) NEC violations are observed in specific examples but it is difficult to prove a really general theorem guaranteeing that FTL travel implies NEC violations [1]. Part of the problem arises in even defining what we mean by FTL, and recent progress in this regard is reported in [1,8].

CP493, *General Relativity and Relativistic Astrophysics*, edited by C. P. Burgess and R. C. Myers
© 1999 American Institute of Physics 1-56396-905-X/99/$15.00

In this note we shall (for pedagogical reasons) restrict attention to weak gravitational fields and work perturbatively around flat Minkowski spacetime. One advantage of doing so is that the background Minkowski spacetime provides an unambiguous definition of FTL travel. A second advantage is that the linearized Einstein equations are simply (if formally) solved via the gravitational Liénard–Wiechert potentials. The resulting expression for the metric perturbation provides information about the manner in which light cones are perturbed.

LINEARIZED GRAVITY

For a weak gravitational field, linearized around flat Minkowski spacetime, we can in the usual fashion write the metric as [4,9,10]

$$g_{\mu\nu} = \eta_{\mu\nu} + h_{\mu\nu}, \tag{1}$$

with $h_{\mu\nu} \ll 1$. Then adopting the Hilbert–Lorentz gauge (*aka* Einstein gauge, harmonic gauge, de Donder gauge, Fock gauge)

$$\partial_\nu \left[h^{\mu\nu} - \frac{1}{2}\eta^{\mu\nu}h \right] = 0, \tag{2}$$

the linearized Einstein equations are [4,9,10]

$$\Delta h_{\mu\nu} = -16\pi G \left[T_{\mu\nu} - \frac{1}{2}\eta_{\mu\nu}T \right]. \tag{3}$$

This has the formal solution [4,9,10]

$$h_{\mu\nu}(\vec{x},t) = 16\pi G \int d^3y \, \frac{\left[T_{\mu\nu}(\vec{y},\tilde{t}) - \frac{1}{2}\eta_{\mu\nu}T(\vec{y},\tilde{t}) \right]}{|\vec{x} - \vec{y}|}, \tag{4}$$

where $\tilde{t}$ is the retarded time $\tilde{t} = t - |\vec{x} - \vec{y}|$. These are the gravitational analog of the Liénard–Wiechert potentials of ordinary electromagnetism, and the integral has support on the unperturbed backward light cone from the point $\vec{x}$.

In writing down this formal solution we have tacitly assumed that there is no incoming gravitational radiation. We have also assumed that the *global geometry* of spacetime is approximately Minkowski, a somewhat more stringent condition than merely assuming that the metric is *locally* approximately Minkowski. Finally note that the fact that we have been able to completely gauge-fix Einstein gravity in a canonical manner is essential to argument. That we can locally gauge-fix to the Hilbert–Lorentz gauge is automatic. By the assumption of asymptotic flatness implicit in linearized Einstein gravity, we can apply this gauge at spatial infinity where the only remaining ambiguity, after we have excluded gravitational radiation, is that of the Poincare group. (That is: Solutions of the Hilbert–Lorentz gauge condition, which can be rewritten as $\nabla^2 x^\mu = 0$, are under these conditions unique up to Poincare transformations.) We now extend the gauge condition inward to cover the entire spacetime, the only obstructions to doing so globally coming from black holes or wormholes, which are excluded by definition. Thus adopting the Hilbert–Lorentz gauge in linearized gravity allows us to assign a *canonical* flat Minkowski metric to the entire spacetime, and it is the existence of this canonical flat metric that permits us to make the comparisons (between two different metrics on the same spacetime) that are at the heart of the argument that follows.

Now consider a vector k^μ which we take to be a null vector of the *unperturbed* Minkowski spacetime

$$\eta_{\mu\nu} \, k^\mu k^\nu = 0. \tag{5}$$

In terms of the full perturbed geometry this vector has a norm

$$||k||^2 \equiv g_{\mu\nu} \, k^\mu k^\nu \tag{6}$$

$$= h_{\mu\nu} \, k^\mu k^\nu \tag{7}$$

$$= 16\pi G \int d^3y \, \frac{T_{\mu\nu}(\vec{y}, \tilde{t}) \, k^\mu k^\nu}{|\vec{x} - \vec{y}|}. \tag{8}$$

Now assume the NEC

$$T_{\mu\nu} \, k^\mu k^\nu \geq 0, \tag{9}$$

and note that the kernel $|\vec{x} - \vec{y}|^{-1}$ is positive definite. Using the fact that the integral of a everywhere positive integrand is also positive, we deduce $g_{\mu\nu} \, k^\mu k^\nu \geq 0$. Barring degenerate cases, such as a completely empty spacetime, the integrand will be positive definite so that

$$g_{\mu\nu} \, k^\mu k^\nu > 0. \tag{10}$$

That is, a vector that is null in the Minkowski metric will be spacelike in the full perturbed metric. Thus the null cone of the perturbed metric must everywhere lie inside the null cone of the unperturbed Minkowski metric.

Because the light cones contract, the *coordinate* speed of light must everywhere decrease. (Not the *physical* speed of light as measured by local observers, as always in Einstein gravity, that is of course a constant.) This does however mean that the time required for a light ray to get from one spatial point to another must always increase compared to the time required in flat Minkowski space. This is the well-known Shapiro time delay, and we see two important points: (1) to even define the delay (delay with respect to what?) we need to use the flat Minkowski metric as a background, (2) the fact that in the solar system it is always a delay, never an advance, is due to the fact that everyday bulk matter satisfies the NEC.

(We mention in passing that the strong energy condition [SEC] provides a somewhat stronger result: If the SEC holds then the proper time interval between any two timelike separated events in the presence of the gravitational field is always larger than the proper time interval between these two events as measured in the background Minkowski spacetime.)

Now subtle quantum-based violations of the NEC are known to occur [11], but they are always small and are in fact tightly constrained by the Ford–Roman quantum inequalities [12,13]. There are also *classical* NEC violations that arise from non-minimally coupled scalar fields [14], but these NEC violations require Planck-scale expectation values for the scalar field. NEC violations are never appreciable in a solar system or galactic setting. (SEC violations are on the other hand relatively common. For example: cosmological inflation, classical massive scalar fields, etc.)

From the point of view of warp drive physics, this analysis is complementary to that of [8], (and also to the comments by Coule [15], regarding energy condition violations and "opening out" the light cones). Though the present analysis is perturbative around Minkowski space, it has the advantage of establishing a direct and immediate *physical* connection between FTL travel and NEC violations. Generalizing this result beyond the weak field perturbative regime is somewhat tricky [1], and we have addressed this issue elsewhere. To even define effective FTL one will need to compare two metrics. (Just to be able to ask the question "FTL with respect to what?").

Even if we simply work perturbatively around a general metric, instead of perturbatively around the Minkowski metric, the complications are immense: (1) the Laplacian in the linearized gravitational equations must be replaced by the Lichnerowicz operator; (2) the Green function for the Lichnerowicz operator need no longer be concentrated *on* the past light cone [physically, there can be back-scattering from the background gravitational field, and so the Green function can have additional support from *within* the backward light cone]; and (3) the Green function need no longer be positive definite.

For example, even for perturbations around a Friedman–Robertson–Walker (FRW) cosmology, the analysis is not easy [16]. Because linearized gravity is *not* conformally coupled to the background the full history of the spacetime back to the Big Bang must be specified to derive the Green function. From the astrophysical literature concerning gravitational lensing it is known that *voids* (as opposed to over-densities) can sometimes lead to a Shapiro time *advance* [17–19]. This is not in conflict with the present analysis and is not evidence for astrophysical NEC violations. Rather, because those calculations compare a inhomogeneous universe with a void to a homogeneous FRW universe, the existence of a time advance is related to a suppression of the density below that of the homogeneous FRW cosmology. The local speed of light is determined by the local gravitational potential relative to the FRW background. Voids cause an increase of the speed of photons relative to the homogeneous background. The total time delay along a particular geodesic is, however, affected by two factors: the gravitational potential effect on the speed of propagation and the geometric effect due to the change in path of the photon (lensing) which may make the total path length longer. Thus traveling through a void doesn't *necessarily* imply an advance relative to the background geometry.

DISCUSSION

This note argues that any form of FTL travel requires violations of the NEC. The perturbative analysis presented here is very useful in that it demonstrates that it is already extremely difficult to even get even started: Any perturbation of flat space that exhibits even the slightest amount of FTL (defined as widening of the light cones) must violate the NEC. The perturbative analysis also serves to focus attention on the Shapiro time delay as a diagnostic for FTL, and it is this feature of the perturbative analysis we have extended elsewhere to the non-perturbative regime to provide both a non-perturbative definition of FTL [1], and a non-perturbative theorem regarding superluminal censorship.

REFERENCES

[†] Electronic mail: visser@kiwi.wustl.edu

[¶] Electronic mail: bruce@thphys.ox.ac.uk

[‡] Electronic mail: liberati@sissa.it

1. M. Visser, B. Bassett, and S. Liberati, *Superluminal censorship*, gr-qc/9810026.

2. M.S. Morris and K.S. Thorne, Am. J. Phys. **56**, 395 (1988).

3. M.S. Morris, K.S. Thorne, and U. Yurtsever, Phys. Rev. Lett, **61**, 1446 (1988).

4. M. Visser, *Lorentzian wormholes*, (AIP Press, New York, 1995).

5. D. Hochberg and M. Visser, Phys. Rev. Lett. **81**, 746 (1998); Phys. Rev. **D57**, 044021 (1998).

6. M. Alcubierre, Class. Quantum Grav. **11**, L73 (1994).

7. S.V. Krasnikov, Phys. Rev. **D57**, 4760 (1998).

8. K. Olum, Phys. Rev. Lett. **81**, 3567 (1998)

9. C.W. Misner, K.S. Thorne, and J.A. Wheeler, *Gravitation*, (Freeman, San Francisco, 1973).

10. R.M. Wald, *General Relativity*, (Chicago University Press, 1984).

11. M. Visser, Phys. Lett. **B349**, 443 (1995); Phys. Rev. **D54**, 5103 (1996); **D54**, 5116 (1996); **D54**, 5123 (1996); **D56**, 936 (1997).

12. L.H. Ford and T.A. Roman, Phys. Rev. **D51**, 4277 (1995); **D53**, 5496 (1996)

13. M.J. Pfenning and L.H. Ford, Class. Quantum Grav. **14**, 1743 (1997);

14. E.E. Flanagan and R.M. Wald, Phys. Rev. **54**, 6233 (1996).

15. D.H. Coule, Class. Quantum Grav. **15**, 2523 (1998).

16. M.J. Pfenning and L.H. Ford, Phys. Rev. **D55**, 4813, (1997).

17. P. Schneider, J. Ehlers, and E.E. Falco, *Gravitational Lenses*, (Springer–Verlag, Berlin, 1992).

18. N. Mustapha, B.A. Bassett, C. Hellaby, and G.F.R. Ellis, Class. Quantum Grav. **15**, 2363 (1998).

19. G.F.R. Ellis and D.M. Solomons, Class. Quantum Grav. **15** 2381 (1998).

Brill Wave Initial-Value Problem

David Hobill and Paul Webster

The University of Calgary, Department of Physics and Astronomy, 2500 University Drive N.W., Calgary, Alberta, Canada T2N 1N4.

I INTRODUCTION

The initial value problem in general relativity is of interest because from a consistent three dimensional (constant time) hypersurface, or slice, we can, using the 3+1 formulation of general relativity, reconstruct the full four dimensional spacetime.

This paper examines the properties of the initial data set of an axisymmetric vacuum spacetime with a gravitational wave whose peak is located at different radial locations. The initial data represents a time symmetric wave. In particular we look at embedding diagrams and locations of any apparent horizons.

II SOLVING THE IVP

We use the three dimensional spatial metric defined by the line element

$$ds^2 = \psi^4(af_{,\eta}^2 d\eta^2 + bf^2 d\theta^2 + cff_{,\eta} d\eta d\theta + df^2 \sin^2 \theta d\phi^2). \tag{1}$$

to represent the spatial geometry on that slice. Where $f = f(\eta)$ is a radial scale function of a logarithmic radial coordinate, η.

The specific form of wave we use is the *Brill wave*, first constructed in 1959 by Brill. The "ground state" (zero Brill wave amplitude) is the Minkowski spacetime.

Brill showed that by choosing the 3-metric, γ_{ab}, so that the spatial line element is of the form

$$ds^2 = \psi^4[e^{2q}(d\rho^2 + \rho^2 d\theta^2) + \rho^2 \sin^2 \theta d\phi^2],$$

where $q = 0$ along the axis and falls off sufficiently quickly radially, the mass of the hyperfurface is well defined.

We choose $q(\eta, \theta)$ to be of the form

$$q = ag_1(\theta)g_2(\eta),$$

where $a =$const and the radial component, g_2, is chosen to be the inversion symmetric gaussian

$$g_2(\eta) = e^{-(g_+)^2} + e^{-(g_-)^2},$$

CP493, *General Relativity and Relativistic Astrophysics*, edited by C. P. Burgess and R. C. Myers
© 1999 American Institute of Physics 1-56396-905-X/99/$15.00

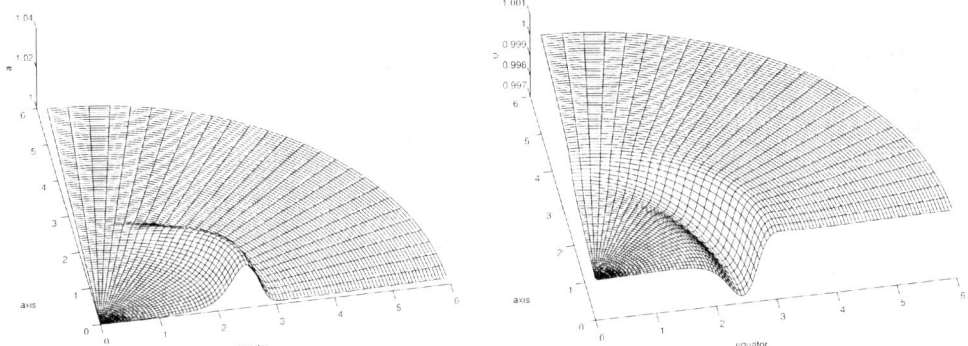

FIGURE 1. Left shows an example of the initial Brill wave in the metric component a, and right shows the associated conformal factor. The data set is $(0.01,1.0,3.0)$.

where $g_{\pm} = \frac{\eta \pm r_0}{\sigma}$. The angular dependence is given by the function g_1, which is chosen to be $g_1(\theta) = \sin^n \theta$, where n is an even integer, usually 2 or 4. The three independent parameters of q which specify the amplitude, initial radial position, and width, are referred to by (a, r_0, σ).

The Hamiltonian constraint,

$$R + (Tr\,K)^2 - K^{ab}K_{ab} = 0,$$

where R is the 3-dimensional scalar curvature, K_{ab}, the extrinsic curvature and $K = K_a^a$, needs to be solved in order to determine the properties of the initial slice (the momentum constraint is solved trivially due to the time symmetric nature of the wave). With the above metric, the Hamiltonian constraint becomes an elliptic second order PDE for the conformal factor, ψ:

$$\psi_{,\eta\eta} + \frac{f_{,\eta}^2}{f^2}\psi_{,\theta\theta} - \psi_{,\eta}\left(\frac{f_{,\eta\eta}}{f_{,\eta}} - \frac{2f_{,\eta}}{f}\right) + \psi_{,\theta}\frac{f_{,\eta}^2}{f^2}\cot\theta + \frac{\psi}{4}\left(q_{,\eta\eta} + \frac{q_{,\theta\theta}f_{,\eta}^2}{f^2} - q_{,\eta}\left[\frac{f_{,\eta\eta}}{f_{,\eta}} - \frac{f_{,\eta}}{f}\right]\right) = 0.$$

The solution to this equation is obtained numerically, using boundary conditions appropriate to a vacuum spacetime with no black hole.

An example of the metric component a and the associated solution of the Initial Value Problem, ψ, is shown in Figure 1.

III   EMBEDDINGS

We can embed γ_{11} in the equatorial plane and an example of this is shown in Figure 2. If the wave is sufficiently large, apparent horizons are formed, indicating the inevitable presence of a black hole.

We can also simultaneously embed γ_{11} and γ_{22}, which is also shown in Figure 2. It can be seen that the radial line along the equator of the γ_{11}, γ_{22} embeddings corresponds to the single line given by the γ_{11} equatorial embeddings.

The metric component γ_{22} can also be embedded for constant radial (η) surfaces, as well as the apparent horizons themselves shown in Figure 3.

Another measure of the geometry of the spacetime is the ratio of the polar circumference to the equatorial circumference, and this is shown in Figure 4 for constant radial (η) surfaces. If the ratio is positive, this indicates the surface is prolate and if it is negative, the surface is oblate. If the ratio is less that $\frac{\pi}{2}$, then the surface is 'flatter that a pancake', and cannot be embedded.

IV   APPARENT HORIZONS

The Maximum Radiation Loss (MRL) is defined by

$$MRL = \frac{M_{ADM} - M_{AH}}{M_{ADM}}, \tag{2}$$

where the apparent horizon mass is given by

$$M_{AH} \equiv \sqrt{\frac{area\ of\ apparent\ horizon}{16\pi}}, \tag{3}$$

and M_{ADM} is the ADM mass measured at the outer bounary of the grid. Here we show how the MRL of the apparent horizons varies with amplitude. The outer apparent horizon should always have a positive MRL, and the inner one having a negative MRL indicates the inability to measure local masses within a black hole. The MRL is shown for the two horizons formed with a varying Brill wave amplitude in Figure 2.

V   CONCLUSIONS

A pure Brill wave spacetime can have interesting and complex structure, with highly oblate and prolate surfaces forming around the wave. Far from the interaction region the spacetime behaves much like a Schwarzschild spacetime but the fine structure is quite complex. Although it appears that a 'bag of gold' type naked singularity might form as the 'size' of the wave increases, the minimal surface is at an increasingly large areal radius, and will never close off. Thus in this context violations of Cosmic Censorship seem unlikely.

The presence of apparent horizons in the vacuum Brill wave space time is an indication that a Brill wave might be evolved from a space time with no black hole to that of a black hole. The outer apparent horizon never attains a mass greater than the ADM mass, again meaning Cosmic Cenorship remains valid.

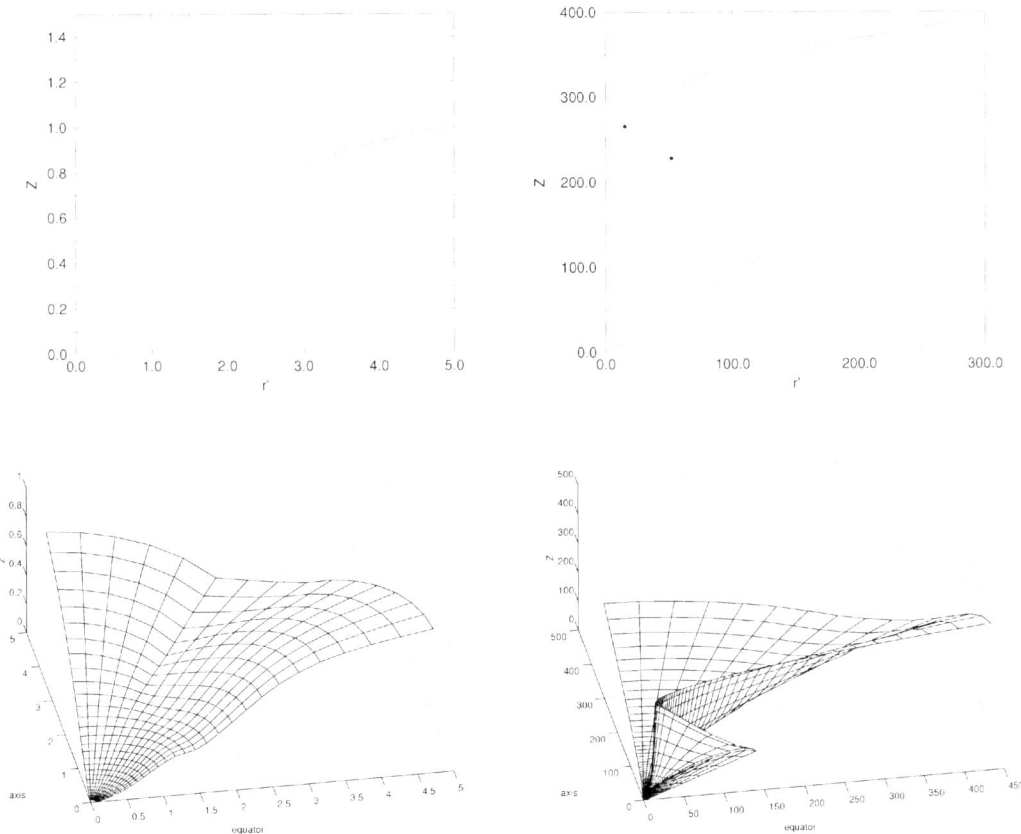

FIGURE 2. The upper left and right graphs show embedding diagrams of γ_{11} along the equator for a Brill wave spacetime with data sets $(0.1,1.0,1.0)$, and $(5.0,1.0,1.0)$. Apparent horizon locations are marked with a dot. The lower left and right graphs show embedding diagrams of γ_{11} and γ_{22} for a Brill wave spacetime with the same data sets as above.

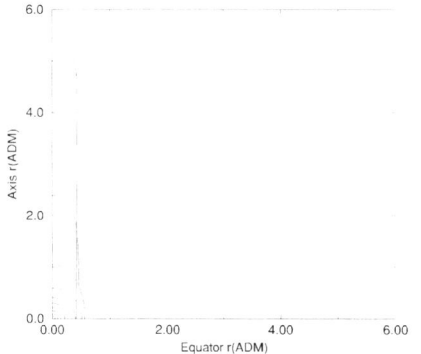
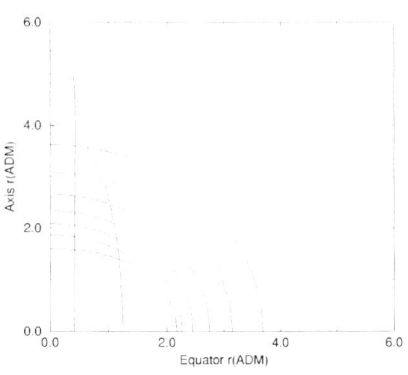

FIGURE 3. Embedding of a series of $\eta =$const surfaces. The initial data set is (3.0,1.0,3.0), and the surfaces (where they intersect the axis from bottom to top) are at $\eta = 0.25$, $\eta = 0.5$, $\eta = 0.75$, $\eta = 1.0$, $\eta = 1.25$, $\eta = 1.5$, $\eta = 1.75$, $\eta = 2.0$, $\eta = 2.25$, and $\eta = 2.5$. The right graph shows the continuation of these surfaces, with the exception of $\eta = 3.0$ which was not embeddable. The surfaces (where they intersect the equator from left to right) are at $\eta = 2.5$, $\eta = 2.75$, $\eta = 3.25$, $\eta = 3.5$, $\eta = 3.75$, $\eta = 4.0$, $\eta = 4.25$, $\eta = 4.5$, and $\eta = 4.75$

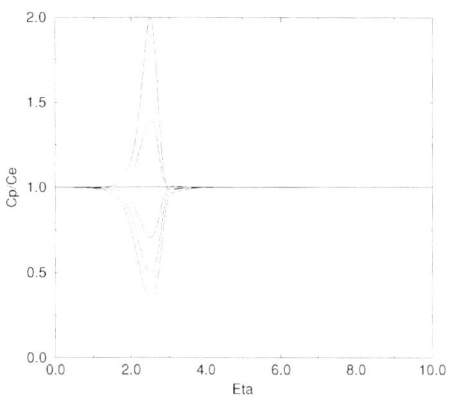

FIGURE 4. The left graph shows the C_p/C_e ratio for initial data set (a,3.0,1.0), where a is (from top to bottom) 1.0, 0.5, 0.0, -0.5, -1.0, -1.5. On the right the MRL for the apparent horizons is shown as a function of the Brill wave amplitude, with $\sigma = 1.5$, and $r_0 = 1.5$.

Dynamic Brill Wave Spacetimes

David Hobill and Paul Webster

*The University of Calgary, Department of Physics and Astronomy, 2500 University Drive N.W.,
Calgary, Alberta, Canada T2N 1N4.*

I INTRODUCTION

In this paper we present the current state of a code designed to study the dynamics of a Brill wave in an axisymmetric spacetime. The code is still under development, and these are preliminary results. In 1959 Brill specified an initial data set that would ensure the ADM mass of the spacelike hyperfurface associated with the initial data set was well defined. Here we take this initial data and attempt to evolve it though time to recreate the full four dimensional solution to Einstein's Equations. This is done by means of the 3+1 formalism, the details of which are given elsewhere [1]. Here we use the convention that $c = G = 1$, and all indices run from 1 to 3.

II FORM OF THE METRIC

A non rotating axisymmetric spacetime has one spatial symmetry about the azimuthal axis, so that in spherical polar coordinates, the metric is invariant under rotations about the coordinate ϕ. This leads to the spatial line element in spherical polar coordinates being given by

$$ds^2 = \psi^4(Ad\rho^2 + B\rho^2 d\theta^2 + C\rho d\rho d\theta + D\rho^2 \sin^2\theta d\phi^2), \tag{1}$$

where A, B, C, D are functions of ρ, θ, and t, and $\psi = \psi(\rho, \theta)$. Combining axisymmetry with equatorial reflection symmetry makes the metric symmetric about the $\theta = \frac{\pi}{2}$ plane, and thus, the independent range of θ, is $0 < \theta < \frac{\pi}{2}$. To facilitate the numerical problems associated with representing a large spatial volume, a new radial coordinate, η, is chosen so that we can represent large radial distances together with a fine grid near the centre of the system, where we expect most of the dynamics to occur

$$\rho = f(\eta). \tag{2}$$

For the duration of this talk, we choose $f(\eta) = \sinh(\eta)$, which has the properties that $\rho = 0$ corresponds to $\eta = 0$, it is symmetric across $\rho = 0$ and a large physical

CP493, *General Relativity and Relativistic Astrophysics*, edited by C. P. Burgess and R. C. Myers
© 1999 American Institute of Physics 1-56396-905-X/99/$15.00

volume can be represented by a relatively small grid in η. This leads to the line element

$$ds^2 = \psi^4(af_{,\eta}^2 d\eta^2 + bf^2 d\theta^2 + cff_{,\eta}d\eta d\theta + df^2 \sin^2 \theta d\phi^2). \tag{3}$$

We also need to define an extrinsic curvature, or second fundemental form,

$$K_{ij} = \psi^4 \begin{pmatrix} H_a f_{,\eta}^2 & H_c ff_{,\eta} & 0 \\ H_c ff_{,\eta} & H_b f^2 & 0 \\ 0 & 0 & H_d f^2 \sin^2 \theta \end{pmatrix}, \tag{4}$$

to determine the embedding of the three dimensional hypersurface in the full four dimensional spacetime.

Here the dynamical quantities are $a, b, c, d, H_a, H_b, H_c, H_d$ depend on t, η, and θ, while $\psi(\eta, \theta)$ is determined by the solution to the initial value problem.

The covariance of general relativity allows us to choose our coordinate system on each slice as we evolve the dynamic quantities, and this is expressed in the 3+1 formalism by choosing the lapse, $\alpha = \sqrt{g_{00}}$, which determines the proper time elapsed between two successive slices at each point, and the shift vector $\beta_i = g_{0i}$, which determines any boosts or rotations of the coordinates with respect to the previous slice. Here we have two non trivial components of the shift vector, the third, β_3 being zero due the axisymmetry of the system. So this gives us three kinematic quantities.

In the 3+1 formalism, the Einstein Equations become a set of ten equations, the scalar Hamiltonian constraint and the three momentum constraints

$$R - K_{ij}K^{ij} + (TrK)^2 = 0, \tag{5}$$

$$D_j(K_i^j - \gamma_i^j K) = 0 \tag{6}$$

and six (with a symmetric metric) evolution equations

$$\partial_t K_{ab} = -D_a D_b \alpha + \alpha[R_{ab} + (trK)K_{ab} - 2K_{ac}K_b^c] + \beta^c D_c K_{ab} + K_{ac}D_b\beta^c + K_{cb}D_a\beta^c. \tag{7}$$

The extrinsic curvature may be thought of as the 'velocity' of the metric in the full four dimensional spacetime, and this can be seen in the relationship between it and the metric

$$\partial_t \gamma_{ab} = -2\alpha K_{ab} + D_a\beta_b + D_b\beta_a. \tag{8}$$

312

III EVOLUTION

Here we assume a time symmetric Brill wave has been constructed, and the associated initial value problem has been solved. The time symmetry of the initial data means the momentum constraints will be identically zero, and this leaves the Hamiltonian constraint, which becomes a linear second order elliptic partial differential equation (PDE), and is solved numerically for ψ. See our other contribution to this volume for a further discussion on the Initial Value Problem.

In order to go about performing the evolution, we need to specify our gauge quantities, α, and β_i. We choose the maximal slicing condition ($TrK = 0$) to determine the lapse. This condition has had considerable success in the past [2], [3], and has the ability to penetrate apparent horizons, if they form. The two components of the shift vector are chosen such that

1. $c = 0$. This simplifies the equations to be solved considerably, and prevents any shear at the axis, $\theta = 0$. Since initially $c = 0$, this means we must enforce

$$\dot{\gamma}_{12} = 0, \qquad (9)$$

   throughout the evolution.

2. $a = b$. Known as the Isothermal gauge, this simplifies the equations further, and has had success in the past [2]. There are also indications that is *must* be used if we are to try to perform a stable evolution with maximal slicing [4]. Initially, $a = b$, so we need to choose our two components of the shift vector such that

$$\dot{\gamma}_{11} = \dot{\gamma}_{22}. \qquad (10)$$

The two conditions 9 and 10 give two coupled fist order PDEs to be solved. To decouple them, we must make them second order, and then once one of these second order PDEs is solved for β_2, one of the original first order equations can be used to solve for β_1 for numerical speed and accuracy.

IV NUMERICAL TECHNIQUES

The evolution through time is performed using a staggered leapfrog method, with the second order elliptic PDEs being solved using a biconjugate gradient method that is described in Press et al. [5]. The main problem in constructing a numerical solution to Einstein's equations for a system of this type is the coordinate singularity at $\rho = 0$, which causes instabilities at the origin very quickly in an evolution. Here we use an interpolation method to guess at a solution in the region that the evolution equations cannot be trusted, provided the guess keeps the constraint equations solved to a required accuracy. If the quantity in question is known to vanish at $\rho = 0$, then an odd order polynomial interpolation is used, and if the quantity is symmetric about $\rho = 0$, then an even order polynomial is used, and the results can be dramatic as shown in Figure 1.

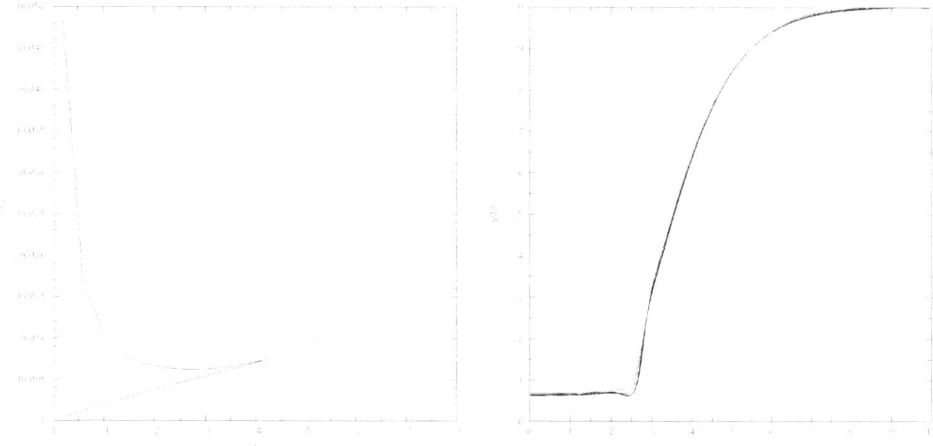

FIGURE 1. In the Left graph, the upper curve is R_{12} calculated directly from the evolution equations, and the lower curve has been interpolated from about $\eta = 0.4$ to $\eta = 0$, given that R_{12} should pass through 0 at $\eta = 0$. The right graph shows the rapid collapse of the lapse with a large Brill wave present.

V   RESULTS

The code actually only evolves for a very short time, due to an error in the slicing condition, which although does not affect the accuracy of the code since the slicing condition is merely a choice of coordinates, does make the lapse collapse much faster than maximal slicing would, effectively halting the evolution of the code very quickly, see Figure 1. Consequently a given Brill wave will only evolve for a short proper time. If the initial wave is, however chosen to be large, but with no apparent horizons on the initial slice present, an apparent horizon is seen to form during the evolution. Throughout this evolution the constraints are held satisfactorily as shown in Figure 2. As yet, due to the short duration of the evolution, it has not been possible to carry out a detailed analysis of the subsequent spacetime, other than to note the apparent horizon oscillates between two spatial locations, appears to be trying to settle down to a more spherical state as the evolution stops. This appears to resemble the 'ringing' of the black hole that has been observed in other numerical evolution's of black holes [3]. The horizon is very oblate upon its formation and is not embedable, and Figure 2 shows the location of the horizon in η coordinates as a function of angle for a number of different times.

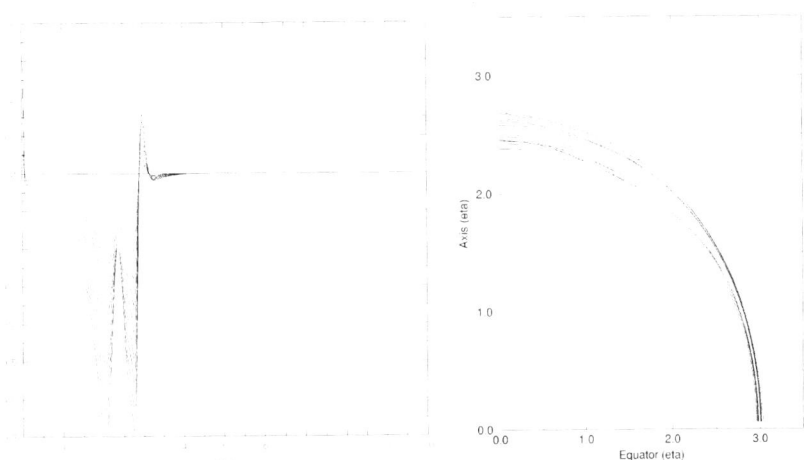

FIGURE 2. The left graph shows one of the momentum constraints at different times for the duration of the evolution, and the right graph shows the 'ringing' of the apparent horizon formed as it evolves to a more to less oblate form.

VI   CONCLUSIONS

A code has been developed that will evolve a Brill wave for a short proper time. If the wave is large enough, an apparent horizon forms towards the end of the evolution, thus, showing that given smooth vacuum initial data in the form of a Brill wave that it is possible to evolve this data consistently with Einstein's equations to a situation where a black is inevitable [6] in a finite proper time.

Unfortunately, the collapse of the lapse does not enable a detailed analysis of the subsequent spacetime to be carried out. Future work will concentrate on developing a true maximal slicing code that will enable a detailed analysis of black hole spacetimes created from a pure Brill wave.

REFERENCES

1. J. York, in *Sources of Gravitational Radiation*, edited by L. Smarr (Cambridge University Press, Cambridge, England, 1979).
2. C. Evans, *NATO A.R.W. on Astrophysical Radiation Hydrodynamics* (Kluwer, Dordrecht, Holland, 1982).
3. D. Bernstein, D. Hobill, E, L. Smarr and J. Towns, Phys. Rev. D **50** 5000 (1994).
4. M. Bardeen, T. Piran, Phys. Rep.**96** 205 (1983).
5. W. Press, B. Flannery, S. Teukolsky, and W. Vetterling *Numerical Recipies*, (Cambridge University Press, Cambridge, England, 1986).
6. R. Penrose, Ann. N.Y. Acad. Sci. **224** 125 (1973).

GRLite and GRTensorJ: Graphical User Interfaces to the Computer Algebra System GRTensorII

Mustapha Ishak[1], Peter Musgrave, John Mourra, Jonathan Stern
and Kayll Lake[2]

Department of Physics, Queen's University, Kingston, Ontario, Canada

Abstract. Two open graphical user interfaces to the computer algebra system GRTensorII are described. Each can be run locally or via the Internet. These provide students and researchers in the area of General Relativity and related fields with advanced truly portable tools that reduce many complex calculations to elementary functions.

INTRODUCTION

GRLite [1] and GRTensorJ are first- and second- generation graphical user interfaces to the computer algebra system GRTensorII [2]. They allow students and researchers in the area of General Relativity and related fields to perform symbolic calculations either locally or through the Web. GRLite is a calculator-style tool for evaluating more common tensors and scalars, reducing them to elementary functions. GRTensorJ provides fully customizable symbolic procedures without recompilation. These interfaces, which are open source, are written in Java and will run on any platform with browser support for JDK1.1.

I DESCRIPTION

Any user can initiate a GRLite or GRTensorJ session remotely on the Web by logging onto the GRTensorII home page and following the links. All that is needed is a browser with support for JDK1.1. The user initiates a session by clicking on the Java applet that opens the GRLite or GRTensorJ graphical user interfaces. They appear as in figures 1 and 2. Behind the scenes, a computer algebra session in GRTensorII is started automatically. GRLite and GRTensorJ use MapleV [3]

---

[1] ishak@astro.queensu.ca
[2] lake@astro.queensu.ca

CP493, *General Relativity and Relativistic Astrophysics,* edited by C. P. Burgess and R. C. Myers
© 1999 American Institute of Physics 1-56396-905-X/99/$15.00

as the algebraic engine. They are compatible with any engine that can output an ASCII stream.

A   GRLite

GRLite is perhaps most appropriate for students. At the moment, it is restricted to classical tensor analysis. The commands are performed through a predefined menu and selection of buttons. The first step is to select a space-time. This choice automatically displays the components of the metric tensor. The second step is to simply click on the object to be calculated. GRLite includes the 35 pre-defined functions as shown in figure 1. After the calculation is displayed the user can apply to it eight simplification procedures from the menu. There are also some support, like buffer clearing and a help system. GRLite comes with a set of pre-defined space-times but also offers a graphical sub-interface for entering new space-times defined for a given session.

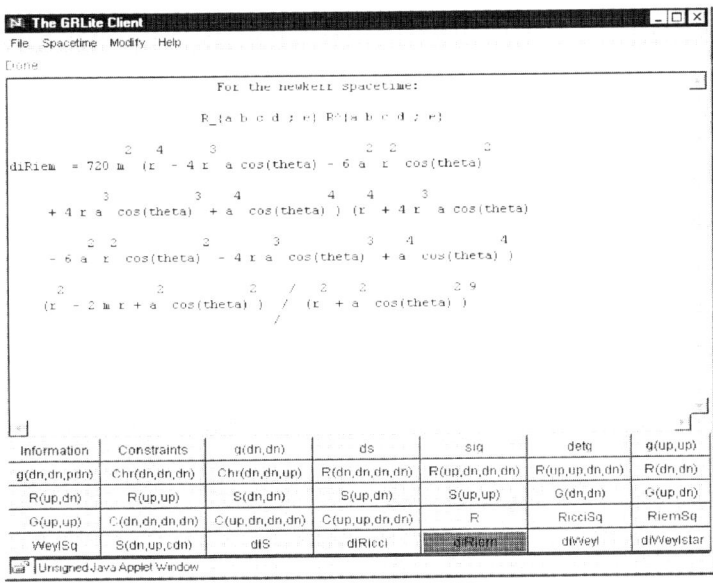

FIGURE 1. GRLite Graphical User Interface.   Calculation of the differential invariant $R_{abcd;e}R^{abcd;e}$ for the Kerr metric. At the time of writing, this calculation executes in about one second on a contemporary PC with MapleV Release 5.1 and GRTensorII 1.74.

B   GRTensorJ

GRTensorJ provides all the functionality of GRLite. In addition, it has a different architecture that allows it to be expanded and programmed by the user. The commands are accessed through menu and sub-menu selections. The first step is to select a space-time in coordinates or in tetrads. Then, the user can select the object(s) to be calculated from the corresponding menus. After the result of the calculation is displayed the user can apply to it any simplification procedure supported by the engine. The help system is built into the menu system as shown in figure 2. GRTensorJ comes with a selection of space-times and a set of pre-defined commands. Further, the user can define new metrics, tetrads and procedures through the GRTensorII definition facilities. These are entered through a sub-interface with the option of saving them on the users disk space. When a session begins, GRTensorJ reads a directory on the server named TextSheets (not to be confused with "worksheets") and builds the menu and sub-menus for the interface from the underlying structure. All sub-directories to TextSheets will appear as primary menu bar items. The names of ASCII files contained within these sub-directories will be displayed as menu selections. These files contain a sequence of commands written in the syntax of the computer algebra engine being used [4]. By selecting a menu item the user sends these commands to the engine. Items to be displayed in the interface window are distinguished simply by an asterisk in the file. In other words, creating new menu items and calculation commands is as simple as creating and editing simple ASCII files. Yet, the file, and resultant menu item, can be the equivalent of an entire worksheet with only the result chosen for display.

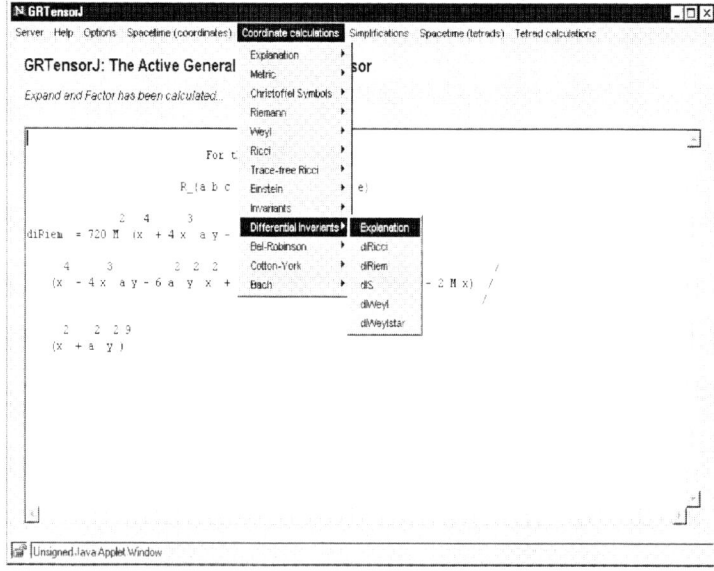

FIGURE 2. GRTensorJ Graphical User Interface. An example of the embedded help system is show. By the nature of the interface, the "help" system is more of the nature of an information system.

318

II INTERNAL DESIGN OVERVIEW

GRLite and GRTensorJ are written in Java and have been designed using an object-oriented approach. In addition, all communications between the user and the server are object-based. GRTensorJ has a multi-layer architecture that allows the generic functionality described above. This architecture is outlined below.

GUI (Graphical User Interface)
User Functional Interface
User ICM Handler
Interchange Modules (ICMs)
Server ICM Handler
Server Functional Interface
Algebraic Engine - Server Structure

III FUTURE DEVELOPMENT

A number of useful features are easily added to GRTensorJ, for example, automatic Latex output [5]. More involved would be the development of a dynamic database of solutions to Einstein's field equations. Such a database would have simple searches, classifications and calculations online. The database would communicate directly with a computer algebra system. A noteworthy feature is that the database would be designed to accept space-times by users through the Web. The object-oriented modularity of such a database would allow links to tools like GRtensorJ.

REFERENCES

1. Lake,K 1997 "Algebraic Computing" in *"Gravitation and Relativity:At the turn of the Millennium" Proceedings of GR-15*, IUCAA, Pune, Edited by Naresh Dadhich and Jayant Narlikar pp 421-427 (gr-qc/9803072)
2. GRTensorII is a package that runs within MapleV. It is entirely distinct from packages distributed with MapleV and must be obtained independently. The GRTensorII software and documentation is distributed freely on the World-Wide-Web from the address http://www.astro.queensu.ca/~grtensor/
3. MapleV is copyright Waterloo Maple Software and use of MapleV through GRLite and GRTensorJ on the Web is by special arrangement with Waterloo Maple Software.
4. For example, qload(test) in GRTensorII under MapleV is equivalent to qload[test] in GRTensorM under Mathematica.
5. A latex call in MapleV produces output suitable for use with LaTeX 2e that appears in the applet window. We are serious when we say that you could, for example, go to the Web, do your calculation, and paste the answer into the paper you are writing!

The CCGRRA Program

Thursday June 10:

Morning session:

08:50 : Opening Remarks: Luc Vinet (Director, CRM)

09:00 : Peter Saulson
"What Will We Learn from Gravitational Wave Detection?"

09:50 : Vicky Kaspi
"Binary Radio Pulsar Timing and General Relativity"

10:40 : Coffee Break

11:10 : Jeffrey Winicour
"The Characteristics of Colliding Black Holes"

12:00 : Lunch

Afternoon session:

13:30 : Sharon Morsink
"Surprises from Rotating Neutron Stars"

14:20 : (**Move to Parallel Sessions**)

    14:30 : Parallel Sessions A

15:50 : Coffee Break

    16:10 : Parallel Sessions B

17:30 : (Parallel Sessions End)

19:00 : <u>Banquet</u>

Friday June 11:

Morning session:

09:00 : Ted Jacobson
"Entropy and Gravity: horizons, entanglement and the holographic Bound"

09:50 : Abhay Ashtekar
"Quantum Geometry and Black Hole Entropy"

10:40 : Coffee Break

11:10 : Leonard Susskind
"The Holographic Principle"

12:00 : Lunch

Afternoon session:

13:30 : Gilles Fontaine
"White Dwarf Stars as Potential Contributors to Baryonic Dark Matter"

14:20 : (**Move to Parallel Sessions**)

14:30 : Parallel Sessions A

15:50 : Coffee Break

16:10 : Parallel Sessions B

17:30 : (Parallel Sessions End)

Saturday June 12:

Morning session:

09:00 : Lev Kofman
"Preheating After Inflation"

09:50 : Bill Unruh
"The effect of second order perturbations on the expansion of the universe"

10:40 : Coffee Break

11:10 : Don Page
"Can Quantum Cosmology Give Observational Consequences of Many-Worlds Quantum Theory?"

12:00 : Closing Remarks
(**Conference Ends**)

Parallel Sessions:

THURSDAY JUNE 10

Parallel session 1A: (Room 1020)

14:30 : Matt Visser
"Superluminal Censorship"
14:50 : Valeri Frolov
"Scattering and Capture of Cosmic Strings by a Black Hole"

15:10 : Yutaka Hosotani
"Dyon solutions in the Einstein-Yang-Mills theory in the asymptotically anti-de Sitter space"
15:30 : Jonathan Oppenheim
"Measurement of Time of Arrival in Quantum Mechanics"

Parallel session 1B: (Room 1020)

16:10 : Ivan Booth
"Naked Black Holes and Quasilocal Energy"
16:30 : Christopher Beetle
"A Hamiltonian Approach to the Mass of Isolated Black Holes"
16:50 : Don Witt
"Topological Censorship and Black Hole Topologies"
17:10 : Munawar Karim
"Casimir force in curved space"

Parallel session 2A: (Room 1050)

14:30 : Thomas Baumgarte
"The "Hydro Without Hydro" Approach to the Intermediate Binary Neutron Star Inspiral"
14:50 : Tzu T. Chia
"The constraints on rates of orbital angular momentum loss, orbital period variation and mass exchange in mass-conservative circular binary systems evolving by Roche lobe overflows"
15:10 : Karl Martel
"Detecting eccentric binaries with LIGO/VIRGO using circular templates"
15:30 : Edward Glass
"The Vaidya metric, the Bach tensor, and radiated power"

Parallel session 2B: (Room 1050)

16:10 : Andrei Frolov
"Critical collapse of the scalar field beyound spherical symmetry"
16:30 : Martin Goliath
"The state space and physical interpretation of self-similar spherically symmetric perfect-fluid models"
16:50 : Eric Poisson
"Radiative falloff in black-hole spacetimes -- Part I"
17:10 : Bill Laarakkers
"Radiative falloff in black-hole spacetimes -- Part II"

FRIDAY JUNE 11

Parallel session 1A: (Room 1020)

14:30 : Ramzi Khuri
"Black Holes, Thermodynamics and Polymers"
14:50 : Daniel Kabat
"Horizons and causality from gauge theory"

15:10 : Don Marolf
"Some brane-theoretic no-hair results (and their field theory duals)"
15:30 : David Lowe
"Holography, Cosmology and the Second Law of Thermodynamics"

Parallel session 1B: (Room 1020)

16:10 : Viqar Husain
"Phase transition in quantum gravity"
16:30 : Shinji Mukohyama
"Entanglement entropy and black hole"
16:50 : Kristin Schleich
"Exotic Spaces in Quantum Gravity"
17:10 : Jack Gegenberg
"Boundary Dynamics of Higher Dimensional Black Holes"
17:30 : Andrew Billyard
"The Stability of Scaling Solutions"

Parallel session 2A: (Room 1050)

14:30 : Eric Woolgar
"Lower Bounds for Horizon Areas of Higher Genus Black Holes"
14:50 : Martin Rainer
"Topological classifying spaces for algebras and geometries"
15:10 : Kayll Lake
"Algebraic computing in GR: The next level"
15:30 : W.R. Wood
"Conformal Symmetry"

Parallel session 2B: (Room 1050)

16:10 : Patrick Sutton
"Dimensional Reduction of the Effective Action and the Multiplicative Anomaly"
16:30 : Manasse Mbonye
"Evolution Of Evaporating Black Holes in a Higher-Dimensional Inflationary Universe"
16:50 : Ali Gaber Aeid
"On the Crossing of Thin Shells"
17:10 : Chris Vuille
"Schrodinger's Equation in General Relativity"
17:30 : Paul Webster
"Dynamic Brill Wave Spacetimes"

CONFERENCE PARTICIPANTS

1. **Shahriar Afshar**
   Omnigenesis Institute, P.O. Box 1281, Brookline, MA 02446, USA

2. **Michael Ashley**
   Dept. of Math. and Comp. Sci., St.Louis University, 221 N Grand Blvd, St.Louis , MO 63103, USA

3. **Abhay Ashtekar**
   Dept. of Physics, Pennsylvania State University, Center for Gravitational Physics & Geometry, 104 Davey Lab., University Park, PA 16802-6300, USA

4. **Claude Barrabes**
   UPRES A 6083, Universite de Tours (France), Faculté des sciences et technique, Parc Grandmont, 37200 Tours, FRANCE

5. **Thomas Baumgarte**
   237D Loomis Lab, University of Illinoois at Urbana-Champaign, 1110 W. Green Street, Urbana, IL 61801, USA

6. **Christopher Beetle**
   Department of Physics, Penn State University, 104 Davey Laboratory, Box 250, University Park, PA 16802, USA

7. **Andrew Billyard**
   Department of Physics, Dalhousie University, Halifax (Nova Scotia), B3H 3J5 CANADA

8. **Jefferson Bjoraker**
   Tate Laboratory of Physics and Astronomy, University of Minnesota, 116 Church St., Minneapolis, MN 55455, USA

9. **Ivan Booth**
   Department of Physics, University of Waterloo, Waterloo (Ontario), N2L 3G1 CANADA

10. **Paul Bracken**
    Centre de recherches mathématiques, Université de Montréal, C.P. 6128, Succ. Centre-ville, Montréal (Qu ébec) H3C 3J7 CANADA

11. **Aaron Bruce**
    105 Mt Hope St., Kitchener (Ontario) N2G 2J6 CANADA

12. **Cliff Burgess**
    Physics Department, McGill University, 3600 University St., Montreal (Quebec) H3A 2T8 CANADA

13. **Daniel Cartin**
Dept. of Physics, Pennsylvania State University, Davey Building, University Park, PA 16802, USA

14. **Chia Tzu Tit**
Physics Department, National University of Singapore, Kent Ridge, 119260 SINGAPORE

15. **Alan Coley**
AARMS Institute, Dept. of Maths, Stats & C. S., Dalhousie University, Halifax (Nova Scotia) B3H 3J5 CANADA

16. **Serge Droz**
Institute for Theoretical Physics, Winterthurerstr. 190, 8052 Zuerich, SWITZERLAND

17. **Ariel Edery**
Physics Department, McGill University, 3600 University St., Montreal (Quebec) H3A 2T8 CANADA

18. **Stephen Fairhurst**
Physics Department, Penn State University, 104 Davey Laboratory, University Park, PA 16802, USA

19. **Gilles Fontaine**
Dep. de physique, Université de Montréal, C.P. 6128, succ. Centre-ville, Montréal (Québec) H3C 3J7 CANADA

20. **Andrei Frolov**
Physics Department, University of Alberta, Edmonton (Alberta) T6G 2J1 CANADA

21. **Valeri Frolov**
Dept. of Physics, University of Alberta, Edmonton (Alberta) T6G 2J1 CANADA

22. **Claude Gauthier**
Dep. de mathématiques et de statistique, Université de Moncton, Moncton (Nouveau-Brunswick) E1A 3E9 CANADA

23. **Jack Gegenberg**
Dept. of Mathematics and Statistics, University of New Brunswick, Fredericton (New Brunswick) E3B 5A3 CANADA

24. **Edward Glass**
Department of Physics, University of Windsor, Windsor (Ontario) N9B 3P4 CANADA

25. **Martin Goliath**
Department of Physics, Stockholm University, Box 6730 S-11385, Stockholm
SWEDEN

26. **Pierre Gravel**
Dep. de mathématiques et informatique, College militaire royal du Canada,
B.P. 17000 Stn. Forces, Kingston (Ontario) K7K 7B4 CANADA

27. **Patrick Greene**
Department of Physics, University of Toronto, 60 St. George Street, Toronto
(Ontario) M5S 1A7 CANADA

28. **Marcus Grisaru**
Physics Department, Brandeis University, Waltham, MA 02454 USA

29. **Eduardo Guendelman**
Physics Department, Ben Gurion University, Beer Sheva 84105 ISRAEL

30. **Guo Wanwen**
Department of Physics and Astronomy, University of Calgary, 2500 University
Drive, N.W., Calgary (Alberta) T2N 1N4 CANADA

31. **Sameer Gupta**
Department of Physics, Pennsylvania State University, 104 Davey Lab., University Park, PA 16803 USA

32. **Yuri Gusev**
Dept. of Physics, University of Alberta, 412 Avadh Bhatia Physics Laboratory,
Edmonton (Alberta) T6G 2J1 CANADA

33. **Tina Harriott**
Mathematical Department, Mount Saint Vincent University, 166 Bedford
Highway, Halifax (Nova Scotia) B3M 2J6 CANADA

34. **David Hobill**
Dept. of Physics & Astronomy, University of Calgary, 2500 University DR.
NW, Calgary (Alberta) T2N 1N4 CANADA

35. **Stephen Horwat**
Physics Department, McGill University, 3600 University St., Montréal
(Québec) H3A 2T8 CANADA

36. **Yutaka Hosotani**
School of Physics and Astronomy, University of Minnesota, Minneapolis MN
55455 USA

37. **Viqar Husain**
Department of Physics, University of British Columbia, 6224 Agricultural
Road, Vancouver (British Columbia) V6H 1B2 CANADA

38. **Boushaki Mustapha Ishak**
   Physics Department, Queens' University, Kingston (Ontario) K7L 3N6 CANADA

39. **Werner Israel**
   Department of Physics & Astronomy, University of Victoria, Box 3055 Victoria (British Columbia) V8W 3P6 CANADA

40. **Ted A. Jacobson**
   Dept. of Physics, University of Maryland, College Park, MD 24742-4111 USA

41. **Daniel Kabat**
   School of Natural Sciences, Institute for Advanced Study, Olden Lane, Princeton, NJ 08540 USA

42. **Munawar Karim**
   St. John Fisher College, 3690 East Ave., Rochester, NY 14618 USA

43. **Vicky Kaspi**
   Dept. of Physics, 37-621 Massachusetts Institute of Technology, 70 Vassar Street, Cambridge, MA 02139 USA

44. **Ramzi Khuri**
   Dept. of Natural Sciences, Baruch College CUNY, 17 Lexington Avenue, Box A-0506 New York, NY 10010 USA

45. **Marcia E. Knutt**
   Physics Dept., McGill University, 3600 University St., Montréal (Québec) H3A 2T8 CANADA

46. **Lev Kofman**
   Canadian Inst for Theor. Astrophysics (CITA), University of Toronto, 60 St George St., Toronto (Ontario) M5S 1A1 CANADA

47. **Gabor Kunstatter**
   Physics Department, University of Winnipeg, Winnipeg (Manitoba) R3B 2E9 CANADA

48. **Hans-Peter Kunzle**
   Dept. of Mathematical Sciences, University of Alberta, Edmonton (Alberta) T6G 2G1 CANADA

49. **Bill Laarakkers**
   Department of Physics, University of Guelph, McNaughton Building, Gordon Street, Guelph (Ontario) N1G 2W1 CANADA

50. **René Lafrance**
   Cégep de St-Hyacinthe, 300 rue Boulle, St-Hyacinthe (Québec) H1G 2J6 CANADA

51. **Kayll Lake**
Department of Physics, Queen's University, Kingston (Ontario) K7L 3N6 CANADA

52. **Harry Lam**
Physics Department, McGill University, 3600 University St., Montréal, (Québec) H3A 2T8 CANADA

53. **Jason LeBlanc**
Department of Mathematics, University of New Brunswick, P.O. Box 4400, Fredericton (New Brunswick) E3B 5A3 CANADA

54. **Julian Le**
Korea Institute for Advanced Study, 207-43 Chungryangri-dong, Dongdaemoon-gu, Seoul 130-12 KOREA

55. **David Lowe**
Physics Department, Brown University, Box 1843, Providence, RI 02912 USA

56. **Lu Kau U.**
Department of Mathematics, California State University, Long Beach, Long Beach, CA 90840 USA

57. **Maxim Lyutikov**
McLennan Labs, Canadian Institute for Theoretical Astrophysics, 60 St George Street, Toronto (Ontario) M5S 3H8 CANADA

58. **Robb Mann**
Department of Physics, University of Waterloo, 200 University Ave., Waterloo (Ontario) N2T 2H5 CANADA

59. **Donald Marolf**
Syracuse University, 201 Physics Building, Syracuse, NY 13224 USA

60. **Karl Martel**
Department of Physics, University of Guelph, 50 Stone Road, suite 158, Guelph (Ontario) N1G 2W1 CANADA

61. **Manasse Mbonye**
Physics Department, University of Michigan, Ann Arbor, MI 48109 USA

62. **Ray G. McLenaghan**
Dept. of Applied Mathematics, University of Waterloo, 200 University Ave W. Waterloo (Ontario) N2L 3G1 CANADA

63. **Jeff McPhail**
Mathematical Institute, Oxford University, 24-29 St. Giles' St., Oxford, OX1 3LB UNITED KINGDOM

64. **Tom Merrall**
Physics Department, Queen's University, Stirling Hall, Kingston (Ontario) K7L 3N6 CANADA

65. **André Allan Methot**
Laboratoire René J.A. Levesque, Université de Montréal, 2955, Edouard-Montpetit, Montréal (Québec) H3C 3J7 CANADA

66. **Sharon Morsink**
Department of Physics, University of Alberta, Edmonton (Alberta) T6G 2J1 CANADA

67. **Jeffrey Morto**
Department of Mathematics, McGill University, Burnside Hall, 845 ouest, rue Sherbrooke, Montréal (Québec) H3A 2T5 CANADA

68. **Shinji Mukohyama**
Yukawa Institute for Theoretical Physics, Kyoto University, Kyoto 606-8052 JAPAN

69. **Robert Myers**
Physics Department, McGill University, 3600 University St., Montréal (Québec) H3A 2T8 CANADA

70. **Michel Olivier**
Department of Physics, University of British Columbia, 6224 Agricultural Road, Vancouver (British Columbia) V6T 1Z1 CANADA

71. **Jonathan Oppenheim**
Dept. of Physics & Astronomy, University of British Columbia, Vancouver, British Columbia, V6T 1Z1 CANADA

72. **Don N. Page**
Dept. of Physics, University of Alberta, Edmonton (Alberta) T6G 2J1 CANADA

73. **Manu Paranjape**
Lab. de physique nucléaire, Université de Montréal, C.P. 6128, Succ. Centre-ville, Montréal (Québec) H3C 3J7 CANADA

74. **Renaud Parentani**
UPRES A 6083, Universite de Tours (France), Faculté des sciences, Parc Grandmont, 37200 Tours FRANCE

75. **Vincent Pelletier**
Physics Department, McGill University, 3600 University St., Montréal (Québec) H3A 2T8 CANADA

76. **Eric Poisson**
Department of Physics, University of Guelph, McNaughton Building, Gordon Street Guelph (Ontario) N1G 2W1 CANADA

77. **Frans Pretorius**
Department of Physics and Astronomy, University of Victoria, P.O. Box 3055, Stn. CSC, Victoria (British Columbia) V8W 3P6 CANADA

78. **Simon Ross**
Department of Physics, University of California, Santa Barbara, CA 93106 USA

79. **William Sajko**
Department of Physics, University of Waterloo, Waterloo (Ontario) N2L 3G1 CANADA

80. **Peter Saulson**
Dept. of Physics, Syracuse University, Syracuse, NY 13244-1130 USA

81. **Kristin Schleich**
Dept. of Physics and Astronomy, University of British Columbia, 6224 Agricultural Rd., Vancouver (British Columbia) V6T 1Z1 CANADA

82. **Sanjeev Seahra**
Department of Physics, University of Waterloo, Waterloo (Ontario) N2L 3G1 CANADA

83. **Tetsuya Shiromizu**
DAMTP University of Cambridge, Silver Street, Cambridge CB3 9EW UNITED KINGDOM

84. **Lenny Susskind**
Dept. of Physics, Stanford University, Stanford, CA 94305-4060 USA

85. **Patrick J. Sutton**
Department of Physics, University of Alberta, Edmonton (Alberta) T6G 2J1 CANADA

86. **Tomoko Uesugi**
Institute of Cosmic Ray Research, University of Tokyo, Midori-cho, Tanashi Tokyo 188-8502 JAPAN

87. **William G. Unruh**
Dept. of Physics and Astronomy, University of British Columbia, Vancouver (British Columbia) V6T 1Z1 CANADA

88. **Robert van den Hoogen**
Dept. of Mathematics, Statistics and CS, Saint Francis Xavier University, Antigonish (Nova Scotia, B2G 2W0 CANADA

89. **Dwight Vincent**

Department of Physics, University of Winnipeg, 515 Portage Ave., Winnipeg, (Manitoba) R3B 2E9

90. **Matt Visse**

Physics Department, Washington University in Saint Louis, CB 1105 Saint Louis, Missouri 63130-4899 USA

91. **Chris Vuill**

Department of Physical Sciences, Embry-Riddle Aeronautical University, 600 S. Clyde Morris Blvd., Daytona Beach, FL 32114 USA

92. **Paul Webster**

Dept. of Physics and Astronomy, University of Calgary, 2500 University Drive, N.W. Calgary (Alberta) T2N 1N4 CANADA

93. **Jeff Williams**

Dept. of Mathematics & Computer Science, Brandon University, 270 - 18th Street, Brandon (Manitoba) R7A 6A9 CANADA

94. **Jeffrey Winicour**

Dept. of Physics & Astronomy, Univ. of Pittsburgh, 221A Allen Hall, 3941 O'Hara Street, Pittsburgh, PA 15260 USA

95. **David Winters**

Physics Department, McGill University, 3600 University St., Montréal (Québec) H3A 2T8 CANADA

96. **Don Witt**

Dept. of Physics and Astronomy, University of British Columbia, 6224 Agricultural Rd., Vancouver (British Columbia) V6T 1Z1 CANADA

97. **Robert Wood**

Department of Mathematical Science, Trinity Western University, 7600 Glover Road, Langley (British Columbia) V2Y 1Y1 CANADA

98. **Eric Woolgar**

Dept. of Mathematical Sciences, University of Alberta, Edmonton (Alberta) T6G 2G1 CANADA

99. **Haider Zaidi**

Department of Physics, University of New Brunswick, Fredericton (New Brunswick) E3B 5A3 CANADA

100. **Andrei Zelnikov**

Department of Physics, University of Alberta, Edmonton (Alberta) T6G 2J1 CANADA

AUTHOR INDEX

A

Aoki, M., 291

B

Barrabès, C., 167
Bassett, B., 301
Baumgarte, T. W., 53
Beetle, C., 174
Bjoraker, J., 285
Booth, I., 182

C

Chia, T. T., 63, 68

D

Duley, W. W., 219

E

Edery, A., 208, 275

F

Fairhurst, S., 174
Frolov, A. V., 141
Frolov, V., 167

G

Gauthier, C., 213
Gen, U., 72
Glass, E. N., 77
Goliath, M., 146
Guendelman, E. I., 201

H

Harriott, T. A., 296
Hobill, D., 306, 311
Hosotani, Y., 285
Hughes, S. A., 53
Husain, V., 238

I

Ipser, J., 60
Ishak, M., 316

J

Jacobson, T., 85
Jaimungal, S., 238

K

Kaspi, V. M., 3
Khuri, R. R., 113
Kofman, L., 191

L

Laarakkers, W. G., 156
Lake, K., 316
Liberati, S., 301

M

Mann, R., 182
Marolf, D., 123
Martel, K., 48
Mbonye, M. R., 161
Morsink, S. M., 15
Mourra, J., 316
Mukohyama, S., 118
Musgrave, P., 316

333

O

Oppenheim, J., 247
Overduin, J. M., 219

P

Page, D. N., 225
Paranjape, M. B., 275
Parentani, R., 167
Poisson, E., 151

R

Rezzolla, L., 53

S

Sajko, W. N., 262, 267
Saulson, P. R., 25
Schleich, K., 132, 233
Seahra, S. S., 219
Shapiro, S. L., 53

Shibata, M., 53
Shiromizu, T., 72, 252, 291
Stern, J., 316
Susskind, L., 98
Sutton, P. J., 257

U

Uesugi, T., 291

V

Visser, M., 301
Vuille, C., 60, 243

W

Webster, P., 306, 311
Wesson, P. S., 219, 262, 267
Williams, J. G., 296
Winicour, J., 35
Witt, D. M., 132, 233
Wood, W. R., 280
Woolgar, E., 137